THE NEUROPSYCHOLOGY OF SMELL AND TASTE

Smell and taste are our most misunderstood senses. Given a choice between losing our senses of smell and taste, or our senses of sight and hearing, most people nominate the former. Yet our senses of smell and taste have the power to stir up memories, alter our mood and even influence our behaviour.

In *The Neuropsychology of Smell and Taste* G. Neil Martin provides a comprehensive, critical analysis of the role of the brain in gustation and olfaction. In his accessible and characteristic style he shows why our senses of smell and taste do not simply perform basic and intermittent functions, but lie at the very centre of our perception of the world around us. Through an exploration of the physiology, anatomy and neuropsychology of the senses, the neurophysiological causes of smell and taste disorders, and their function in physical and mental illness, G. Neil Martin provides an accessible and up-to-date overview of the processes of gustation and olfaction.

The Neuropsychology of Smell and Taste provides a state-of-the-art overview of current research in olfactory and gustatory perception. With sections describing the effect of odour and taste on our behaviour, and evaluating the contribution current neuroimaging technology has made to our understanding of the senses, the book will be of interest to researchers and students of neuropsychology and neuroscience, and anybody with an interest in olfaction and gustation.

G. Neil Martin is a Chartered Scientist and a Fellow of the Royal Society of Arts. As Director of the Human Olfaction Laboratory at Middlesex University, his research covers human olfaction and the effect of ambient odour on behaviour. He has written over a dozen books on psychology and teaches courses in neuropsychology, biological psychology, forensic psychology, health psychology and integrative medicine. He received his PhD in psychophysiology from the University of Warwick.

Brain, Behaviour and Cognition

Series editors: Glyn W. Humphreys and Chris Code

Published titles:

Milestones in the History of Aphasia
Theories and protagonists
Juergen Tesak and Chris Code

Anomia
Theoretical and clinical aspects
Matti Laine and Nadine Martin

Neuropsychology of Art
Neurological, cognitive and evolutionary perspectives
Dahlia W. Zaidel

Classic Cases in Neuropsychology, Volume II
Chris Code, Yves Joanette, Andre Roch Lecours and Claus-W. Wallesch

Category Specificity in Brain and Mind
Emer Forde and Glyn Humphreys

Neurobehavioural Disability and Social Handicap Following Traumatic Brain Injury
Rodger Ll. Wood and Tom McMillan

Developmental Neuropsychology
A clinical approach
Vicki Anderson, Julie Hendy, Elisabeth Northam and Jacquie Wrennall

Developmental Disorders of the Frontostriatal System
Neuropsychological, neuropsychiatric and evolutionary perspectives
John L. Bradshaw

Clinical and Neuropsychological Aspects of Closed Head Injury
Dr J. Richardson

Communication Disorders Following Traumatic Brain Injury
Skye McDonald, Chris Code and Leanne Togher

Transcortical Aphasias
Marcelo L. Berthier

Spatial Neglect
A clinical handbook for diagnosis and treatment
Ian H. Robertson and Peter W. Halligan

THE NEUROPSYCHOLOGY OF SMELL AND TASTE

G. Neil Martin

Psychology Press
Taylor & Francis Group
LONDON AND NEW YORK

First published 2013
by Psychology Press
27 Church Road, Hove, East Sussex BN3 2FA

Simultaneously published in the USA and Canada
by Psychology Press
711 Third Avenue, New York, NY 10017

Psychology Press is an imprint of the Taylor & Francis Group, an informa business

British Library Cataloguing in Publication Data
A catalogue record for this book is available from the British Library

Library of Congress Cataloguing in Publication Data
Library of Congress Cataloging-in-Publication Data
Martin, G. Neil.
The neuropsychology of smell and taste / G. Neil Martin.
pages cm.
Includes bibliographical references and index.
1. Neuropsychology. 2. Smell. 3. Taste. I. Title.
QP360.M3516 2013
612.8--dc23
2012030467

ISBN: 978-1-84872-100-5 (hbk)
ISBN: 978-1-84872-137-1 (pbk)
ISBN: 978-0-20307-014-7 (ebk)

Typeset in Times by
GreenGate Publishing Services, Tonbridge, Kent

MIX
Paper from responsible sources
FSC www.fsc.org FSC® C013056

Printed and bound in Great Britain by
TJ International Ltd, Padstow, Cornwall

For Niki, louloudi mou

CONTENTS

ILLUSTRATIONS

Figures

Table

SERIES PREFACE

From being an area primarily on the periphery of mainstream behavioural and cognitive science, neuropsychology has developed in recent years into an area of central concern for a range of disciplines. We are witnessing not only a revolution in the way in which brain–behaviour–cognition relationships are viewed, but also a widening of interest concerning developments in neuropsychology on the part of a range of workers in a variety of fields. Major advances in brain-imaging techniques and the cognitive modelling of the impairments following brain injury promise a wider understanding of the nature of the representation of cognition and behaviour in the damaged and undamaged brain.

Neuropsychology is now centrally important for those working with brain-damaged people, but the very rate of expansion in the area makes it difficult to keep up with findings from the current research. The aim of the *Brain, Behaviour and Cognition* series is to publish a wide range of books that present comprehensive and up-to-date overviews of current developments in specific areas of interest.

These books will be of particular interest to those working with the brain-damaged. It is the editors' intention that undergraduates, postgraduates, clinicians and researchers in psychology, speech pathology and medicine will find this series a useful source of information on important current developments. The authors and editors of the books in the series are experts in their respective fields, working at the forefront of contemporary research. They have produced texts that are accessible and scholarly. We thank them for their contribution and their hard work in fulfilling the aims of the series.

Chris Code and Glyn W. Humphreys
University of Exeter, UK and University of Birmingham, UK
Series Editors

PREFACE

In 1892, Henry's arguably more creative brother, William James, wrote this withering assessment of two of our oldest senses and other physiological phenomena: 'Taste, smell, as well as hunger, thirst, nausea and other so-called "common" sensations need not be touched on… as almost nothing of psychological interest is known concerning them' (William James, 1892, *Briefer Course Psychology*).

Picking up the baton, Sir Victor Negus in 1958, not to be outdone in terms of complete collapse of motivation and interest, wrote: 'the human mind is an inadequate agent with which to study olfaction, for the reason that in Man the sense of smell is relatively feeble and not of great significance'.

Given this largely unreceptive and, frankly, arctic view of these senses, even the most charitable soul would not view it as a positive augur for a book on both. For a start, there might not be enough material to fill it. Second, the material may be of epiphenomenal interest only. Smell and taste, it might be argued, are minor senses, of occasional sensory interest constituting a pleasurable distraction but performing intermittent basic, perfunctory functions that are too elementary and quotidian to warrant sophisticated neuropsychological inspection. Some of this is true. Given a choice between losing the senses of smell and taste, or the sense of sight or hearing, people would normally nominate the former, rather than the latter. We rely on the dominant senses more; this is why they are dominant. We are no longer quadripedal and do not rely on our chemosenses to navigate, to mate, to make ingestive decisions, to influence behaviour and so on, in the way we did before we became bipedal, serendipitously noticing the air was fresher and cleaner above ground (odours are heavy; they like the ground or bottom floor).

However, a life without either is like sight without colour or like somatosensation experienced through rubber gloves. These senses are considered unimportant because we take them for granted and we do rely on them to an extraordinary casual degree – the reason we do this is because they very rarely go wrong (one noteworthy example notwithstanding). Myopic or presbyopic individuals can wear glasses to correct their degenerating lenses, the tone-deaf can cope with this inconvenience, a person with a headache will find the slightest noise an irritation that tips into unbearability. But because the senses of smell and taste – the invisible senses – perform so well day to day, we only notice major impairments in their function when these impairments affect us badly. The most ersatz example is the common cold where individuals famously misattribute the failure to perceive food flavour to the inability to taste (rather than the inability to smell, which is what occurs). Food flavour is predominantly olfactory, not gustatory.

We also underestimate how efficient these senses are. Our sense of smell is more effective than a smoke detector. According to Engen (1982), we can recognize odours within 0–3 seconds of encountering them, and at a distance of between 1–2 m (one of the reasons why olfaction is more productively and creatively studied by psychologists than gustation. The Japanese Sanitation Centre noted that we (humans) can detect the malodorous isoamyl mercaptan (a variant of which is added to odourless propane gas to make it pungent) at 0.77 parts per trillion (Nagata and Takeuchi, 1990). Cain (1977) concluded that our noses are more sensitive than a chromatograph. We can probably detect ethyl mercaptan (which is added to gas) at around 1 part per billion (Whisman *et al.*, 1978), the equivalent, as Yeshurun and Sobel note (2010), of three drops in an Olympic Swimming Pool. It is probably not on the basis of little understanding that Brillat-Savarin had argued that the nose 'acts as the first sentinel, crying out, "Who goes there?"' We also have very low detection thresholds for certain odours such as d-limolene and ozone (Cain *et al.*, 2007), a phenomenon considered in more detail in Chapter 1. In short, our sense of smell – for a sense we appear not to rely on or which we regard as being of little significance – is very effective. Less gloriously, the external agent of this sense bequeathed to psychology one if its more colourful legacies. Emma Eckstein, a patient of Wilhelm Fliess – the otolaryngologist of orgon machine fame – suffered severe nasal bleeding after the removal of a nasal pack that had been left in her nasal cavity after surgery (the surgery itself was spuriously recommended for the interruption of nasal neurotic reflexes, whatever they might have been). Following this, Fliess dreamed about the 'after care', a somnolence that subsequently led to Freud's musings on the nature of dreams.

Smell and taste serve a vital purpose: they are essential for stopping us from killing ourselves, not only by detecting noxious odours, a mephitic alarm that saves us, but by preventing us from ingesting material that can harm us (rotten or spoiled food). These senses also interact with the largest cranial nerve, the trigeminus, which adds another dimension to behavioural life – this somatosensory nerve mediates the heat-delivery of a chilli, the tear-evocation of an onion and the jolting assault of ammonia. It adds another survival dimension to the panoply of chemosensation – the stimulation of the trigeminus activates the same fibres and substances implicated in the experience of pain. Combined with this survival role is another that allows us to derive feelings of pleasure or disgust from objects, environments and people. Scent is an effective person-repellant and attractor. As McBurney (1986), echoing Brillat-Savarin earlier, put it when describing the utility of the chemosenses:

> I argue that smell and taste are first and foremost our gatekeepers for ingestion and monitors of social behaviour. Their principal task, therefore, is to answer these questions: 'What is that stuff in my mouth (or about to go in my mouth)?', 'Do I like it?', and (for odour) 'Who is that and what do I want to do about him or her?'

At a psychological, but no less interesting level, scents in the environment affect our behaviour in ways of which we are barely conscious. Exposure to pleasant odour is associated with increases in charitable behaviour (Baron, 1997), better anagram formation (Baron and Thomley, 1994), increased emotional experience when reading literature (Cupchik and Phillips, 2005), changes in brain electrical activity associated with attention (Martin, 1998), reduced visual vigilance (Gould and Martin, 2001), improved verbal recall (Herz, 1997), reduced anxiety in women waiting for dental surgery (Lehrner *et al.*, 2000), increases in pain perception (Martin, 2006) and increased accurate recall of memories of events experienced years previously (Aggleton and Waskett, 1999), amongst other effects.

The psychological effects of exposure to odour have also been demonstrated outside the laboratory, in applied contexts such as the workplace (Sakamoto, *et al.*, 2005). Mental concentration levels have been reported to be higher during exposure to the odour of lavender but not jasmine (Sakamatoto *et al.*, 2005) whereas participants who slept in the presence of jasmine odour performed cognitive tasks more rapidly and reported being more alert after waking (Raudenbush *et al.*, 2003). The odours of peppermint, jasmine, ylang-ylang 1, 8-cineole and menthol appear to have no beneficial effect on reaction time (Ilmberger *et al.*, 2001), but respondents' ratings of these odours as positive or negative did influence reaction time: odours rated as positive were associated with faster reaction times (Ilmberger *et al.*, 2001).

The behavioural effects of malodour are more stereotypically predictable. Malodour has strong historical and medical associations with ill-health, especially disease and infection ('malaria', for example, literally means 'bad air' and was the name given to conditions arising from the inhalation of noxious fumes emanating from Roman marshes; Martin, 1996). Malodour is also an environmental hazard: it is the major source of public complaints to local government authorities in the US and Europe (Nicell, 2009). Exposure to it is associated with significant increases in ill-health and psychological annoyance, leading to a seriously impaired quality of life, stress, insomnia, eye irritation, nausea, headaches, irrational behaviour and anorexia (Nicell, 2009; Sucker *et al.*, 2009). Laboratory studies have found that exposure to unpleasant odours increases pain perception (Martin, 2006; Villemure *et al.*, 2006) and the stereochemical, androstenone (described in Chapter 1), has also been associated with increased perception of pain, especially in women (Villemure and Bushnell, 2007). Frequency, intensity, duration, offensiveness and location are all important factors that determine the strength of people's responses to environmental malodour.

Increases in an individual's negative mood are significantly associated with exposure to a foul-smelling odour. Across their lifespan, people remember unpleasant odours better than pleasant ones (Larsson *et al.*, 2006) and these odours are detected more quickly than pleasant odours (Boesveldt *et al.*, 2010). The likeability of faces decreases in the presence of malodour administered at below-threshold levels (Li *et al.*, 2007). Participants exposed to an unpleasant smell are more inclined to rate strangers that are similar to themselves more positively than they would dissimilar strangers (Rotton *et al.*, 1978). Rotton (1983) reported that women who rated paintings and black and white photographs in a room polluted with ethyl mercaptan gave significantly lower scores of 'well-being' to the photographs and judged the pictures to be less professional and less worthy (but no less tasteful) than did participants in an unpolluted room. These participants also reported lower feelings of pleasure and levels of arousal than participants in the air-conditioned room. The longer the exposure to the malodour, the less the pleasure taken in completing the task. Participants detect fewer proof-reading errors in a polluted room but detect more when moved to an unpolluted room. Participants taken to an unpolluted room and asked to solve a series of puzzles, the first and second of which were insoluble (a measure of frustration) attempted fewer puzzles after a previous, 30-minute exposure to the malodour. Malodour, therefore, is a highly instrumental sensory stimulus, capable of directing emotion, thought and behaviour.

The behavioural effects of scent have been more comprehensively studied than the effects of taste because taste is the briefer sensation and is generated only for one purpose: the detection and appreciation of food inside the mouth. While we do habituate quickly to scent, we have, to some degree, more control over and exposure to its spatial distribution and its uses – we use it to deodorize, to make us attractive, to repel, to freshen, to relax and so on. (None of these could apply to taste specifically, the measurement of responses to which have been mainly physiological in nature or limited to the study of psychophysics and valence/pleasure.) The maximum field of influence of scent, therefore, is greater.

Unlike the naming of tastes (where we are moderately competent), our naming of odours is notoriously bad. Of the tens of thousands of odours we can detect, we can name very few accurately. As chemicals, these odours should not surprise us with their problematic linguistics. When sitting in the garden, we do not comment on the fragrance of the phenyl ethyl alcohol, while admiring the verdant source of the isobutyl methoxypyrazine as we nibble our peeled amyl acetate, stopping occasionally to sip thiourine with ethanol, while avoiding the scent of isovaleric acid from discarded footwear. Instead, we use descriptors. We say that something smells or tastes like a referent, even though this referent is nothing more than chemistry and chemistry that interacts with our olfactory and gustatory systems to provide the unitary percept we are familiar with. So, we enjoy the scent of a rose, and of freshly mown grass, of a banana, and the taste of gin with tonic, and avoid malodorous feet – the names, the psychology, we give to the chemistry.

This book aims to describe, review and discuss the contribution of psychology and neuropsychology to our understanding of smell and taste. Chapter 1 provides an introduction to the senses and describes the ways in which psychology and chemosensory science measures our responses to these stimuli. An understanding of how olfaction and taste is measured is important (not least because the subsequent chapters refer to the specific functions measured and the tests and techniques used). Equipped with this information, the bulk of the book will be better understood.

Chapter 2 considers some of the individual differences that exist in these senses, particularly sex and age. There are also modality-specific individual differences (such as supertasting and specific anosmia for androstenone). As these have implications for our understanding of both senses, these specific eccentricities are described there.

The book assumes very little in the way of prior knowledge of the brain and neurophysiology, apart from the basics. Chapters 3 and 4 describe and review the current knowledge of the neurophysiology and anatomy of chemosensation (Chapter 3), and how electrophysiology and neuroimaging have helped advance the understanding of the cerebral basis of olfactory and gustatory perception (Chapter 4). The emphasis in the book is on human neuropsychology but, as with all areas of psychology, it is informed by animal research – and much of what we understand of the initial stages of olfactory processes has been derived almost exclusively from animal work. Therefore, animal work is cited to inform understanding of human function. Given the significance of odour and taste to humans – compared with other species – however, these data are described circumspectly.

Chapter 5 continues the pure neuropsychological theme by describing disorders of smell and taste, the effect of degenerative disease on olfactory function and the effect of particular psychiatric disorders on smell and taste perception.

Finally, Chapter 6 brings together the material in the previous chapters to examine the neuropsychological basis of flavour – one of the largest remaining challenges for neuropsychology and chemosensation. It is, possibly, an impossible challenge. This chapter also examines the interaction between smell and taste and other modalities, such as vision, somatosensation and audition. Flavour is the sum of this interaction.

This book would not have been possible without the support of several people. First, and importantly, Chris Code and Glyn Humphreys, who were gracious enough to see the merit of this entry to their *Brain, Behaviour and Cognition* series and commission the manuscript; George Mather and Jamie Ward, who were kind enough to comment on the original proposal for Psychology Press; and Phil Jerrod, Laura Elllis and Becci Edmondson at Psychology Press for their advice, patience and good humour. For permissions to use figures, their enormous help and their encouragement for the book, a Brobdignagian thank you to Jessica Albrecht, Ivan Araujo, Thomas Bitter, Douglas Braaten, Warrick Brewer, Jelena Djordevic, Richard Doty, Diego Luis Garcia Gonzalez, Chris Hawkes, David Kareken, Don Katz, Alan Mackay-Sim, Paul Moberg, Claire Murphy and Bruce

Turetsky. More proximally, my thanks to Paul Assoul and all at Coffee Bean and the staff at Costa High Barnet for providing a temporary office in the Summer, Autumn and Winter of 2011 when the book was being researched and written. In addition to thinking I know a bit about smell and taste, I am further equipped with the knowledge of how to grill an Emmenthal and ham ciabatta and prepare a small Americano with hot milk which professionally, if you work in smell and taste, this is something of an advantage. Finally, but it could never be finally, thanks to Niki, the salt and pepper of my life: efharisto.

If you would like to write to me with comments and feedback about the content of the book, my email address is n.martin@mdx.ac.uk (or tweet @thatneilmartin).

ABBREVIATIONS

ACC	anterior cingulate cortex
AD	Alzheimer's Disease
AML	ascending method of limits
AMTL	anteromedial temporal lobe
AON	anterior olfactory nucleus
B-SIT	Brief Smell Identification Test
CN	cranial nerve
CNS	central nervous system
DAT	Dementia of the Alzheimer Type
DBS	deep brain stimulation
dlPFC	dorsolateral prefrontal cortex
DT	dopamine transporter
EEG	electroencephalography
EP	evoked potential
EPI	Eysenck Personality Inventory
ERP	event-related potential
fMRI	functional magnetic resonance imaging
fNIS	functional near-infrared spectroscopy
ISI	inter-stimulus interval
KS	Kallmann's syndrome
LGN	lateral geniculate nucleus
LOT	lateral olfactory tract
LPOFC	lateral prefrontal orbitofrontal cortex

MCI	mild cognitive impairment
MEG	magnetoencephalography
MRI	magnetic resonance imaging
MTLE	medial temporal lobe epilepsy
NA	nucleus accumbens
NST	nucleus of the solitary tract
OB	olfactory bulb
OCD	obsessive–compulsive disorder
OE	olfactory epithelium
OEP	olfactory evoked potential
OFC	orbitofrontal cortex
PBN	parabrachial nucleus
PD	Parkinson's Disease
PET	positron emission tomography
PFC	prefrontal cortex
PMC	primary motor cortex
POC	primary olfactory cortex
PTA	primary taste area
PTC	primary taste cortex
rTMS	repetitive transcranial magnetic stimulation
SOC	secondary olfactory cortex
SS	single staircase
SSS	sensory-specific satiety
TBI	traumatic brain injury
UPSIT	University of Pennsylvania Smell Identification Test
VBM	voxel-based morphometry
VMPN	ventroposteromedial nucleus

1

SMELL AND TASTE

An introduction to the psychology of chemosensation

1.1 Unique features of smell and taste

Smell and taste are chemosenses, that is, they are sensory systems that respond to chemical stimulation and whose chemical stimuli bind to receptors to create a sensation. Both are two of the most neglected and unusual in the sensory panoply in that each exhibits features that uniquely and dramatically separate it from the dominant senses of vision and audition, and even somatosensation. For example:

- Olfaction is the only sense with receptors directly exposed to the environment;
- Humans have an ability to detect hundreds, if not thousands, of different odours but only five or so tastes. However, the same humans are notoriously poor at identifying such odours;
- There is no agreed classification system for odour; there is more agreement for taste;
- Unlike vision, hearing and touch there is no olfactory dimension that relates stimuli to sensation; it has no predictable frequencies nor limited spectra (Mackay-Sim and Royet, 2006);
- Also unlike vision and audition, the olfactory system requires a third of the genome; vision requires three genes; audition requires a structure that develops from genes coding for other aspects of development (Mackay-Sim and Royet, 2006);
- The olfactory cortex has three layers, unlike most other sensory cortices;
- Taste and smell receptors regenerate every sixty days – thus, our current chemoreceptors did not exist two months ago;
- Smell is probably the most manipulable and confusable of the senses: people can be convinced that an odourless substance is odorous or that they are smelling something they are not;
- Taste is invariably confused with smell although smell provides the greatest contribution to food flavour;

- Olfactory dysfunction may be a better marker of risk of degenerative disease (e.g. Alzheimer's Disease (AD)) than more conventional neurophysiological or clinical measures.

Formally, smell is known as olfaction and taste as gustation and the chemicals that stimulate each sense are called odorants or tastants. In the case of gustation, sensation is produced by the interaction between the tastant on the tongue and the depolarization that occurs in the taste bud field it stimulates. In the case of odour, the molecules are inhaled via the external nares (nostrils) with the air that carries them, and are processed, via transduction, by the olfactory apparatus at the top of the nose and beyond (Chapter 3 describes this pathway and process in detail). An odorant is an odour compound which means that it is volatile (and, therefore, evaporates quickly) and hydrophilic and lipophilic (it can dissolve in oil and water). However, the sense of smell also delivers olfactory information from another source of respiration, other than external: from inside the mouth.

1.2 Orthonasal and retronasal breathing

Typical olfactory perception involves two types of breathing – orthonasal and retronasal. With orthonasal breathing, odour molecules enter the anterior or external nares, travel through the nasal cavity and are transported to the olfactory apparatus at the top of the nose and onward to the cortex. Retronasal breathing occurs in the oral/buccal cavity where odorants stimulate the posterior or interior nares of the pharynx (the receptors here are called nasopharyngeal receptors and are supplied by two cranial nerves (CNs), neither of which are olfactory), and travel to the olfactory apparatus and the olfactory receptors at the top of the nose (Burdach and Doty, 1987). It is this process that creates food flavour.

A fruit juice inserted into the mouth and rolled on the tongue while the nose is pinched shut, will be almost impossible to identify, although the identification of the juice's taste (that it is sweet or sour) will be relatively unaffected. However, if the nostrils are released, identification of the juice will be almost immediate because the internal nasopharyngeal receptors have been stimulated by the odour molecules released by the tastant and these molecules have stimulated the epithelium retronasally, via the back of the mouth. The failure of retronasal perception is the reason why when individuals suffer colds and the 'flu, they claim to be unable to taste food. What they actually mean is that they cannot smell the food – they can easily determine whether the food tastes salty, sweet and so on if pressed. What they are unable to do is identify the flavour (and, therefore, the food). The reason for this inability is the impairment in retronasal perception of odour. In this sense, therefore, the sense of smell is both an exteroreceptor and an interoreceptor. Unless we are ingesting, the exteroreceptive function is the most common and important, 'giving warning of enemies and other noxious things and guiding the animal to mates, food and other desirables' (Herrick, 1933). There are also psychophysical differences between orthonasal and retronasal breathing. Thresholds for odours are lower in the former and odours may be perceived more intensely (Voirol and Daget, 1986), especially when coupled with tastants administered to the tongue (Gillan, 1983), and identified more accurately (Pierce and Halpern, 1996).

Odour is the primary determinant of food flavour – providing a crucial interoceptive function – but it has important interactions with taste (and other sensations) to create the flavour ensemble (discussed in detail in Chapter 6). Retronasal breathing increases the judged intensity of taste whereas orthonasal breathing has little effect on these judgements. As Brillat-Savarin wrote in his *Physiology of Taste*,

> I am not only convinced that without the co-operation of smell there can be no complete degustation, but I am also tempted to believe that smell and taste are in fact a single sense, whose laboratory is the mouth and whose chimney is the nose…When smell is intercepted, taste is paralysed.

Moncrieff (1967) had observed that if the eyes and nose were closed, pureed apple and onion would be perceived as being the same olfactorily. That is, they would be unidentified. However, they would be discriminated on the basis of non-olfactory cues: the apple would taste more sour and the onion sweeter or sharper. Other pairings that would be impossible to identify if olfaction was impaired include red wine and unsweetened black coffee, raw apple and potato, tomato and orange, diluted raspberry and sweetened milk, sweetened milk and sweetened milk with vanilla, and peach and apricot. These might be discriminated from each other reasonably well – red wine is lighter and less bitter than coffee, raw apple is sweeter/sourer than potato – but not correctly identified, thus demonstrating the importance of olfaction to the identification of food flavour.

In view of the above, it is not difficult to empathize with the refrain and lament of Hollingworth and Poffenberger (1917) in their *The Sense of Taste*: 'Why should it be the rule that, since the taste and smell qualities are to be confused, smell should so commonly sacrifice its claim, so that odours are called tastes rather than vice versa?' One answer to this conundrum, they suggest, is the illusion of predominance created by the cutaneous stimulation in taste. A curious phenomenon, illustrated by the von Skramlik test, is that odours can be perceived through the mouth. If a person pinches their nose and inhales and exhales through the mouth, a sensation will emerge (Mozell *et al.*, 1969). That all food-related identification appears to involve the tongue, however, a physical structure that senses food and its various characteristics from temperature and texture to its actual taste, may account for the dispropportionate importance placed on it when we perceive flavour.

Smell is the first chemosensory custodian of survival: we can sniff food for signs of spoiling or decomposition. Off-milk, meat and vegetables, if their appearance provides no indication or harm or rotting, can be detected in this way and are prevented from ingestion. If its scent and appearance indicate that food is ingestible, taste is the final custodian and its most important role is to prevent gastrointestinal risk. Thus bitter substances are normally perceived as less pleasant than are sweet or salty (and sour) tastants – most poisons are bitter-tasting, suggesting an evolutionary role for the development of receptors allowing us to detect bitterness. Preference for tastes with a strong bitter component tends to develop during young to late adulthood – thus a liking for coffee, green vegetables, whisky, gin and other bitter-tasting foods is virtually absent in children (not that you would normally feed children whisky and gin). Unfamiliar food also elicits withdrawal behaviour or curiosity short of eating (a gustatory neophobia). Bitter substances or spoiled, contaminated food elicit a disgust reaction, evidenced by changes in physiognomy, feelings of revulsion and behavioural withdrawal (Rozin and Fallon, 1987). At the core of food/taste disgust are feelings of danger, distaste and knowledge of the nature or origin of the food (Rozin and Fallon, 1987). Analysis of the features of food that people find disgusting suggests that there are two dimensions: textural properties (unpleasant) and reminders of 'livingness' or 'animalness' (Martins and Pliner, 2006). Predictors of disgust include culture, history of exposure and the ability to detect disgust-eliciting taste. There is also evidence that this disgust can extend to moral disgust. Recent work in social psychology suggests that people in clean-scented environments engage more in charity work, express reciprocal trust, enagage with unknown people more and show more interest in voluntary work (Lilenquist *et al.*, 2010). Eskine *et al.* (2011) found that people who were given a teaspoon of a bitter substance to taste expressed greater moral disgust when judging controversial topics such as incest or the acceptability of eating a dead dog. The more conservative respondents responded even more robustly after tasting the substance. Of course, this effect could be due to intensity, rather than bitterness, and a useful experiment would alter the intensities of this and other tastes.

1.3 Smell and taste: basic features and assumptions

Phylogenetically, smell and taste are two of the oldest senses, if not the oldest. The subcortical structures we now identify as the limbic system – structures involved in our most basic behaviours such as hunger, thirst, sexual drive, homeostasis, aggression – were originally described as the rhinencephalon or 'smell–brain', a term given to these regions and structures by Paul Broca. They are evolutionary contemporaries of thermosensation and nociception (pain).

Smell and taste are also both near senses, or 'short distance sensory modalities', to use Sherrington's (1906) term. That is, the stimuli that result in transduction of sensory signals are close to the site of transduction (the nose and mouth). This contrasts with audition and vision, which are far senses, or 'long distance sensory modalities' (you can identify a building or a sound from ten miles or 10m away).

But despite this similarity, olfaction is the more flexible sense. Taste requires molecules to make contact with receptors on the tongue, located inside the mouth – which is normally shut – and this involves the conscious act of introducing stimuli into a cavity and making direct contact with the stimulus producing the sensation. This contrasts starkly with its partner which is tireless and open – the process of respiration means that the olfactory system is continuously working, inhaling air which carries molecules that comprise odorants. This is a feature it also shares with audition. Taste, on the other hand, is similar to vision and more like somatosensation – in that while receptors are theoretically available to be stimulated, there needs to be an agentic action in order to create stimulation in taste and touch (and the eyelid needs to be opened for vision – although a closed lid, of course, does not prevent visual stimulation). The sense of smell is automatic but the degree of response depends on the degree of automaticity. The most obvious illustration of this is that sniffing may be more effective than natural breathing in allowing olfactory perception.

There is also some debate over whether smell and taste are analytic or synthetic senses. The prototypical synthetic sense is vision. Yellow combined with red will not result in a bit of yellow and a bit of red but orange: seeing orange means we are none the wiser as to whether it is orange or a combination of two wavelengths. Audition, conversely, is considered to be the prototypical analytic sense; it does not blend sounds and we can distinguish the auditory components that contribute to sound combinations (such as musical instruments in a song). Olfaction is probably, on balance, a synthetic sense because we can discriminate, at best, three or four odours in an odour mixture – no more (Laing and Glemarec, 1992) – and odours can be blended like colour frequencies. Taste is considered an analytic sense because the qualities of taste do not combine to form new tastes.

A sniff is shorter and more vigorous than a breath and it assists olfactory perception. LeMagnen (1945/1946) found that participants' sensitivity to eucalyptus improved when the number of inhalations increased from twelve to ninety-six per minute. Similarly, Rehn (1978) found a threefold increase in sensitivity to the odour of pyridine sensitivity when inhalations increased from 4.5 to sixty per minute (the odour of pyridine is distinctive of the top notes of Oxo or coffee). However, there are negative findings – Teghtsoonian *et al.* (1978) found no improvement in intensity perception when sniffs increased from 15.2 to 32.4 per minute. Laing (1983) observed that the average human inhales at a rate of thirty breaths per minute, with a volume of 200cm^3 and a duration of 0.4 seconds per inhalation but also reported that varying the number of sniffs, the size of sniffs and the interval between sniffs does not improve odour perception. What Laing suggests is that sniffing confirms a person's perception of the odour, rather than determines it.

To test the efficiency of sniffing and scent tracking in humans, Porter *et al.* (2007) asked thirty-two participants to follow a 10m trail of chocolate essential oil in open grass while participants were kneeling and blindfolded. Two-thirds were able to do this effectively. With increased training – three times a day, three days a week for two weeks – the amount of deviation from the scent trail reduced and speed along the trail increased. Sniffing also increased over three days and this sniffing increased with increasing speed along the trail. In terms of sniffing velocity, the right naris was found to have a velocity of 0.45msecs^{-1}; in the right, this was 0.30msecs^{-1}. The right nostril also had an advantage in terms of spatial reach – the maximum spatial distance at which an odour could be detected. Its reach was 1.5–2cm to the right; the left was 1–1.5cm to the left. In a final experiment,

Porter *et al.* compared monorhinal with birhinal tracking – sniffing with one nostril resulted in decreased accuracy (30 per cent compared with 66 per cent) and increased slowness (20 per cent slower). This asymmetrical airflow and its psychological consequences is described in Chapter 3.

1.4 Classification of smell and taste

There is more agreement on the classification of tastes than odours. In 1755, Polycorpe Poncelet proposed the existence of seven distinct tastes which he called *acide*, *doux*, *amer*, *piquant*, *fade*, *austere* and *aigre-doux* (sweet-sour). The topology did not surive, however, although some of these descriptors are considered basic tastes. There are five agreed basic tastes: salty, sweet, bitter, sour and umami. There is evidence that these tastes can be distinguished at the molecular and chemical level (and this evidence is discussed in Chapter 3). This fifth taste, umami, was first reported in 1909 by Ikeda, who noted a sensation different to the classical tastes was elicited by various foods such as meat, mushrooms, some fish, cheese and tomatoes (as well as human breast milk). Terms describing this gustatory sensation include 'brothy', 'meaty' or 'savoury'. The chemicals glutamate and 5'-nucleotides are probably the components in these foods that make up the umami taste. Other, chemical, examples of stimuli that create the umami taste are: monosodium L-glutamate (MSG), guano monophosphate (GMP) and inopsine 5'-monophosphate (IMP). It was once thought that umami worked by enhancing the flavour of other tastes but there is no consistent evidence that adding the taste of umami to a variety of foods lowers or increases the thresholds of other tastes: if umami is added to food, salty tastes are not perceived as more or less salty.

Taste is the easier of the chemosenses to classify. Over 300 years, various interested philosophers, bien pensants and scientists have attempted to make sense of odour by delineating classifications of odour quality, grouping scents according to type of chemical or psychological quality. Thus, topologies (or, even more appropriately, nosologies) have relied on either grouping scents by molecular structure/family or sensory quality (fruity, sweet, faecal and so on). In view of recent developments in molecular genetics, traditional forms of classifying odours may be a forlorn and fruitless exercise but it is useful to examine the history of this approach to understanding our responses to odour.

The first attempt at classifying odour was that of von Linne in 1756 who proposed that there were seven odour qualities, followed by Albrecht von Haller in 1763 who suggested three and Rimmel in 1868 who identified eighteen. Perhaps the two most influential classification systems were proposed by Henning (in 1916 and 1924) and Amoore (in 1962) and the latter had dominated thinking in this field. Henning proposed that there were six salient odour qualities, based on their psychological characteristics, forming an analogue to his taste tetrahedron (a model of taste quality). These six qualities occupied points on a smell prism and qualities shifted as you moved from one point to the next (see Figure 1.1). He argued that mixing two basic odours would not give you an odour that was somewhere in between.

Amoore hypothesized that, based on his examination of over 600 odorants, stimuli could be classed according to how they bind to the microvilli of olfactory cells – odours of different shapes would bind with similarly shaped cells, in a lock-and-key fashion. Thus this stereochemical theory proposed that there were five or seven stereochemical binding sites and each was responsible for the sensation of each of seven primary odours. These primary odours were identified as camphoraceous, musk, floral, peppermint, ethereal, pungent and putrid. What unifies each type of primary odour is its common molecular structure. However, not all odours that smell the same have the same molecular, chemical structure and similar chemical structures can produce very different odours. The most famous example of the latter is l (aevo)-carvone and d (extro)-carvone. Almost mirror images of each other chemically, one elicits the sensation of caraway, the other spearmint. Amoore moved his attention away from this model in his later work to focus on discovering basic odours based on anosmia – the inability to detect an odour. On the basis of this work, he proposed that there were thirty basic odours (Amoore, 1970; Amoore *et al.*, 1975). Two other models, which have fallen by the wayside, were those of Davies (1953) and Mozell (1970). Davies proposed that odour quality was produced by the rate of diffusion of odour through the receptor

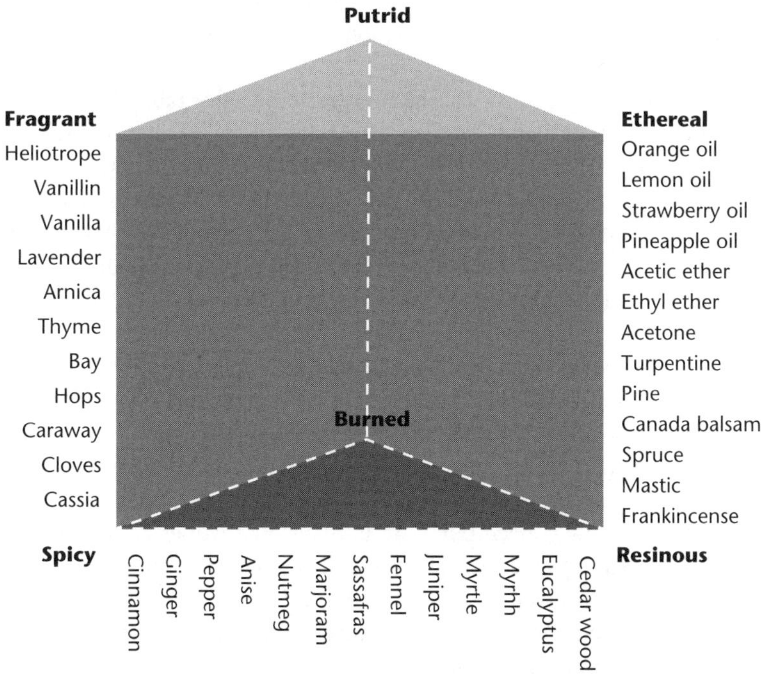

FIGURE 1.1 Henning's odour prism, still decorating textbooks

cells and the resistance of the cell membrane to being punctured by the odour. Mozell argued that quality was determined by the pattern of activation/spread of molecules across the olfactory epithelium (OE). While these models were ambitious and based on a great deal of intellectual industry, it is difficult to argue with McBurney (1986) who concluded that their 'greatest contribution was in decorating textbooks' (p. 122).

There is greater consistency of agreement that, at the general level of perception, odours can be classified into three types based on their mechanical effects on neurophysiology. Thus, odorants can be regarded as pure odours, which stimulate the olfactory nerve exclusively, as trigeminal odorants, which stimulate the trigeminal nerve primarily, and bimodal odours, which stimulate both CNs. The basis on which these distinctions were made are discussed in the psychophysiology section in Chapter 3 but, in brief, it has been found that trigeminal stimuli can elicit electrical potentials in people who have lost their sense of smell but not odours (Kobal and Hummel, 1988). Patients with anosmia are unable to detect the odours of phenyl ethyl alcohol and vanillin. Trigeminal stimulation – discussed in more detail in Chapter 3 – is somatosensory in nature. The painful 'smell' of ammonia, the lachrymal response to peeling onions, the perception of CO_2 and the burning sensation to acetic/formic acid are all sensations mediated by, and the responsibility of, the trigeminal nerve. It is responsible for the sensation of chilli burn and other pain sensations in the mouth and face as well as tactile perception of odour and thermoreception. Some odours, such as eucalyptol, have a bimodal effect, producing that characteristic menthol scent (mediated by the olfactory nerve) but also a sensation of cooling when inhaled (mediated by the trigeminal nerve).

The classification models described above, especially those based on psychological characteristics and, to some extent those topologizing on the basis of chemistry because these models group odours according to categories that invite particular psychological responses, use psychological dimensions as guiding principles. The most obvious is pleasantness. In fact, Engen (1982) has concluded that 'it is clearly the hedonic meaning of odour that dominates odour perception'. Evidence consistently demonstrates that when people respond to odour, the initial response is hedonic – the individual decides whether they like it or not (Khan *et al.*, 2007). Pleasantness is also a useful way to help us discriminate between odours and groups of odours (Berglund *et al.*, 1973; Schiffman, 1974); odours of different pleasantness are evaluated at different speeds (Bensafi *et al.*, 2002).

One important, perhaps the most important, determinant of odour pleasantness is intensity. Intensity describes the number of molecules in an odorant presented at a given time. A low concentration of an odour that would conventionally be regarded as unpleasant (faeces, for example, or the flavouring Devil's dung/asafoetida) can be perceived neutrally or positively. Some fixatives in perfumes are not the sorts of odorants you would expect to find on the shelves in Harvey Nichols but they appear in colognes and scents at such low concentrations as to be barely detectable. Conversely, a normally positively regarded odour can be judged negatively if its concentration is very strong.

Related to intensity is adaptation or habituation. We are all familiar with walking into someone's home and detecting a distinctive odour permeating that environment but that, after a while, becomes unnoticeable. This process is adaptation or habituation and it is useful in that it maintains olfactory equilibrium, making us accept (adapt to) a new environment that is not harmful and, in so doing, makes the appearance of any other odour more detectable. Repeated stimulation can lead to a decrease in sensitivity but this recovers eventually if there is no further stimulation (Dalton *et al.*, 2002). Dalton and Wysocki (1996) examined the effect of long-term odour exposure on detection ability and intensity rating. As most habituation studies in olfaction use exposure times lasting minutes or hours rather than days or weeks, Dalton and Wysocki exposed participants to one of two odours continuously in their homes for two weeks. They found that thresholds for detecting the exposed odour were higher after two weeks compared with the participants' baseline response and they also found weaker concentrations less intense. The detection decrement was seen even up to two weeks following the final exposure.

1.5 Measuring olfaction

Alexander Graham Bell, amongst his many other sterling observations on the way the world worked, had this to say about scent:

> Have you an ambition to found a new science? Why not measure a smell? Can you measure a smell? Can you measure the difference between one smell and another? … Odours are becoming more and more important in the world of science and medicine – and the need of more knowledge, as surely as the sun shines.

The measurement of scent has developed dramatically since Graham Bell's observation. Measurement of olfactory responses involves the administration of odour, clearly, but what is required of a test dictates the nature of the stimulus used. For example, a detection test will involve a different procedure to an identification or discrimination test. The use of odour in psychophysiological and neuroimaging research also employs specific techniques and methods that are unique to this sense. The preceding comments, in fact, broadly describe the two types of olfactory testing employed in psychological and neuropsychological work – one involves the exposure to an odour and the making of a judgement; the other involves the delivery of an odour in a specialized way and the person either perceives passively or makes a judgement of some kind (e.g., on intensity, valence). Some illustrations of olfactory methods can be seen in Figure 1.2.

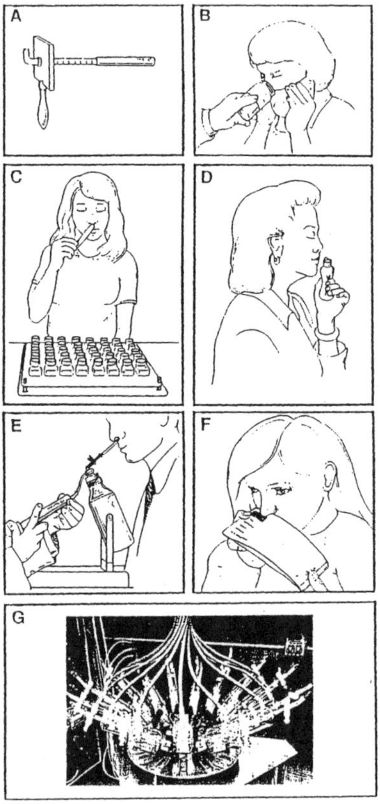

FIGURE 1.2 Some of the techniques and methods used to study the sense of smell empirically

1.6 Tests of olfactory function and ability

Our understanding of the ability to detect, perceive and identify odours depends on an understanding of relatively sophisticated psychophysical tests. Tests of detection, and some tests of discrimination, are solely based on psychophysics and utilize the concept of just-noticeable-difference embodied in Weber's Law, the smallest increase in concentration ΔI necessary to discriminate between two stimulus concentrations. Thus, participants are asked to determine whether a previous odour was stronger than a currently presented one when the target differs only minutely. These tests make use of different concentrations of odours; the lowest detectable concentration would represent the participant's olfactory detection threshold.

Current tests that measure olfactory ability explicitly are listed in Table 1.1.

TABLE 1.1 Some common measures of olfactory ability

Takagi and Takagi olfactometer test
San Diego Odour Identification Test
Smell Threshold Test
Alcohol Sniff Test (Davidson and Murphy, 1997)
Three-odour pocket Smell Test
12-Odour Brief Smell Identification Test/Cross-cultural Smell Identification Test
12-item Odour Memory Test (Bromley and Doty, 1995)
The University of Pennsylvania Smell Identification Test (UPSIT)
Odour Confusion Matrix Test
Scandinavian Odour Identification Test
Sniffin' Sticks (Hummel *et al.*, 1997)
Viennese Olfactory Test Battery (Lehrner *et al.*, 1999)
Jet Stream Olfactometer Test (Onoda and Ikeda, 1999)
Eight-Odour Identification Test

Of the tests in Table 1.1, the most widely used are the UPSIT, Sniffin' Sticks and the San Diego OIT largely because their creators are supremely prolific researchers and publish work extensively on them and partly because they are reliable and well validated.

As the introduction suggests, the method of presentation in these odour studies is important. The discipline's history has utilized various types of procedure, methods and techniques from the early pioneering draw tube olfactometer of Zwaardemeker to more modern olfactometers (there is more on olfactometry on p. 16), glass bottles, glass rods, felt-tipped pens, scratch and sniff cards (microcapsules), squeezy bottles, perfumers' strips, alcohol pads and blotting paper. These do not include the cruder methods of presentation used in behavioural studies such as impregnated paper, aroma diffusers such as the AromaCube, impregnated necklaces, aerosols and others. The degree of experimental control in these studies is restricted but then these experiments are designed only to ensure the isointense, suprathreshold delivery of an odour – that it can be smelled. The direction of smelling has also detained some authors – whether odour administration is delivered monorhinally to one nostril or birhinally to both and some of these studies are reviewed in Chapter 3.

Psychophysical tests fall into three general categories: (i) detection/recognition threshold tests, (ii) tests of discrimination and (iii) tests of identification.

1.6.1 Detection threshold tests

Tests of detection threshold aim to determine the lowest possible concentration at which a person can detect an odour. Doty and Laing (2003) distinguish between two types of threshold: absolute and recognition. Absolute threshold is the 'lowest concentration where such a presence is reliably detected'; recognition threshold is 'the lowest concentration where odour quality is reliably discerned'. Both thresholds are normally measured by forced-choice, i.e. participants are asked which of two stimuli smells the strongest or which of two odours has a particular quality.

At the psychophysical level, tests adopt one of two procedures: the ascending method of limits (AML) or single staircase method (SS). In the AML, the participant is presented with a series of odours from low to high concentration and the point at which a participant claims to be able to detect the odorant is taken as the threshold. In the SS method, which is similar, the concentration of an odorant is increased after a trial in which the participant claims not to be able to detect an odour and is decreased following a positive reaction. Both have been used in academic research and in clinical work. Thus Cain (1982) used the AML to determine thresholds in forty-three patients with olfactory dysfunction and noted impairments in both nostrils.

As Doty and Laing note, however, these thresholds are never truly absolute and they fluctuate according to the methods used, participant characteristics and even time of day. They note that some variables that affect threshold measurement include the method of dilution used in creating the odorants, the volume of inhalation, the type of task and the number of trials administered. Pressure and humidity can affect the sense of smell, with hypobaric pressure affecting thresholds for n-butanol to a worse extent than hyperbaric pressure (Kuehn *et al.*, 2008): thresholds were lower under humid than dry conditions. Stevens *et al.* (1988) observed that the variability in responses across time to three odours when testing occurred over thirty days was as great as between-subject variability on the same day. That is, the same person's threshold varied over the testing period. The testing also relies on correct responses being given – that is, responses depend on the integrity of the participant, a perennial problem in psychophysics. Therefore, the reliability of threshold measures is low with the AML method showing less reliability than the SS method.

The ability to detect odour can improve with training. This has been found with typical odours (Engen and Bosack, 1969; Cain and Gent, 1991) and, more controversially, with the stereochemical, androstenone. Androstenone is described in more detail in Chapter 2 but is thought to be the commonest scent that cannot be detected. The anosmia can appear in 50 per cent of the population. Wysocki *et al.* (1989), however, found that repeated exposure to androstenone could improve recognition in participants who claimed not be be able to detect it, a finding that has been replicated (Mainland *et al.*, 2002). A putative receptor for androstenone detection has been reported (Keller *et al.*, 2007).

Related tests measure the magnitude of differences between odours. The suprathreshold scaling procedure is one such method. The intensity of an odour is probably its most important psychological characteristic, as noted in the preface. Intensity ratings also depend on context. If a moderately strong odour is presented to a participant along with either a weak or a strong odour, it is perceived as stronger when paired with a weak rather than a strong odour (Eyman *et al.*, 1975). Even objective tests, therefore, highlight the subjectivity of our response to smell.

In magnitude tests, a number or anchor is assigned to a particular concentration of an odorant (sixty, say) so that a concentration four times as strong would be rated 240 (4 × 60), a concentration half as strong would be thirty and so on. In some studies an anchor is provided; in others, the participant is allowed to choose (Doty and Laing, 2003).

1.6.2 Tests of discrimination

Tests of discrimination examine a person's ability to distinguish between two or more odorants on the basis of some quality, whether a category – flower or food – or a psychometric dimension such as strong/weak, pleasant/unpleasant, familiar/unfamiliar, etc. Typically, a person is presented with two pairs of odours – one pair is the same, one pair is different – and are asked which pair is the same/different (O'Mahony, 1979). In a variant of this method – the triangle method – the participant discriminates one odour from a series of odorants (Frijters *et al.*, 1980).

The first published study of olfactory sensitivity was probably Gamble's (1898) which observed that two concentrations of an odour could be detected if they differed by 25–33 per cent. Cain (1977) later reported that humans could discriminate between two concentrations of odours that differed by 7 per cent. Odour familiarity, however, is important because people are less able to discriminate between unfamiliar than familiar odours, a phenomenon that is probably largely explained by the cognitive assistance provided by labelling (Mingo and Stevenson, 2007).

The most sophisticated form of discrimination analysis is multi-dimensional scaling. This statistical technique takes data from discrimination studies, creates a correlation matrix that is subjected to an algorithm which then plots and groups these odours in a two-dimensional space (Schiffman *et al.*, 1981). The advantage of the technique is that it can allow the grouping of odours that would not otherwise have been grouped in this way because there would have been no a priori reason for doing so. It is, therefore, highly data driven and free from experimenter bias.

Like detection, discrimination ability can improve with training, experience and practice, even if the odour is unfamiliar (Jehl *et al.*, 1995). Expert professionals who are familiar with their stimuli – perfume counter workers and wine tasters – show better discrimination ability than non-experts (Bende and Nordin, 1997; Hummel *et al.*, 2004). Stimuli that are personally meaningful are more effectively discriminated from other stimuli. For example, people have been able to discriminate between their own t-shirts and those of 100 others, based on scent alone (Lord and

Kasprzakc, 1989), can discriminate between the odour of their own dog and others' (Wells and Hepper, 2000) and can discriminate between the odour of their own baby and others' babies (Porter, Cernoch and McLaughlin, 1983). Five to eight-year-old children can discriminate the odour of their younger siblings and nine-year-old children can identify their friends from scent alone (Porter and Moore, 1981; Mallet and Schaal, 1998). Individuals are able to do this despite their self-reported lack of confidence in the discrimination decisions they make (Lundstrom *et al.*, 2008).

1.6.3 Tests of identification

Tests of odour identification require participants to either (i) name an odour presented, (ii) agree or disagree with a name given to an odour or (iii) choose a name (or object) from a range of names (or objects) for an odour they have been presented with. Studies suggest that we are notoriously poor at identifying odours, despite being very efficient at detecting them. Thus, we cannot name around 50 per cent of odours associated with household items (Cain, 1979; de Wijk and Cain, 1994) and we have problems in matching an odour to its name and vice versa (Olsson and Jonsson, 2008). Practice, however, appears to improve identification performance as does the familiarity of the odour (Cain, 1979; Homewood and Stevenson, 2001).

Possibly, the most widely used test of identification is the UPSIT, developed by Richard Doty and colleagues at the University of Pennsylvania. In this test, forty microencapsulated odours are scratched and sniffed and participants are given four possible labels for each (Doty *et al.*, 1984b; Doty, 1995). The measure of identification ability is the number of odours correctly identified. Data from the test collected from hundreds of individuals and norms are thus available for sex, age and other variables. The test is also available in French, German and Spanish, and in short form which is also designed for cross-cultural research (Doty *et al.*, 1996). This Brief Smell Identification Test (B-SIT) features twelve odours including menthol, clove, leather, strawberry, lilac, pineapple, smoke, soap, natural gas and lemon. There are some cultural differences that emerge on the test – Indian and Chinese graduate students and children's UPSIT scores, for example, are lower than those of young US adults (Frank *et al.*, 2004). In contrast, scores are comparable on another smell test, the Sniff Magnitude Test, thereby suggesting that there may be a language component on the UPSIT that can impair performance for non-olfactory reasons. The full form of the test has good test–retest reliability; the shorter form shows less reliability (Doty *et al.*, 1984a, 1984b). Analyses of the test–retest reliability of the most common tests (see Table 1.1) have found correlations of 0.9 for UPSIT, 0.72 for the Scandivanian Odour Identification Test, 0.73 for the short, twelve-item version of this test, and 0.73 for Sniffin' Sticks (Doty *et al.*, 1989; Hummel *et al.*, 1997; Nordin *et al.*, 1998).

The latter, Sniffin' Sticks, is another, well-researched instrument, developed by Thomas Hummel, which tests the three components of olfactory function: discrimination, threshold and identification. The results from each component are combined to create the respondent's composite TDI score (threshold, discrimination, identification). In the threshold test, sixteen dilutions of the odorants n-butanol and phenyl ethyl alcohol are administered using the SS method at increasing concentrations. The lowest concentration detected is the threshold. The discrimination component involves the presentation of sixteen sets of three pens. Two of the pens have the same odorant, one has a different odorant and the task is to discriminate which is different. Finally, identification is measured by a person's ability to choose the right name for sixteen odours from four descriptors. A score below sixteen indicates anosmia. The subtests individually are not as reliable as the measure combined. Test–retest reliability of the discrimination component has been found to be 0.8, the identification component 0.88 and the threshold component 0.92 (Haehner *et al.*, 2009). The test has been applied successfully cross-culturally (Shu *et al.*, 2007).

An interesting issue in olfactory testing is to what extent these tests load on a similar factor. That is, are the components – threshold, discrimination, identification – truly separable or do they load on a common, unitary factor. Threshold measures, for example, load on the same factor as UPSIT (Doty *et al.*, 1994). Doty *et al.* (1994) argue that a single component may measure a common source of variance. However, studies of brain injury (see Chapter 2) suggest that thresholds are unimpaired but identification scores are affected by injury, indicating that the components may be separable and do not measure a common source.

One other study suggests that tests may measure one generic factor. Lotsch *et al.* (2008) examined the reliability of the components of the Sniffin' Sticks test in 2,076 controls and 102 individuals with Parkinson's Disease (PD; see Chapter 5). The components were able to diagnose the illness with moderate success. For example, threshold performance predicted 64 per cent of PD, discrimination 56 per cent and identification 47 per cent. In a theme returned to in Chapter 5, olfaction was impaired in 99 per cent of PD patients. A principal components analysis of the subcomponents produced two factors – all tests loaded on one factor, the second factor accounted for threshold.

1.7 Discriminating and identifying odours in mixtures

We are highly sensitive to odours that do not seem to belong in an odour mixture (Rabin, 1988) and are better at detecting this interloping odour if it is familiar rather than unfamiliar, and unpleasant rather than pleasant. We are not particularly effective at detecting and identifying more than three odours in a mixture, however. In fact, three appears to be our limit (Berglund, 1974). Even perfumers cannot name more than four components in an odour mixture accurately (Berglund, 1974) and training over three weeks does not appear to improve this ability (Laing and Francis, 1989; Livermore and Laing, 1996).

Our stubborn inability to identify more than three elements applies even when odours are 'non-blenders' – these are odours that do not mix well with other odours at all. We also tend to underestimate the number of odours in a mixture, probably for this reason (Schiet and Frijters, 1988). The ultimate reason for this is unknown but may be due to some inhibitory action at the olfactory bulb (OB) (Doty and Laing, 2003), or the ability of some odours to reach receptor cells more quickly than others, two propositons considered in Chapter 3. When two odours are present in a mixture, we tend to rate the intensity of the mixture as lower than when we rate both odours separately, a finding that was reported in the nineteenth century: the odour of camphor was found to cancel out the aroma of gasoline and juniper (Aronsohn, 1886).

1.8 Measurement of the neural response to odour: olfactometry

To be able to measure and record the brain's response to an odour, we can use electroencephalography (EEG) to examine olfactory effects on spontaneous electrical activity, or neuroimaging to observe the effects on processes surrounding neuronal activation, or blood flow. However, the EEG, a measure of the brain's electrical activity, is chaotic, messy and non-linear. Recording EEG and then averaging each period of data representing exposure to odour, leads to distinctive peaks and troughs appearing at various times following stimulus onset. These are evoked potentials (EPs) or event-related potentials (ERPs) – so called because a potential is evoked by or is related to a stimulus sensed or perceived in the participant's environment – and have been recorded in non-olfactory modalities for decades.

However, olfaction is a special sense in that its stimuli are invisible (unlike visual stimuli) and can only be detected by the person exposed to them when they are administered in a proximal environment (unlike a sound – such as a voice saying 'Hello to Jason Isaacs' next to a person's ear at 60dB which can be heard by the person and the person around them at approximately the same time; an odour at the weakest detectable concentration would only be detectable by the participant). The way in which an odour is delivered, therefore, needs to be accurate and controlled. A method that allows this is olfactometry, and it is a technique used in olfactory ERP research and neuroimaging studies. In olfactometry, a gaseous flow of air can be combined with an odorant contained in a bottle or flask which is passed through a tube placed near the entrance of the nostrils. The flow is computer-controlled which means that the odour delivery can be precisely timed and the velocity of airflow can be controlled. The device that delivers the odorant is an olfactometer (Figure 1.3).

There are two methods of stimulus delivery via olfactometer: the flow method and the pulse/blast method. The flow method is considered to be far superior to the pulse/blast method – in which the air or odour is delivered, as the name suggests, as short blasts or pulses rather than as a continuous flow – because even odourless air delivered via the blast method produces ERPs (in the ERP jargon, the shape of the wave during chemical stimulation using the flow method should be a 'square wave'). The temperature and humidity of the flow is also important and relative humidity is recommended to be 50 per cent. It is also recommended that the flow is warmed to 35–37°C (Evans *et al.*, 1993). To prevent nasal respiration, a breathing

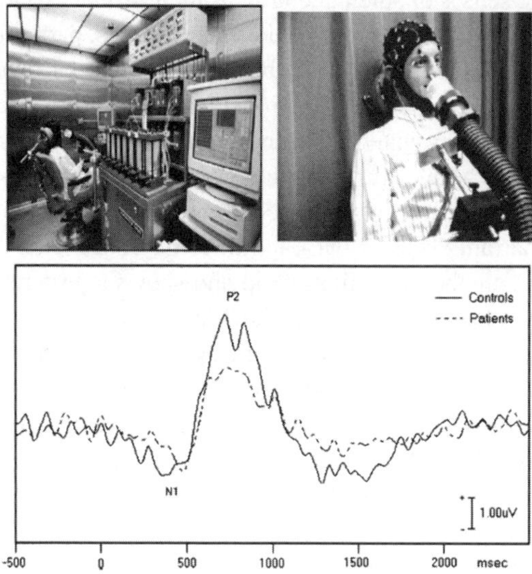

FIGURE 1.3 A modern olfactometer and the olfactory evoked potentials produced by this method (a control group compared with a patient group)

technique called velopharyngeal closure has been developed which avoids this (Kobal, 1981). Chapter 4 describes some of the principal studies that have shed light on how the human brain processes odour, using olfactometry.

1.9 Development of olfactory perception

Like the parts of the brain responsible for basic functions such as respiration and movement, those parts devoted to the sub/cortical functioning of gustation and olfaction are a vestige of our evolutionary history. Perhaps a vestige of the past when olfaction and gustation played a greater role in our survival – before we began to live bipedally – these regions subserving scent, and the number of receptors allowing this subservice, are greater in other quadrupedal mammals that rely more on these senses. Dogs such as German Shepards, for example, have between 95–169cm^2 of OE. Boxers have approximately 120cm^2 and even Pekinese have three times that of humans (Issel-Tarver and Rine, 1997). As Issel-Tarver and Rine suggest, there is no strict correlation between the amount of anatomy devoted to the sense of smell and olfactory acuity but those mammals with extensive olfactory epithelia do tend to use their sense of smell more extensively, and have dramatically greater olfactory acuity than do humans. Vision is our dominant sense and we are, compared to other mammals, microsmatic rather than macrosmatic. We rely less on the olfactory system than do other mammals. When asked which sense and organ of sense they would relinquish if they had to, people nominate the senses of smell and taste as their primary candidates (Martin *et al.*, 2001a). The poor-relation status of the chemosenses, particularly olfaction, has led to the sense of smell being described by the cliché of choice in medical circles, the Cinderella of the Senses (The *BMJ* once published a Christmas edition in which it listed all the physical organs and processes described by medicine as Cinderella. Medicine is a field stuffed with disadvantaged relatives).

The sense of smell is functional in the foetus and appears to be functioning reasonably well post-partum. Thus, a two- and three-day old foetus can distinguish between its own and another's amniotic fluid and shows a preference for its own mother's milk rather than another woman's (Marlier *et al.*, 1998; Schaal *et al.*, 1998), orients more toward a breast pad worn by its own mother than a stranger (Schaal, 1988) and prefers an unwashed breast to a washed one (Varendi *et al.*, 1994). Mothers and infants can also recognize their own odours (Cernoch and Porter, 1985). The reason for this orienting and learning is the cross-over between the mother's diet and the amniotic fluid (see below). Pregnant mothers asked to ingest garlic or a placebo forty-five minutes before amniotic fluid was taken from the body had fluid that was stronger and smelling more like garlic if they ate the food (Mennella and Beauchamp, 1993). This phenomenon has been found with various foods – mint alcohol and vanilla are others – suggesting that what the mother eats percolates through to her milk (Mennella and Beauchamp, 1991b, 1998, 1999). (This amniotic fluid is inhaled as well as swallowed by the foetus (Pritchard, 1965).) Mothers fed a diet of carrot produced children who preferred carrot-flavoured cereal during infancy; control children did not show this preference (see Forestell and Mennella, 2007).

During breast-feeding, an infant will feed for longer and consume more breast milk if the mother has consumed vanilla (Mennella and Beauchamp, 1996). Infants exposed to chamomile found this more attractive two to three days later than those who had not (Allam *et al.*, 2006). Alcohol also increases breast milk's intensity, even half an hour after the mother has ingested alcohol (Mennella and Beauchamp, 1998). However, one reason why infants may not suckle as much from a mother who has ingested alcohol may be that the alcohol inhibits milk production and, therefore, there is not much to suckle (Mennella and Beauchamp, 1991b).

The respiratory rate of infants during exposure to formula milk is greater in bottle-fed infants and to breast milk for breast-fed infants (Soussignan *et al.*, 1997). Breast-fed infants will also prefer the breast milk of an unfamiliar mother to formula milk (Marlier and Schaal, 2005), indicating that the complex nature of breast milk and its odour has a special, biological significance for infants (or may be a function of familiarity – they preferentially attend to colostrum, breast milk and sebum, for example; Marlier *et al.*, 1997). Breast-fed infants will also eat more cereal when it is combined with breast milk than without (Mennella and Beauchamp, 1999). The degree of attachment and 'imprinting' that occurs in the first few hours and days of the neonate's life is important psychologically. The mechanism of the odour-learned maternal attachment is unknown but in rats this appears to depend on the release of norepinephrine by the locus coeruleus into the OB (Moriceau and Sullivan, 2004). By the time imprinting ends (ten days post-partum), this release has decreased, suggesting that its role has been fulfilled.

Infants, however, do not respond to odours in a meaningfully adult way. The well-cited work of Steiner (1977) on neonates' responses to tastes – indicating that there are stereotypical facial responses to specific tastes – has not been replicated (it is famously not peer reviewed). Infants' response to odours is not also sufficiently sophisticated to allow the discrimination of odour types by examining physiognomy alone (Soussignan *et al.*, 1995; Soussignan *et al.*, 1997). However, an exception seems to be infants with pervasive developmental disorder – independent, naïve observers can distinguish between these infants' reactions to pleasant and unpleasant smells when watching film footage of the infants' reactions (Soussignan *et al.*, 1995). While infants can learn to tell the difference between pleasant and unpleasant odours, their preferences tend to mimic those of adults by the time they are 5–6 years old (Balogh and Porter, 1986; Schmidt and Beauchamp, 1988).

A technique called functional near-infrared spectroscopy (fNIS) has found that brain activation in babies is different depending on whether the odour is pleasant or unpleasant. When twelve male and eleven female six-hour-old to seven-day-old babies were exposed to the mother's colostrum, vanilla or distilled water, activation was seen in the left orbitofrontal cortex (OFC), the tip of the frontal lobe, of all babies to vanilla odour. For colostrum, then, the older the baby, the greater the decreased activation in the left OFC (the lack of regional-specificity in fNIS may be due to the anatomical limitations of the technique). When responses to the unpleasant odours of detergent and disinfectant were compared, there was greater involvement of the right OFC (Bartocci *et al.*, 2000; Bartocci *et al.*, 2001). An EEG study of neonates and young adults found no change in direction or location of activity with age (Sanders *et al.*, 2002). But these imaging techniques are susceptible to the obvious problem of artifact generation with this particular sample (Dominguez, 2011): children are exemplary fidgeters and the motion artifact can contaminate functional magnetic resonance imaging (fMRI) recording.

The olfactory ability of children as they develop in the younger years is difficult to measure – especially their ability to identify odour – because their responses up until the age of five are more or less unuseable. Doty (1995) reports that the responses of 5–9 year olds are similar to those of 80 year olds, i.e. poor. The performance of 8–14 year olds is not much better (Cain *et al.*, 1995). The ability to identify odour depends on the ability to experience and to develop a sophisticated vocabulary to interpret sensations (Cain *et al.*, 1995). A study of over 200 French 6–10 year olds' uses of odour found that girls were more olfaction-oriented than were boys (especially when the odours concerned themselves, other people or the environment) (Ferdenzi *et al.*, 2008). Ferdenzi *et al.* also found that the children's ability to describe the olfactory elements of their world increased with age. Monnery-Patris *et al.* (2009) similarly reported that olfactory sensitivity and identification increased with age and was better in girls than boys. However, when verbal ability was controlled for, the effect of sex disappeared but the effect of level/year of schooling remained. It is easier to measure hedonic responses in younger children than it is discrimination or identification ability. Thus, children younger than five have a surprisingly neutral attitude to scents that adults find disgusting – odours such as sweat and faeces (Schmidt and Beauchamp, 1988). Identification is better in teenagers (de Wijk and Cain, 1994) and performance seems to begin to improve from the age of 3.5–5 years – the above caveats notwithstanding (Richman *et al.*, 1992).

There are no validated tests for 3–5 year olds that take into account their linguistic inexperience, their intellectual development and their lack of familiarity with some of the odours used in tests validated on adults (Dalton *et al.*, 2009). Thus, Dalton *et al.* have taken steps to trial an age-appropriate olfactory test that can contribute to the 'NIH toolbox', a US-funded project supported by sixteen national health institute centres, which comprises a series of brief tests of cognitive, memory, motor, emotional and sensory function across the age span. One model is the San Diego Odour Identification Test. (In this test, described earlier, grocery items are placed in opaque bottles and eight of these items are sniffed and matched against a line drawing. This takes five minutes to complete.) Work on establishing the Toolbox Pediatric Odour Identification Test is currently underway and is attempting to determine six odours that 3–4 year olds can recognize. Using the Monell-Jefferson Chemosensory Clinical Research Centre odour identification test as an inspiration, the participant's task in the test would be to identify the odour by pointing to a picture, named by the experimenter. There would be three distractors – one very similar to the odour, two dissimilar. Data are, as yet, unavailable.

1.10 Measuring gustation

The components of gustatory testing are similar to those in olfaction but intensity is also measured. Although the threshold and identification measures for odour are correlated, there are no such significant correlations for bitter, sweet, salty and sour (Cowart *et al.*, 1997).

The spatial requirements of olfactory testing – birhinal or monorhinal stimulation – are nothing as compared to taste. The compartmentalisation of the tongue into discrete taste areas is, as Chapter 3 will show, flawed. However, because the tongue receives innervation from different branches of the three CNs (Chapter 3 describes this anatomy in more detail), different parts of the tongue are more sensitive than others. Therefore, tests of gustation investigate whole-mouth perception (Wrobel and Leopold, 2004) or regional perception where a tastant is placed on specific sectors of the tongue. A technique called electrogustometry – where a weak electrical current is passed over the tongue – is also used; this elicits a metallic taste when applied. It can also help identify injury to one of the nerves of the tongue (the chorda tympani).

Historically, whole-mouth testing – 'sip and spit' – had been common in industry testing and the quantities administered are either three drops or eight cups

(Harris and Kalmus, 1949; Henkin *et al.*, 1963). In terms of threshold measurement, the three-drop method employs two drops of water and one drop of tastant. Threshold is reached when the participant can correctly identify the taste three trials in a row. The eight-cup method involves the participant sampling four tastants and four cups of water. Thresholds for NaCl and sucrose are higher for the three-drop method than the eight-cup (Weiffenbach *et al.*, 1983), probably because the cup method covers a greater area of the tongue. In the cup method, over 1ml of fluid are administered over the tongue; in the drop method, the quantity is less than 0.1ml.

When a tastant is applied to the same area of the tongue and consistently, the intensity rating of that tastant declines and eventually disappears (Bujas *et al.*, 1991), although not totally because mouth movement and swallowing do not allow this. The application of water restores the intensity rating, thus demonstrating a form of adaptation in the gustatory system (Bartoshuk, 1991). There is also a relationship between intensity and degree of taste. Thus, the intensity of a sweet tastant increases with increasing sweetness (Pangborn and Giovanni, 1984). Intermediate sweet levels tend to be the best preferred.

Modern tests of chemogustometry involve the application of small (5mm diameter) saturated filter paper disks on the tongue, the use of saturated cotton swabs or the administration of a tastant via a pipette (Tomita *et al.*, 1986; Bartoshuk, 1989b; Deems *et al.*, 1991). Thickeners such as cellulose or gelatine are sometimes used which means the tastant can be delivered as small cubes that remain fairly stable on the tongue (Delwiche *et al.*, 2000). The nature of these stimuli means that fairly precise regions of the tongue can be stimulated (Kobayakawa *et al.*, 1999).

Like olfactometry, there are techniques employed in neuroimaging studies of taste that ensure that the tastant is delivered in a controlled way. Thus, de Araujo *et al.* (2003a, 2003b) devised a technique whereby 0.5–2ml of solution was delivered to the participant's mouth through tubes following visual or auditory cues to the participant to swallow. Kobayashi *et al.* (2004) developed a perfusion method whereby a tube delivered a tastant to the tongue and another removed the liquid via suction, although this could allow for the tastant to escape between the inlet tube and the tongue as tubes were held between the lips. A different tube method (Ogawa *et al.*, 2005) involved the participants holding a tube that had a small hole through which the tastant was deposited. A disadvantage of this method was that only dorsal areas of the tongue could be stimulated. To avoid some of the problems associated with these techniques, Kami *et al.* (2008) devised a technique whereby a cylinder containing a tastant was attached to a mouthpiece. The participant inserted their tongue into this mouthpiece and a solution was run along it at a constant rate. Their fMRI study using three participants found that this method activated the primary taste area (PTA) when sucrose was delivered.

An additional problem with neuroimaging studies of taste is that water or liquid can activate the central nervous system (CNS) so using it as a control is not a way of genuinely controlling for the liquid carrier for the tastant. It can also create after-impressions, post initial tasting (Galindo-Cuspinera *et al.*, 2006). Ways of obviating this include using dried wafers or tablets or using films made from cellulose or starch. However, these involve chewing, thus introducing another confound. Similarly, filter paper has to be discarded because it is not edible and does not dissolve. One way of solving this problem would be to create edible, dissolvable films that can be placed on the tongue and that do not require chewing; Smutzer *et al.* (2008) have developed such a technique using polymer pullulan.

1.11 Development of taste perception

Like olfaction, gustation is functional in the foetus. When born, the neonate appears to respond differently depending on the taste quality it senses. Infants will stop crying when presented with a sweet solution, although this reaction dissipates after six weeks (Barr *et al.*, 1994; Smith and Blass, 1996) and will generate more hand–mouth contact (Barr *et al.*, 1994). This effect occurs with natural sugars (sucrose) and artificial sweeteners (saccharin). There is also evidence that sweet tastants can reduce pain in healthy preterm and term infants (Blass and Watt, 1999) and in older children between 8–11 years old (Miller *et al.*, 1994). This phenomenon is analogous to the ability of familiar odours to reduce crying during distressing medical procedures (Goubet *et al.*, 2003).

In terms of consumption, infants tend to consume more sucrose and MSG than bitter, salty, sour or neutral solutions (Beauchamp and Moran, 1982; Beauchamp and Pearson, 1991) but there is an inconsistent pattern of behaviour observed in the first few months. Thus, facial reactions to tastes are stable in the first month but after this the infant will refuse to open its mouth or push the caregiver's hand away until around the sixth month. Bitter substances will be rejected (Kaijura *et al.*, 1992). Infants and non-human primates generate similar facial expressions to sucrose and their responses to quinine are also similar. There is a greater coherence of response similarity between human infants and great apes than between human infants and new/old world monkeys (Steiner *et al.*, 2011).

Preference for sweet and umami is seen in the first week post-partum (Beauchamp and Pearson, 1991) and bitter and sour tastants are rejected (even if they are dissolved with sucrose; Vasquez *et al.*, 1982; Kaijura *et al.*, 1992). The response to salt shows a stereotypical pattern. For example, the neonate will reject or be indifferent to NaCl but by six months – through to two years of age – the infant will prefer salty food when it is presented in soup (Beauchamp *et al.*, 1986; Beauchamp and Moran, 1984), perhaps due to the maturation/activation of salt receptors. After that, however, children between two years, six months and five years tend to reject salt solutions (Beauchamp *et al.*, 1986), unless it is carried in a vehicle like vegetable soup. Thus, Beauchamp and Cowart (1987) found that preference for the highest concentration of NaCl in this soup was 65 per cent in 3–6 year olds, 78 per cent in 7–10 year olds and 13 per cent in 18–26 year olds. Infants of mothers with morning sickness – and, therefore, prone to vomiting

– show increased preference for salt intake (Crystal and Bernstein, 1998). In a fascinating experiment, Beauchamp and Moran (1982) fed infants sweetened water in the first six months of life and then monitored their preference for sweet foods after this period. They found that these infants preferred highly sweetened foods whereas a control group not fed sweetened water preferred sweet foods but the magnitude of preference in the former group was greater.

In a similar study of 4–5 year olds, Sullivan and Birch (1990) fed infants either sweetened or salted tofu and monitored their preference for these foods after several weeks of exposure. Their later preferences were dictated by the foods they were exposed to: exposure to sweet foods led to preference for sweet foods. The exposure did not generalize to colour or texture. So preference for tofu-like foods such as ricotta did not increase.

In this way, children (and adults) can be trained to like novel foods. Birch and Marlin (1982) and Pliner (1982), for example, found that repeatedly exposing participants to novel foods on ten occasions can lead to increased acceptance of these foods. Wardle and colleagues (Wardle *et al.*, 2003a, 2003b) found that exposing preschoolers or undergraduates over a period of 10–14 days to foods that were novel or not preferred lead to greater acceptance of this food. The effect appears to be selective, however. Thus, this enhanced preference following exposure is found for fruit and sweet substances, but not sour (Liem and de Graaf, 2004).

The development of the gustatory system appears to be complete by puberty but there is an interesting sex difference in that boys' function matures later. Compared to men, women and prepubertal girls, prepubertal boys have higher detection thresholds for sucrose and NaCl (James *et al.*, 1997). Some tastes were even less well detected: boys' thresholds for citric acid, for example, were seven times higher than adults'. In general, children have preferences for stronger NaCl and sucrose tastes than do adults (Desor *et al.*, 1975; Desor and Beauchamp, 1987).

These individual differences – age and sex – exert significant influence over our ability to process odour and taste, from responding to them hedonically, to detecting and identifying them. The pattern of response that occurs across development and between the sexes, as well as that which can result from personality and culture, is described fully in Chapter 2.

2

INDIVIDUAL DIFFERENCES IN SMELL AND TASTE

Age, sex, personality and culture

With senses that are as incompletely understood as smell and taste, it is not surprising that the responses to them tend not to be universal (although some responses are more universal than others, especially gustatory ones). Individual differences exert strong and different effects on each sense and some effects are more remarkable than others. Those that are of greatest importance are age and sex and, to a lesser extent, personality and culture. There are also modality-specific individual differences such as the inability to detect androstenone and its derivatives (a specific anosmia, discussed in Chapter 5) and high reactivity to bitter taste (PROP tasters).

2.1 Age (ing) (olfaction)

As with all the senses, olfaction declines with age (Murphy, 1985; Schiffman, 1997) and it appears to be not a wave goodbye as you pop down the shops but a tearful au revoir as you take flight to another continent. According to the US National Health Institute Survey, chronic loss of olfactory perception is estimated to be 1.42 per cent in the general population (2.7 million), 1.99 per cent in the 55–64 age group, 2.65 per cent in 65–74 age group and 4.6 per cent in the over 75s (Hoffman *et al.*, 1998). Murphy *et al.* (2002) reported impairment estimates of 9.4 per cent in a 53–97-year-old age group but objective data suggest the incidence is twice that (24.5 per cent) (Figures 2.1a and b).

For the elderly, olfactory loss is not seen as a trivial inconvenience manifested in the over-percolation of lavender and excessive use of Estée Lauder. It can have serious implications such as increased risk of danger from failure to detect fire or gas (Chalke and Dewhurst, 1957; Stevens *et al.*, 1987) and of ingesting spoiled food. There is also now considerable, convincing evidence that olfactory impairment is an early risk factor – in one disorder the earliest risk factor

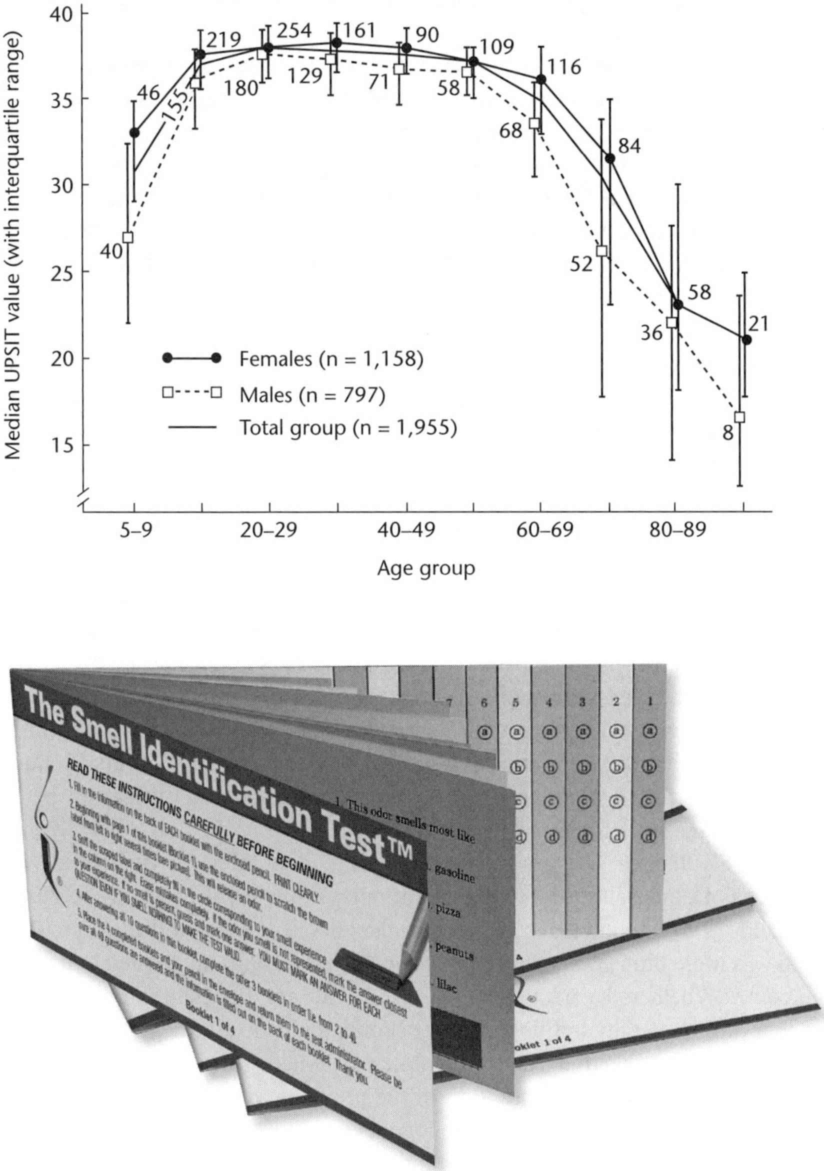

FIGURE 2.1 (a) sex and age differences reported by Doty *et al*. (1984b)
(b) Smell Identification Test

– for degenerative diseases such asPD and AD (this evidence is reviewed in Chapter 5).

There is a large body of historical research into the effect of ageing on olfactory function. This has been complemented by more rigorous contemporary investigations using more robust methodology and very large samples. Thus, one of the most comprehensive studies of its kind (Doty *et al.*, 1984) examined UPSIT performance in 1,955 5–99 year olds and found that performance peaked between the age of 30 and 50 and declined significantly after the age of 70. Half of the 65–80 year olds showed impaired performance; this increased to three-quarters in the over 80s (with smokers performing more poorly than did non-smokers).

The US Beaver Dam Wisconsin Epidemiology of Hearing Loss Study (which, despite its name, also records olfactory function) has examined 2,800 individuals at five- and ten-year stages using the San Diego Odor Identification Test (Schubert *et al.*, 2009, 2011). At the first, five-year follow-up when 2,491 respondents participated, 24 per cent of the sample had impaired olfactory ability (Murphy *et al.*, 2002). This increased to 63 per cent in a subgroup of 80–97 year olds. Men were more severely impaired than women and, as evidence that self-reported ability provides questionable data, only 9 per cent reported olfactory impairment, a percentage that became less accurate with age. Schubert *et al.* (2009, 2011) found some interesting interactions. Age, a history of nasal polyps, a history of a deviated septum and heavy alcohol use were associated with increased risk of olfactory impairment. Exercising once a week was associated with decreased risk. Another substantial (cross-sectional) study of 2,928 individuals aged between 57 and 85 years (Schumm *et al.*, 2009) found a 67 per cent decline in olfactory identification performance on the Sniffin' Sticks test.

Probably the earliest study demonstrating gerontological difficulties with smell was Douglass (1901) who noted atrophy in the nasal tissue of the elderly. Vaschide (1904) found that sensitivity to the odour of camphor declined with age (and was more blunted in men). Studies up until the 1960s all pointed to the same outcome. Whether the odour was faeces, sweat, amyl acetate, almond, coffee, peppermint or tetrahydrothiophene, a chemical added to coal gas to help identify this poison, sensitivity/detection ability declined with age and the older the participant, the greater the loss. Part of the problem with such studies, however, although they were universal in their conclusions, was the variability in the methods used, a lack of adequate controls and the recruitment of varied samples.

Greater experimental control emerged in studies such as Schiffman's (e.g. Schiffman and Pasternak, 1979). A longitudinal study of 150 participants aged 19–95 years followed over three years reported a decline in UPSIT performance with age (Ship *et al.*, 1996). By the time the participants reached their 80s, they had lost one UPSIT point per year. A Swedish version of the National Geographic Smell Survey found a detection and identification decline with age (Larsson *et al.*, 2000) but, interestingly, no sex difference. In a study using Sniffin' Sticks, the more pleasant odours (pineapple, apple, anise, banana, lemon, cinnamon) were less well identified than were unpleasant odours such as fish, garlic and turpentine (Konstantinidis *et al.*, 2006).

Recognition memory for odour is also comparatively poor in the elderly (Larsson, 1997). At around two months following initial exposure, 70 per cent of people can recognize an odour when re-presented (Lawless, 1978). Within two weeks, the elderly are performing at chance; at 6–7 months, the young are performing above chance (Murphy *et al.*, 1991). The result is even less positive when the odours are unfamiliar chemicals. Familiarity is closely allied to successful recognition (Rabin and Cain, 1984). Within 2–3 hours, recognition for unfamiliar odours is poorer in 61–88 year olds compared with 22 year olds (Stevens *et al.*, 1990). Hedner *et al.* (2010) examined the inter-relationships between semantic knowledge, verbal fluency and odour identification ability. They examined olfactory sensitivity, discrimination and identification (using the Sniffin' Sticks), executive function (measured via the digit backward task), verbal fluency (naming as many words beginning with the letter B), episodic memory (recognition of previously seen concrete nouns), general knowledge and the similarities test (what features do a banana and orange share) in 170 30–87 year olds and found that episodic and semantic memory contributed to discrimination and identification. Finally, Wehling *et al.* (2011) administered the Scandinavian Odour Identification Test and a series of cognitive tests (vocabulary, matrices, the Digit Symbol Test, Trail Making, a version of the Stroop test, everyday memory questionnaire and the California Verbal Learning test) to 240 participants aged between 45–79 years. Those who were unaware that they had an olfactory deficit performed more poorly on the verbal learning, memory and attention and processing speed tasks compared to those who thought they had normal function and those who believed they had an impairment.

The response of the elderly is very similar to that of young children. When presented with ten everyday odours and asked to memorize them, younger children, older children and the elderly showed poor identification success (35, 39 and 22 per cent, respectively) compared to 18–30 year olds and 31–57 year olds (60 and 65 per cent, respectively) (Lehrner *et al.*, 1999). After a retention interval of fifteen minutes, correctly named odours showed an effect of age, with the young adults recognizing the odour better. In this study, the poor performance of the younger children is probably attributable to their lack of vocabulary for naming odour. In the elderly, one proposition is that this population also experiences difficulty in accessing odour names, which affects episodic memory (Larsson, 1997). Interestingly, there is some evidence that subjective (non-olfactory) memory complaints are associated with olfactory discrimination and identification impairments, but not threshold problems (Sohrabi *et al.*, 2009).

Although intensity ratings for suprathreshold odours may be lower in the elderly, this appears to have no relevance to retention (Stevens *et al.*, 1990) and some studies note that the age-related impairment in detection threshold, discrimination and identification is not related to intensity (Markovic *et al.*, 2007). In addition, Markovic *et al.* also report that odours (from Sniffin' Sticks) are rated as more pleasant beyond the age of 50, indicating that there is some fragrant lining to a largely mephitic cloud.

The cause of the loss of function in the elderly is not known, although reasonable candidates include a reduction in OB volume or bulbar degeneration due to ageing or disease, reduced neural density, synaptic reduction or hypometabolism in the brain (in addition to cognitive variables such as impaired verbal retrieval which, itself, may be neurally mediated). There may be a cellular or neurophysiological basis to the olfactory loss, given recent research into the precursors of AD and PD. Yousem *et al.* (1999a) examined neural activation during the passive presentation of the odours of eugenol, phenyl ethyl alcohol and hydrogen sulfide in five 73 year olds and five 24 year olds. In both age groups, there was activation in the right inferior frontal regions, and the left and right superior and perisylvian areas. However, activation was greater in the frontal lobe, perisylvian regions and cingulate gyri in the younger sample, especially in the right hemisphere. In a positron emission tomography (PET) study of olfactory performance in five 27 year olds and six 71 year olds, who were matched for olfactory discrimination and identification, Kareken *et al.* (2003) recorded brain activation while participants completed the UPSIT (placed under the nose). They found that olfactory stimulation was associated with bilateral activation in important primary and secondary olfactory cortical areas (described in detail in Chapter 3) while the discrimination element of the UPSIT (which involves the activation of working memory) predictably activated the hippocampus and the identification component, the left inferior frontal region. In the older group, however, there was hypoactivation (a decrease) in the left gyrus rectus and medial orbital gyrus during stimulation; these areas were also activated during discrimination and identification.

This suggests – although the sample in this study was small – that there are areas that either compensate for inactivity in the conventional olfactory regions in the elderly or that are activated because participants are engaging in a different form of cognitive processing. A similar experiment using the UPSIT administered to eleven young (average age of 23 years) and eight old (average age of 66 years) participants was undertaken by Wang *et al.* (2005). While both groups activated the primary olfactory areas (the primary olfactory cortex (POC), the entorhinal cortex, the hippocampus, the OFC, the insula, the parahippocampal gyrus, the thalamus, the hypothalamus and the inferior lateral frontal cortex), activation was lower and less intense in the older group. A reduction in activation in the primary olfactory areas – including the piriform cortex, the amygdala and the OFC – was also seen in Cerf-Ducastel and Murphy's (2003) study. The passive smelling of limonene, methylsalicylate and eugenol by six young and six elderly participants led to activation in the OFC, the hippocampus and thalamus, the anterior insula, inferior postcentral gyrus, inferior and superior temporal gyri and cerebellum in the young. The only significant regions of activation in the elderly were seen in the left inferior temporal gyrus and the primary visual cortex (Suzuki *et al.*, 2001).

The role of possible cognitive differences in explaining olfactory performance was explored in Cerf-Ducastel and Murphy's (2009) fMRI study of ten 66–86 year olds and ten 20–25 year olds. Participants completed an odour recognition task in which sixteen familiar odours were presented to them before scanning and these odours plus foils were presented during scanning in a recognition test. Activation was generally decreased in the older group but there were also regional differences between the groups. In the young, activation was observed in the mesiotemporal lobe, the prefrontal cortex (PFC), the fusiform and parahippocampal gyrus and the lateral and medial parietal cortex during retrieval. In the older sample, there was activation in the left and right claustrum, putamen, the tail of the caudate nucleus, the cingulate cortex, the insula, the left middle and frontal gyrus and the right precentral region. They also made more false alarms and fewer correct rejections. There was also greater activation in the cerebellum which the authors suggest might be related to attention or, as suggested above, compensation. Of interest is Ferdon and Murphy's (2003) study showing that when ten young and ten elderly adults were exposed to amyl acetate and told to press a button when they smelled an odour, activation was lower in the older sample in the superior semilunar lobule (Crus I) and the inferior semilunar lobule (Crus II) but similar activation (but more variable in the elderly) in the posterior quadrangular lobule (Crus VI) of the cerebellum. The authors suggest that this deactivation may be related to the attentional demands of the olfactory task as changes in these regions have been associated with allocating attentional resources to tasks.

One study has examined hedonic preference in old and young participants using fMRI (Reske *et al.*, 2010). Fifteen women aged between 21 and 47 years were exposed to the odours of yeast, vanilla, a neutral/control chemical and odourless air delivered to the right nostril. In general, the yeast was rated as less pleasant, more intense, more arousing and more disgusting and this was reflected in greater OFC and superior temporal cortex activation to this odour. It also activated the anterior cingulate gyrus, the insula and the motor cortex suggesting that the body is preparing itself for withdrawal from a stimulus it finds unpleasant. In the younger women, there were region-specific increases in the dorsolateral OFC and caudate nucleus to this odour, i.e. stronger responses to the aversive odour. While the neuroimaging data are indicative, there is currently not much work available on which a reliable conclusion can be based. Other psychophysiological indices, however, such as the olfactory EP, present a more coherent narrative suggesting a neural sluggishness to odour as we get older. EPs are discussed in Chapter 4.

2.2 Age (ing) (gustation)

In what will seem like a chemoreceptive leitmotif (or cliché, depending on your point of view), the degree of experimental attention directed towards age differences in gustation has fallen behind that of olfaction. The reasons for this include all of the other generic reasons, but one salient and paramount reason may be that the sensory scope of taste is stunted. That is, we can only perceive five (perhaps one or two more) tastes and when people claim hypogeusia what they are actually complaining of is an underlying olfactory deficit (see Chapter 5 for a discussion of this). When tested objectively, people are normally able to distinguish between basic tastes: distinguishing between odours, and detecting and identifying them, is the problem.

Nevertheless, there are a few studies exploring shifts and changes in taste preference and intensity perception across the adult lifespan. Early studies suggested that gustatory dysfunction resulted from anatomical problems such as a decrease in the number of taste buds (Arey *et al.*, 1935), but research since has not been able to confirm the age-related, pathological absence of lingual tissue/ taste buds. Generally, thresholds for all the four major tastes are elevated as humans grow older – there is a reduction in sensitivity at threshold and suprathreshold levels (Corso, 1971; Bartoshuk *et al.*, 1986; Mojet *et al.*, 2001, 2003, 2004). For example, three concentrations of NaCl placed on the tip or anterior tongue were discernible by 20–29 year olds but were detected at chance levels in 70–79 year olds, whatever the concentration or lingual topology (Matsuda and Doty, 1995). Performance improved as concentration increased. The result is consistent with a reported reduction in whole-mouth sensitivity to salt in the elderly (Grzegorczyk *et al.*, 1979; Weiffenbach *et al.*, 1982). The elderly also report that 18µm of citric acid and 180µm of quinine has a weaker taste than do younger tasters (Cowart, 1989).

There is evidence that the elderly prefer stronger tasting food than the young (Murphy and Withee, 1986; deGraaf *et al.*, 1996), that they prefer more salt in vegetable juice that is low in NaCl and more sucrose/citric acid in a lemon drink (Murphy and Withee, 1986), prefer sweeter concentrations of food (Mojet *et al.*, 2004), prefer more sucrose in orange lemonade, strawberry jam and yoghurt, but not chocolate spread or porridge (De Jong *et al.*, 1996), require twice as much salt in a tomato soup to detect the salt (Stevens *et al.*, 1984), detect bitter tastes less well than sweet tastes and sour less than salty (Gilmore and Murphy, 1989; Nordin *et al.*, 2003) and find sour stimuli in water to be less intense (Nordin *et al.*, 2003). That said, there are exceptions – Chauhan (1989) found no age differences for salt preference in chicken soup and one study found a preference for less salt in the same soup (Drewnowski *et al.*, 1986). The method of presentation is an important determinant of taste – if tastes are presented in a food matrix, rather than as single stimulants, the age effects are not as clear (Mojet *et al.*, 2003): when umami is presented alone, the elderly show reduced sensitivity; in a matrix, they do not (Mojet *et al.*, 2003).

The ability to identify tastes remains relatively buoyant after the age of 65 but it continues to be poorer than the ability of adults under 35 years. The former group can identify tastes with 83 per cent accuracy; this is 95 per cent in the under 35s (Deems *et al.*, 1991). Twenty-two year olds perceive citric acid and sucrose placed on the anterior tongue to be five times as strong than do those over 74 years old (Bartoshuk, 1989).

In the olfaction section, Schiffman's experiment with blended food was summarised. Murphy (1985) found that young participants were better able to identify blended food but when nose clips were worn, the superiority disappeared (thereby demonstrating that the superiority was olfactory, not gustatory in nature). However, age differences are infrequently found for identifying different tastes in combination. For example, Mojet *et al.* (2004) dissolved two compounds of each taste at five concentrations in a food solution – sucrose/aspartame (sweet) in iced tea, quinine hydrochloride (bitter) in a chocolate drink, acetic/citric acid (sour) in mayonnaise and NaCl/KCl (salty) in tomato soup, and MSG/inosine 5-monophosphate (umami) in bouillon. Increased concentration affected the perception of other tastes but there were no age-related effects.

Of all the tastes, however, sweet appears to be the one that is least susceptible to changes in its perception with age (Schumm *et al.*, 2009). In contrast, the least recognized taste is sour. In Schumm *et al.*'s experiment, only 39 per cent were able to correctly identify this when a taste strip was placed on the tongue.

Given that the deficits in gustation are not as pronounced as those for olfaction, it is perhaps not surprising that little neuroanatomical or neurophysiological work in humans has sought to explore the cause of the deficit. What is surprising however is the microscopic degree of attention that has been devoted to the endeavour. Only one neuroimaging study to date, for example, has systematically explored gustatory responses in the elderly. Jacobson *et al.* (2010) administered caffeine, citric acid, sucrose and NaCl to a group of young (average = 23 years) and elderly (average = 72 years) participants during periods of hunger and after eating and asked them to rate the pleasantness of each tastant. During hunger, activation was found in the insula in both groups, as was activation in the OFC, hippocampus, amygdala and caudate nucleus. As might be predicted, activation was greater during hunger than after satiety (the effect of eating on fMRI response to taste is considered extensively in Chapter 1). However, activation was much greater in these regions in the elderly group and there was an interesting interaction with type of taste. The age difference was greatest to caffeine (bitter) and smallest to sweet.

Underlying the problems with taste in the elderly is the conventional confusion over the two senses. In a study of 750 patients complaining of chemosensory problems at a North American clinic, two-thirds reported having taste loss and 87 per cent reported anosmia. The actual percentage of patients with a verifiable taste impairment was 4 per cent (Deems *et al.*, 1991).

2.3 Sex (olfaction)

One of the canards of chemoreceptive science is that women have a superior sense of smell to men. However, the canard is not comprehensively correct. It is correct that women outperform men on several aspects of olfactory function and respond differently to the hedonic tone of odour, but the sexual superiority dissipates with other types of olfactory function (and even the superior performance may be explained by other factors).

The earliest known experimental study demonstrating a sex difference was that of Bailey and colleagues (Bailey and Nichols, 1884; Bailey and Powell, 1885) which found that men showed greater olfactory sensitivity than did women. This was followed by the more famous – in chemosensory circles – and widely cited study of Toulouse and Vaschide (1899) which found that women showed greater sensitivity for the scent of camphor. Several other studies followed in the first half of the twentieth century, generally reporting negative results. One exemplar asked participants whether they could smell something and, if so, how strongly (Kloek, 1961). The odours in this study were coffee, nutmeg, cinnamon, steroid sex hormones and cyto-pentadecanolide (Exaltolide) which has a musky odour and which appeared to have some sexual significance (see section 2.4). Studies could not agree on anosmia, with Griffiths and Patterson (1970) finding that 7.6 per cent of women could not detect thibetolide, compared with 44.3 per cent of men, and Whissell-Buechy and Amoore (1973) reporting no sex difference in anosmia.

Then, an interesting series of studies was published showing female superiority in sensitivity to Exaltolide (LeMagnen, 1952; Koelega, 1970), citral (Schneider and Wolf, 1955) and 11-oxahexadecanolide (Koelega, 1970). In LeMagnen's study, women found the odour of Exaltolide to be more intense than did children and men (who were virtually non-responsive). One reason for this was thought to be sex hormones: oestrogen was argued to improve sensitivity whereas androgens lowered sensitivity. Koelega and Koster (1974) studied this more comprehensively in a study of male and female children, adolescents and adult students who smelled four flasks of odorant and were asked which one was different to the other three. The odours included amyl acetate, pyridine, Exaltolide and androstenol. Women were generally better, and the largest sex difference found was for the biological odours. Koelega (1994), however, found sex differences only for some odours (especially amyl acetate) and not for 'biological' odours such as Exaltolide. No difference in odour detection was found in a group of forty-two women and forty-five men who smelled fifty-eight compounds (Punter, 1983).

With the development of validated, standardized and reliable tests of olfactory function (described in Chapter 1), greater consistency emerged. Thus, Wysocki and Gilbert's (1989) National Geographic Smell Survey (which had 1.4 million responses) found female superiority on all measures. Women were found to have superior (lower) detection thresholds, better discrimination ability, better ability to identify and to name, better memory and to show more emotional responses (Hulshoff Pol *et al*., 2000; Choudhury *et al*., 2003). Doty *et al*. (1984b) found that women outperformed men on the UPSIT. Cain (1982) reported significantly better identification ability in women in a task that required participants to identify eighty common odours (such as tuna and crayons). The superior performance even extended to odours such as machine oil and beer. Women were able to identify 92.5 per cent of the odours better. Stevenson and Repacholi (2003) also reported better olfactory naming in women. Using the UPSIT and a large sample (455 men, 742 women), Doty reported that women outperformed men on 90 per cent of the odours administered. The superiority appears to be culture-free (Doty *et al.*, 1985; Barber, 1997) and the sex effect is also seen longitudinally (Ship *et al.*, 1996).

Women also show better odour memory (Lehrner, 1993; Oberg *et al.*, 2002) but they also show better word recall and face recognition generally (Herlitz *et al.*, 1999), thus suggesting that the superiority may be task-general rather than domain-specific. Performance appears to be better for familiar rather than unfamiliar odours, perhaps indicating that olfactory information is processed with elaboration. Larsson *et al.* (2003), however, found a sex difference for unfamiliar odours – recollection was better in women – but no difference for recollection based on familiarity. Even at early ages, females show good olfactory ability. Four-year-old girls have been found to be better at identifying the person they play with, by odour, than are boys (Verron and Gaultier, 1976), a finding replicated in prebubescent girls in relation to their favoured friends (Mallet and Schaal, 1998).

In terms of hedonic response, Kenneth (1927) reported that women preferred the odour of camphor, menthol, citronella and valerian whereas men preferred the odours of cedarwood oil, pine oil and musk. But whether men and women prefer different non-biological odours in a context-independent way is unclear. Seubert *et al.* (2009) found that women gave more positive and less negative hedonic responses to eugenol. In an interesting aside, people tend to associate masculine perfumes with the colours blue and green and feminine ones with the colour pink (Zellner *et al.*, 2008).

Men and women do appear to respond differently to the sex-steroids, androstenone/androstenol, and their derivatives. While sensitivity to this odour is poor at best – it is the clearest example of a specific anosmia – there is evidence that people can be trained to detect it, with repeated exposure (Wysocki *et al.*, 1989). Specific anosmia for the odour in women is thought to range between 2 and 24 per cent, and between 13 and 44 per cent in men (Bremner *et al.*, 2003). Chapter 4 describes some of the neural and behavioural effects of the androstenes.

In his study, LeMagnen (1952) noted that sensitivity to Exaltolide was lowest during menstruation and highest during ovulation. It then declined and plateaued until menstruation. Pregnancy and the menstrual cycle affect detection thresholds, but the effects for the latter are more consistent and better-studied (Hummel *et al.*, 1991). Vierling and Rock (1967) partly replicated LeMagnen's findings in seventy-three student nurses but reported two sensitive periods – 1–17 days before menstruation and during the luteal phase (eight days before). Similar troughs and peaks in sensitivity have been reported by Schneider and Wolf (1955) and Koster (1965, 1968) to the odours citral and m-xylene (a neutral odour). The trough tends to occur fourteen days before menstruation. At this point, sensitivity is highest and declines rapidly thereafter, although the milestones depend on the length of the woman's cycle.

One of the most comprehensive studies of olfaction and the menstrual cycle was that of Doty *et al.* (1981) who recorded olfactory detection performance across seventeen menstrual cycles of women not taking oral contraceptive, and six women taking the contraceptive. Sensitivity peaked mid-cycle, mid-luteal and in the second half of menstruation, regardless of contraceptive status. These olfactory changes across the cycle reflect other preferences for particularly masculine or symmetric faces and particular odours (Gangestad and Thornhill, 1998; Schieb *et al.*, 1999). Dalton *et al.* (2002) found a five-fold increase in sensitivity in women of reproductive age to benzaldehyde, a finding replicated with the lemon/orange scented citralva. Boys and girls' sensitivity did not differ.

The overwhelming evidence in favour of female olfactory detection superiority, whatever its cause, suggests that the ability may be innate and predetermined. Thus, female neonates prefer the odour of the mother to that of a stranger or artificial odour than do boys (Balogh and Porter, 1986; Makin and Porter, 1989), a finding that probably reflects detection ability and learning rather than a hedonic preference. If the ability is innate, it is reasonable to assume that this would be reflected in differences either in olfactory apparatus or its neural/neurophysiological functioning. But while there are well-demonstrated sex differences in volume of particular brain regions and grey matter in general (Allen *et al.*, 2003; Luders *et al.*, 2005), olfactory variation is difficult to determine in humans because the technology is not sensitive enough to detect the small changes that might contribute to this difference. Thus, neuroimaging work with common odours has found no specific regional sex differences in activation (Yousem *et al.*, 1999b; Bengtsson *et al.*, 2001), but has found higher levels of activation in women (Yousem *et al.*, 1999b). Bengtsson *et al.* used PET to compare the responses of twelve women and eleven men in their 20s, to the odours of vanillin, cedar oil, lavender oil, eugenol and butanol presented in glass bottles birhinally (the last four were considered olfactory-trigeminal stimulants). Activation was seen bilaterally in the amygdala, piriform cortex (see Chapter 3) and insula in both sexes, suggesting to the authors that female olfactory superiority is cognitive, rather than perceptual.

Olfactory evoked potential (OEP) (see Chapter 4) amplitude is higher in women (Doty *et al.*, 1985). Interestingly, a study of the external olfactory apparatus itself has found that nasal capacity differs between men and women, with men showing a larger lateral meatus, suggesting that odour is driven away from the OE (Hornung and Leopold, 1999). A comparison of grey matter in olfaction-related brain regions using voxel-based morphometry (VBM) found that concentrations were higher in women in BA10, BA11 and BA25 (the OFC), in the hippocampus bilaterally and in the right amygdala and left basal insula. In men, grey matter was greater in the left entorhinal cortex, dorsal insula and BA25 in the OFC (Garcia-Falgueras *et al.*, 2006).

2.4 'Biologically significant' odours

The animal kingdom is noted for its ability to use semiochemicals to transmit information. 'Pheromones', or chemical signals, may function as sexual attractants, copulation coordinators, alarm signals and territory markers. According to the original progenitors of the term, a pheromone is a specific chemical compound that is secreted by animals and can influence the recipient's physiological and behavioural response when the recipient is of the same species (Karlson and Luscher, 1959). Most research in this domain has revolved around the well-known pig pheromones, androstenone and androstenol – the ketone and alcohol form of the pheromone respectively (Melrose *et al.*, 1971). 5-alpha-en-3-one and its derivatives are delta-16 steroids with a musk-like odour and are secreted in foaming saliva during porcine ruts (Melrose *et al.*, 1971).

Androstenone, specifically, has a urinous odour; androstenol has a sandalwood or musky fragrance. The unusual characteristic of this chemical is that approximately half the population cannot detect it and the half that can tend to describe the odour as pungently and unpleasantly ruinous, 'like a gents' lavatory'. The other half describe it as a 'chemical'-type odour. It is the most well-studied and well-demonstrated specific anosmia (and led to a major theory of olfaction, described in Chapter 1); male anosmia for androstadienone is greater than women's (Hummel *et al.*, 2005b).

McClintock (1971) famously reported menstrual synchrony in a group of 17–22-year-old female residents of a college dormitory, which may have been attributable to semiochemistry. That is, women who shared dorms together experienced synchronized menstrual cycles. Women who spent less than three times a week with male companionship, however, reported longer cycles. McClintock did not advocate that her data indicated a pheromonal effect, only that this was a possibility. Russell *et al.* (1980) applied the sweaty secretions of a woman who had a history of twenty-eight-day cycles and experience of 'driving' (influencing) other women's cycles, on the upper lips of five women, three times a week for four months. The mean difference in cycle onset for the experimental group was 3–9 days before the experiment; 3–4 days during driving. Control (no odour) participants' figures were eight days and 9.2 days, respectively. However, the experiment was not single – or double-blind – and the woman who provided the samples was also one of the experimenters.

Women reported shorter menstrual cycles when compounds from the follicular (late) stage of another woman's cycle were placed on the upper lip; longer cycles were reported when receiving ovulatory compounds (Stern and McClintock, 1998). Chemical substances collected from lactating women increased the 'sexual motivation' of other women (such as sexual desire and fantasy) (Spencer *et al.*, 2004) and women with partners experienced more sexual desire whereas women without partners experienced more sexual fantasies.

Recently, attention has been drawn to more psychologically driven effects of exposure to semiochemistry. Androstadienone, for example, has been found to increase people's perception of pain (Villemure and Bushnell, 2007; an effect also found with non-steroidal odours: Martin, 2006). Men asked to judge the happiness of facial expressions that varied from neutral to happy, were more likely to rate happy faces as less happy when exposed to the sweat of men who had been on a high rope course (an anxiety-provoking exercise) compared with exposure to sweat from men who had been cycling (Zernecke *et al.*, 2011).

2.5 Sex (gustation)

The effects of sex on taste are less pronounced if measurable at all. One of the earliest studies found that men and women showed no differences in their rating of the intensity of sucrose but that men preferred higher concentrations (Enns *et al.*, 1979). Other studies find no sex differences in pleasantness (Leshem *et al.*, 1998).

2.6 Personality (olfaction)

The relationship between personality type/dimension and olfactory threshold and preference is variable at best. One of the earliest – if not the first – quasi-experimental study of its kind was that by Moncrieff (1966). He claimed to find odour preference differences between introverts and extroverts but no statistical testing was undertaken and personality assessment was based on a brief conversation with the experimenter. Subsequent studies have focused on Eysenck's Extraversion-Introversion and Neuroticism personality dimensions, but other personality types and measures have been employed (e.g. shyness, NEO-FFI). Olfactory measures have involved either threshold, detection or preference judgements.

Koelega (1970) was the first to examine the relationship between extraversion, neuroticism and olfactory threshold systematically, using amyl acetate, exaltolide, muscore, butanol, dupical and musk as odours. He found little relationship between personality and threshold, a finding replicated by Filsinger *et al.* (1987), using the Eysenck Personality Inventory (EPI), and Hvastja and Zanuttini (1991). His follow-on study, examining olfactory threshold for butanol, amyl acetate, isovaleric acid, exaltolide and musk (a mixture of pleasant, unpleasant and 'biologically relevant' odours) found an interaction between personality type and sex (Koelega, 1994): neuroticism in women was associated with higher sensitivity for butanol; emotionally stable men had heightened sensitivity for isovaleric acid. This interaction between neuroticism and sex was also seen in a Rovee *et al.* (1973) study in which women high in trait anxiety had higher butanol thresholds than did low-anxiety women. Pause *et al.* (1998) examined olfactory sensitivity to linalool, androstenone and isoamyl acetate in individuals who completed the German Personality Inventory which measures twelve personality dimensions. They found increased sensitivity to odours as neuroticism scores increased. Neuroticism is also implicated in odour naming: one study has show better naming performance in people high in neuroticism and those less open to experience; high impulsivity was associated with low identification ability (Larsson *et al.*, 2000).

Shyness in men has been associated with differences in threshold performance: shy men have lower olfactory thresholds (Herbener *et al.*, 1989). Specifically, shy blue-eyed men had lower thresholds than did sociable brown-eyed men, suggesting a degree of higher sensitivity, attention or vigilance in shy men although the result has not been replicated (effects of neuroticism notwithstanding). Emotional awareness has been associated with better odour detection. For example, female room-mates (twenty-two pairs) who were asked to identify their cohabitee on the basis of their body odour (carried via worn a t-shirt), were better able to do this if they scored highly on a measure of emotional awareness (Zhou and Chen, 2009).

A recent, comprehensive study of personality and threshold sensitivity measures for gustation, olfaction, trigeminal perception and pain, using Sniffin' Sticks as the olfactory measure and the NEO-FFI as the personality measure, found higher sensitivity in those scoring high in agreeableness and higher trigeminal sensitivity in those scoring high in neuroticism, but no other personality effects (Croy *et al.*, 2011).

Two experimental studies shed some additional light on a fairly dimly-lit area. Fiore (1992) asked ninety-two female college students to indicate the type of personality they would associate with three different perfumes – Samsara, Coco and White Shoulders (representing cyphre, oriental and floral). Wearers of the floral perfume were regarded as being less uninhibited and less traditionally male (masculine). They were perceived as less impulsive, aggressive, assertive, dynamic, confident, sophisticated and outgoing but more timid than cyphre and oriental perfume wearers. The latter perfumes were perceived as uninhibited and traditionally male. Interestingly, Fiore notes that the perfume used in Baron's (1983) experiment in which formally dressed and perfumed female job applicants were rated less favourably and as more unapproachable than informally dressed candidates by male interviewers, was floral. She speculates that wearing an oriental odour might have led to even more dire employment consequences.

Chen and Dalton (2005) exposed participants to film clips designed to arouse happiness, sadness, hostility or no emotion and presented pleasant (lemon/orange), unpleasant (methylindole, a faecal odour) or neutral (rubbing alcohol) odour into the environment. Women detected the pleasant odour faster than they did the neutral. Those scoring high in neuroticism and anxiety detected the emotional odours more quickly. Emotional state induced by the films was associated with increased olfactory intensity in men. High anxiety women perceived emotional odours as more intense and high anxiety and neurotic men detected the emotional odours more quickly. Cherinskii *et al.* (2010) also found that introverts showed increases in EEG activation (see Chapter 4) to essential oils, with smaller increases seen in extroverts. What this means is open to debate.

2.7 Personality (gustation)

If the research on personality in olfaction is rare, the research on gustation is virtually blue. An early study of personality in individuals with high or low quinine taste sensitivity found that the more sensitive types were more 'intuitive' (Corlis *et al.*, 1967). A study of taste detection and EPI responses found no relationship between personality and taste (Zverev and Mipando, 2008). Similarly, Croy *et al.* (2011) found no relationship between NEO-FFI scores and gustatory threshold performance.

In terms of taste preference, there is evidence that salty and sweet foods are preferred by those high in neuroticism (Kikuchi and Watanabe, 2000) and that intense sweetness is preferred by those who are more outgoing (Stone and Pangborn, 1990). Elfhag and Erlanson-Albertsson (2006), using the Swedish Universities Scales of Personality to measure personality (and fat and sweet preference) in obesity, found that preference for strong, sweet taste was associated with increased neuroticism, lack of assertiveness and embitterment. Preference for fat taste was related to fewer attempts at restricting and controlling food intake.

2.8 Culture (olfaction)

In general, there are cross-cultural similarities in people's responses to odour: there is a universal correlation between odour familiarity and pleasantness, for example. The more familiar a person thinks an odour is, the more likeable it is judged. A study of Japanese and German participants found that the number of memories evoked by pleasant and unpleasant odours was similar (Schleidt *et al.*, 1988). But different cultures do rate the pleasantness of some odours differently and there is more agreement on what is unpleasant than what is pleasant.

In one study, Haller *et al.* (1999) found that exposure to vanilla in childhood affected a German participant's food preferences later in life (Germans, at one time, received bottled milk flavoured with vanilla). When German participants were asked to rate ketchup or ketchup scented with vanilla, those who had been bottle-fed preferred the vanilla ketchup, compared with those who were breast-fed.

More directly, Doty *et al.* (1985) compared the ability of American Korean, Caucasian, African American and Japanese participants to identify odours on the UPSIT. The Koreans were better at identification than were the Caucasians and African Americans; the last two groups were better than the Japanese, probably because the USA-validated odours were more familiar to the Westerners than to the Japanese. A study of odour detection thresholds (the lowest concentration at which a person can detect an odour) reported lower detection thresholds for Japanese ink and aniseed (Hübener *et al.*, in press).

Cultures also differ in the way they classify odour. For example, when Japanese and Sherpa people were asked to classify twenty artificial scents into perceptually similar categories, there was agreement on most but the Japanese classified some odours as 'fishy': Sherpa are not used to eating fish but the Japanese are famously fish-friendly (Ueno, 1993). Americans and French people are more likely to describe fruit odorants as sweets or flowers and flower odorants as cleaning products than are Vietnamese raters (Chrea *et al.*, 2004).

One comprehensive study asked a sample of Japanese and German participants to rate the pleasantness and 'edibility' of three classes of odours which the authors described as 'European', 'Japanese' and 'International' (Ayabe-Kanamura *et al.*, 1998). Of the European odours, the Japanese sample rated the odours of church incense, anise and almond as less pleasant than did the Germans, but rated the odour of cheese and pinewood as more pleasant. Of the international odours, the

German sample rated perfume to be more pleasant and the odours of beer and peanuts to be less pleasant than did the Japanese.

When asked whether the substance represented by an odour was edible, the Japanese rated the Japanese food odours to be more edible than did the Germans; the Germans found anise and almond to be more edible. The odours of cheese and peanuts were rated as more edible by the Japanese than the Germans.

A coda, however. The sense of smell is notoriously unreliable at naming familiar or unfamiliar odour, a difficulty characterized by the term 'tip-of-the-nose' phenomenon. In the study above, 25 per cent of the Japanese sample thought that India ink represented an edible substance; 40 per cent of Germans thought that Vick's Vaporub did.

2.9 Individual differences in taste: the case of supertasters

In 1931, a peculiar psychological phenomenon was observed from a chemical reaction. Some people who tasted the crystals of a thiourea called phenylthiocarbamide found the chemical bitter, others found it tasteless (Blakeslee, 1931; Fox, 1932). This serendipitous finding hinted at the first evidence for the genetic basis of taste and for a taste quality with a unique function: to warn of dangerous, unpleasant or poisonous ingestants (whether rancid fat, poison, urea, etc., as discussed in Chapter 1).

Around a third of Europeans and North Americans show heritable lack of taste sensation for phenylthiocarbamide and its related compound, 6-n-propylthiouracil (PROP) (Guo and Reed, 2001). There is no other extensive taste polymorphism of this kind, only for bitterness, and there is evidence that the taste receptors for PROP are different from those for other bitter tastes such as quinine or urea. There may be up to eighty receptors in humans (Adler *et al.*, 2000) and the hT2R-4 bitter taste receptor in humans reacts to PROP (Chandrashekar *et al.*, 2000). Current thinking suggests that the contribution is multilocus and multi-allele (Olson *et al.*, 1989) and two genetic loci have been identified in PROP tasters – 5p15 and Chr7 (Guo and Reed, 2001). In non-tasters with a specific ageusia – an inability to taste – for PROP, the responsible gene seems to be TAS2R38 (Mennella *et al.*, 2005).

Bartoshuk (1979, 1993) noted that the PROP thresholds of non-tasters were well in excess of tasters (whose threshold was below 0.1μmol per litre). Bartoshuk argued further, based on threshold data, there were three groups of tasters – these were non-tasters, medium tasters and supertasters (the middle with one dominant allele and the last with two) (Bartoshuk *et al.*, 1994). She found in these supertasters that 3.2μmol per litre was perceived as more bitter and more intense than 1.0μmol/l of NaCl was salty. In the middle group, the two tastes were perceived as equal in taste and quality. Based on this work, it was suggested that 16 per cent of people were non-tasters, 56 per cent were medium tasters and 28 per cent were supertasters (Figure 2.2).

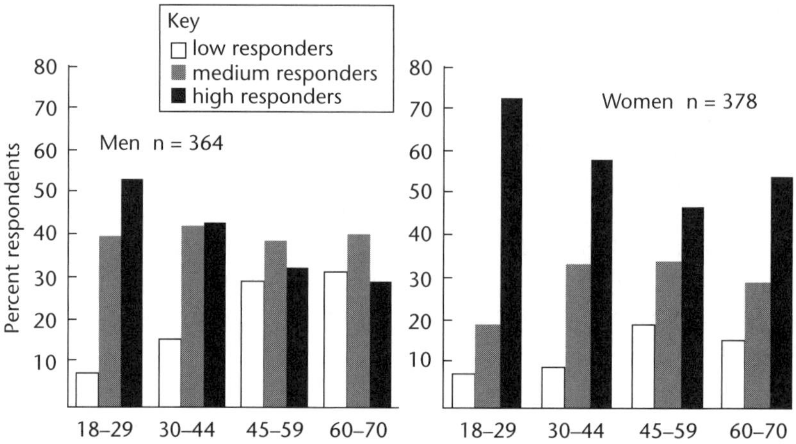

FIGURE 2.2 Some sex and age differences in PROP status

There is evidence for specific receptor types for other tastes, sweet being the most obvious as Chapter 3 will discuss. However, the unique polymorphism for bitter appears to be just that – unique. What is interesting from a psychological perspective are the behavioural consequences of this genetic anomaly, such as food choice. For example, there is growing evidence that non-tasters prefer highly fatty and fat-flavoured foods such as high-fat milk, salad dressing and fatty sweets (Hayes and Duffy, 2008). Coupled with interesting new research demonstrating that the non-taster phenotype is associated with a mutation in the gene that regulates the protein gustin (which helps determine taste bud development) (Padiglia *et al.*, 2010), this behaviour pattern can present a health risk. PROP tasters have also been found to show activation to bitter taste in the posterior left dorsolateral and left and right ventrolateral PFC, when compared to non-tasters (Bembich *et al.*, 2010).

This, recently observed, neurofunctional difference in groups to specific tastes is supplemented by a large number of neuroimaging studies of smell and taste in healthy and unusual populations and these are described and discussed in Chapter 4. The next chapter considers some of the more fundamental aspects of the neuropsychology of smell and taste: the mechanics of the organs of chemosensation, how they work and how the brain supports them.

3

SMELL AND TASTE

Anatomy, development, neuroanatomy and neurophysiology

3.1 Peripheral mechanisms in olfactory processing

Olfactory life begins at the nose sometimes with a sniff, but usually with just a breath. Odour molecules progress through the two nasal cavities called nares (singular = naris), commonly known as the nostrils, and into the two nasal vaults, which are separated by a fleshy bit of cartillage, the nasal septum. The nose warms the air and also acts as a filter, keeping out unprepossessing organisms. It also constricts and dilates asymmetrically which leads to changes in airflow to, and mucal concentration in, the different nostrils (a phenomenon possibly under the control of the hypothalamus which is described in a later section). Sniffing, which increases the rate of airflow and, therefore, the transport of more molecules, also increases the response of the olfactory system (Mozell *et al.*, 1991). The percentage of the air we breathe that reaches the olfactory cleft is estimated to be 10–15 per cent (Hahn *et al.*, 1993).

The surface area of the nasal vault is increased by a process of folding and these projecting folds are called conchae. At the top of these conchae is mucosa, secreted by Bowman's glands and the goblet cells, which is rich with two types of receptor cells – respiratory cells and sensory cells. Sensory neurons belong to a small olfactory region or area olfactoria which contains the OE. This is located approximately 7cm along the nares and occupies around 1.5–2.5cm^2 in humans and 50cm^2 in dogs. There are thought to be around 20–50 million olfactory neurons in humans located here and the same number of unmyelinated axons projecting to the CNS. The remainder of the mucus membrane contains the respiratory cells (and is called the respiratory epithelium). This occupies most of the mucosa. Sensory cells form bundles and traverse bilaterally through a plate of bone called the cribriform plate found in the anterior cranial fossa, reaching the two OBs that lie underneath the frontal lobe, specifically the olfactory sulcus, the region which separates two gyri in the frontal lobe – the gyrus rectus and the medial orbitofrontal gyrus. The olfactory neurons are unique in being the only neurons in the body to have their surface exposed to the external environment. Because of this, they are particularly susceptible to damage and injury (as explained in greater detail in Chapter 5) (Figure 3.1).

3.2 The development of the olfactory apparatus

One of the most remarkable features of the olfactory system is the ability of the sense's sensory neurons to regenerate in the OE, a phenomenon that can be experimentally demonstrated in the laboratory with the sampling of tissue culture which leads to the expression of proteins of sensory neurons and supporting neurons (Graziadei and Monti Graziadei, 1986). Early studies had shown that the mitotic cells of rodents could produce neurons in the dentate gyrus and in the OE of these animals (Altman, 1969). It is now known that new neurons can be generated in the OB and OE, and may move from core areas to the periphery, becoming interneurons and granule and glomerular cells. Doetsch *et al.* (1997) have observed that most granule cells are produced postnatally but interneurons are found during adult development. This regenerative feature of smell is also seen in taste.

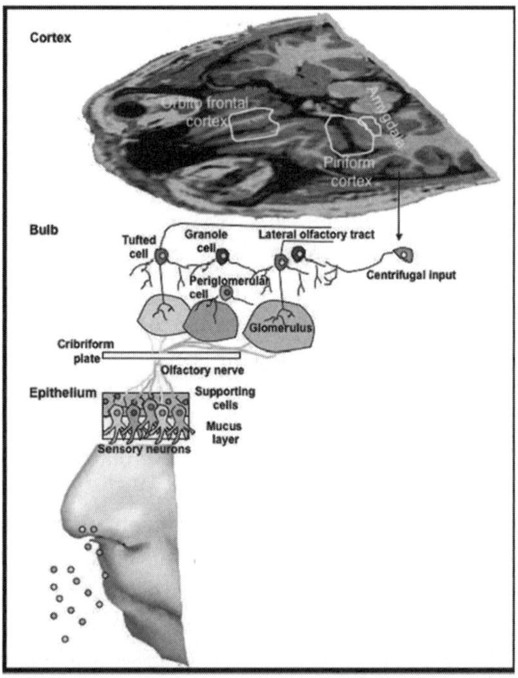

FIGURE 3.1 The pathway of odour molecules from nares to cortex

3.3 The olfactory epithelium (OE)

The OE, also known as the regio olfactoria, is composed of several layers of six types of cells: olfactory receptor neurons, microvillar cells, sustentacular (supporting) cells, globose basal cells, horizontal basal cells and Bowman's glands. The olfactory (and gustatory) basal cells are immature sensory cells. That is, they can develop into olfactory (or gustatory) sensory neurons. Basal cells are the stem cells of chemosensation, involved in neurogenesis of cells postnatally and regenerating receptors after injury. Thus sensory receptor cells regenerate after a time, a process that continues throughout their lifetime. There seems to be a natural homeostasis at work in the olfactory system where injured or damaged cells (or cells that die through apoptosis, programmed cell degeneration) are regenerated thus maintaining a similar number of neurons (Schwob *et al.*, 1992). Microvillar cells regulate the composition of mucus.

The cells most important to olfaction are the sensory receptors that are found on bipolar neurons, with short peripheral processes (dendrites) and a long central process (axons). These are responsible for the detection of odour and the OE is where this process occurs. If the cilia are removed, olfactory recognition is almost impossible (Bronshtein and Minor, 1977). They are the site of sensory transduction in the olfactory system and stimulate the production of adenylyl cyclase (which requires the presence of GTP and thus the involvement of GTP-binding proteins) and cyclic adenosine monophosphate (cAMP). (Increased cAMP leads to depolarisation of the olfactory neurons via direct activation of a cylic, cation-permeable channel; Nakamura and Gold, 1987.) There are around 10–15 million receptors (Moran *et al.*, 1982) and, unlike taste cells, olfactory cells are primary neurons as they send a single axon projection directly to the OB and CNS. The dendritic short process of the sensory cell projects to the surface of the mucosa and gives rise to cilia which form a dense mass and which are about 15–200 micrometres in diameter and 0.2 micrometres thick. Odorant molecules need to penetrate the mucus, be absorbed by it and reach these cilia for olfactory sensation to occur. At the cilia, odour molecules bind to specific recognition sites – the cilia contain seven transmembrane receptor types that interact with incoming molecules, activating G-proteins. This binding is thought to result in the opening of Na^+ and K^+ ion channels thus creating depolarization, which leads to an action potential.

It is thought that each of the olfactory receptor cells expresses only a single odorant receptor gene. Of the 300,000 genes in the mouse genome, 1,000 are olfactory and 350 receptor classes have been identified (Buck and Axel, 1991; Glusman *et al.*, 2001). Buck and Axel, the Nobel prize winners of this work, were able to clone eighteen different members of a multigene family that encodes the seven transmembrane proteins and where expression is limited to the OE. This family encodes individual odorant receptors but Buck and Axel noted that although we can detect around 10,000 odours we do not have that number of olfactory receptors – instead, structurally similar odours may activate the same receptors and subfamilies of receptor may recognize variations in a particular group of odorants (note that Amoore's topology of odour included seven psychophysically distinct classes of odour). They also note that the size of this multigene family suggests that the olfactory system uses more receptors, expressed by a smaller number of neurons, than does the visual system.

The longer process of the olfactory neuron is an unmyelinated axon that synapses with the OB, the next major structure in the olfactory system. When these axons are bundled together and are surrounded by Schwann cells, they form the olfactory nerve (CNI). Before it reaches the OB, the nerve travels through the pitted cribriform plate of the ethmoid bone. Free nerve endings of the ethmoid branch of the trigeminal nerve (described below) also reach the epithelium. The nerve then terminates in the OB in spherical cell masses called the 'glomerular olfactoria' (usually called glomeruli).

These two nerves are, in fact, only two of four systems involved in the transmission of olfactory information. The two others – the accessory olfactory system and the terminal nerve and the former is described below. The accessory olfactory system's role appears to be one of detecting non-olfactory, non-volatile stimuli, at least in non-humans; its role in humans is, even now, unknown (Stockhorst and Pietrowsky, 2004). The terminal nerve (or nervus terminalis, vomeronasal organ or Jacobson's organ) is thought to have a similar, reproductive role but its location in humans has been debated (as has, when it is found, its function).

Central to an understanding of the psychology of olfaction is how this system allows us to distinguish between odours (and produce behavioural reactions to others). The odour of banana has 350 components whereas coffee has 800 but despite the phenomenal complexity of odorants we recognize one as banana and the other as coffee (Stockhorst and Pietrowsky, 2004). There are around 400,000 odours and odour compounds that we can probably detect. We can tell the difference between them but does the chemical composition of these odours lead to this ability to discriminate? Evidence suggests that olfactory receptor neurons might integrate olfactory information – these neurons in rats respond more strongly to binary than unitary odours, for example (Su *et al.*, 2011) and it appears that the number of these neurons, rather than their size, is what is crucial for olfactory sensation (Escada *et al.*, 2009). Current thinking in molecular biochemistry argues that there is a large number of olfactory receptor types and that the features of odour molecules are recognized by families of receptor proteins (G-protein receptors). If olfactory receptors share similar receptor proteins they project to the same area of the OB and form 'epitope maps' (Hudson, 1999). This spatial interpretation of olfactory information transmission is discussed below.

3.4 The olfactory bulb

The OBs are part of the telencephalon, are symmetrical and receive projections from their respective nasal cavities. They provide the first level of analysis in the olfactory system. Evidence suggests that their projections are ipsilateral, as is the projection from the bulb to the brain. The very first synapse in the olfactory system occurs here and this fact – the parsimony of the synaptic connection in a major sense – is unique in the sensory system. It occurs between axons of sensory receptors and second order dendrites in a specific layer of the bulb and with specific cells. The vertebrate OB contains seven layers – the outer layer of this (the olfactory nerve layer) contains olfactory receptor neurons and glial cells. Axons from the neurons project to the deepest layers of the OB (the glomerular layer; see below). The layer with the largest cell bodies is the granular layer and most inputs synapse here or in the glomerular layer. The deepest layer is the subependymal layer.

The bulbs are small and delicate, more so in humans than other animals. Because we are microsmatic creatures, as are all primates, and emphasize the use of the visual sense rather than olfaction (Zhang and Webb, 2003), the bulbs (and the cavity of the OE) are smaller than they are in other mammals (Smith *et al.*, 2004). The first magnetic resonance imaging (MRI) investigation of the OB in healthy individuals was able to identify these structures in seventy out of eighty participants (Susuki *et al.*, 1989). Coronal MRI slices can highlight the bulbs, despite their size – they are 6–14mm long and 3.7mm wide (Castillo and Mukherji, 1996). Sagittal slices make their observation difficult.

Approximately 25,000 receptors synapse on collections of cells called glomeruli (singular: glomerulus) in the OBs, of which there may be ninety or so in the human CNS; the actual figure is unknown (Maresh *et al.*, 2008). Synapses occur with the interneurons that surround each bulb (these represent the juxtaglomerular cells; the largest of them are called periglomerular) (Wilson and Mainen, 2006). The principal cells of the glomeruli are the mitral and tufted cells, both major output neurons, and these project to a region that becomes the lateral olfactory tract (LOT). There are approximately 60,000 LOT fibres in rabbits, 250,000 in cats and 48,000 in rats (Brunjes *et al.*, 2005). Olfactory neurons make excitatory connections with the dendrites of periglomerular cells that are inhibitory – producing the neurotransmitter GABA. Other neurotransmitters involved include glutamate (excitatory) and dopamine. Glutamate is released in the bulb and other regions of the secondary olfactory cortex (SOC) but we have little understanding of the role of neurotransmisson in olfaction. The LOT projects to the cortex, via the olfactory peduncle. It is the projections of these mitral cells to the cortex that allows perception to occur. This has been called the 'amplification step' of olfactory processing because around 5,000 olfactory neuron axons converge on one glomerulus in rats (Wilson and Mainen, 2006).

Early olfactory processing involves the transduction of odour molecules by large families of olfactory receptors, transduction that is then converted into odour maps or images in the OB (specifically, in the glomerular layer) which synaptically extracts features of odour molecules (Mori and Yoshihara, 1995). This, according to Shepherd (2005), 'puts our understanding of early olfaction on par with early vision' and other senses. Odorants from the OE stimulate patches of the glomerulus and the glomeruli are thought to be the source of the spatial code in the olfactory system whereby mitral cell activity is distributed spatially (Mombaerts *et al.*, 1996). Thus particular groups of neurons activate areas of different glomeruli – ketones activate similar areas of the glomerulus whereas acid alcohols activate different patches (Leon and Johnson, 2003).

Current undertanding of this process is that structurally similar molecules activate the same or similar areas of the glomerulus and this activity spreads across the OB, hence the spatial nature of the activation (Johnson *et al.*, 2004, 2005). According to Brunjes *et al.* (2005), the glomeruli 'parse' the information they receive into information streams, 'setting up a map of odor space in the bulb'. This would suggest that activity in the OB is 'chemotypically' organized and provides a spatial map representing the identity of an odour or odour type (Leon and Johnson, 2003; Schaefer and Margrie, 2007). Mitral cells that are near to each other project to similar areas of the glomerulus and appear to act as 'molecular feature detectors' (Mackay-Sim and Royet, 2006): different groups of mitral cells are activated by different functional or structural groups of odours. For example, work with rats has found that optical isomers (enantiomers) that are mirror images of each other, activate different spatial areas of the glomeruli; those that are undistinguishable from each other (l and d-limonene) activate similar areas and are not discriminated by rats (Mackay-Sim and Royet, 2006). Mackay-Sim and Royet conclude that this represents a process very similar to that which occurs in vision, at least, at its end point. There is a topographical representation of an odour that is converted into a chemotopic map by the mitral cells that extract features from the odorant. These features are then assembled later along the cortical pathway. This can be compared with vision where features such as orientation, movement, colour and so on are extracted early on in the visual system and are then reassembled later along this system to provide us with our perception (of a face or a building).

Buck and Axel (1991), in their pioneering paper, put forward two hypotheses for how the odorant receptors might act: either sensory neurons expressing a specific receptor are expressed in one specific part of the OE or that they are distributed randomly but project to specific areas of the OB. The evidence suggests the latter (Mombaerts *et al.*, 1996) and carbon monoxide and stereochemicals appear to excite activation in one glomerulus (Suh *et al.*, 2004; Kurtovic *et al.*, 2007). However, it is currently unclear how this spatial distribution of olfactory information reflects decisions about odour properties (Mainen, 2006; Haddad *et al.*, 2010). In a principal components analysis of twelve studies that examined neural responses to several odorants, Haddad *et al.* (2010) found that two odorant factors/quality accounted for more than half of the variance – approach and withdrawal characteristics in non-human animals and pleasantness of odour in humans, plus odorant toxicity. Haddad *et al.* suggest that these two types of response may occur concurrently and reflect parallel types of olfactory processing. They also caution that the first factor may reflect intensity (rather than explicitly pleasantness) judgements. There is also an argument that as odours are coded by specific groups of olfactory receptors (but there are too many of them), then the system processes olfactory information not just spatially but temporally also (Haddad *et al.*, 2010). Thus spatial patterns may be driven by or co-occur with different temporal disribution of activation. Getchell *et al.* (1984), for example, proposed that odours can differ in the time taken for a receptor cell to become activated. 'Fast' odours could thus be recognized faster, because they reach the OB more quickly. Laing *et al.* (1994) investigated whether which of two odorants or odours in mixtures separated by 50msecs would be detected first. They concluded that the intensity of the odour predicted faster identification.

3.5 Primary olfactory cortex

The next stage of processing occurs after the bulbs project to the cortex, directly to what is known as the 'primary olfactory cortex' (POC). The term POC provides perhaps an overly simplistic and misleadingly unitary description as it comprises various, disparate cortical regions. These are the anterior olfactory nucleus (AON) or region, the prepiriform cortex, the lateral entorhinal cortex, the ventral tenia tecta (this is ribbon-like and covers the corpus callosum), the nucleus of the LOT, the olfactory tubercle and the cortical nucleus of the amygdala. Various researchers will include variants of these regions – e.g. piriform cortex, entorhinal cortex – but this is a comprehensive list of the constituents of the POC. The entorhinal cortex becomes the parasubiculum and subiculum (Mackay-Sim and Royet, 2006) and is found in the ventromedial surface of the temporal lobe (Brodmann's area 28). It merges with the CA1 area of the hippocampus and has twenty-three different fields (Insausti *et al.*, 1995). According to one study, approximately 13 per cent of the total area of the entorhinal cortex is supplied with olfactory input in monkeys; around half receives input in rats (Amaral *et al.*, 1987). Collectively, the POC receives the greatest number of inputs from the OB, hence their 'primary' sensory status.

Some have suggested that the piriform cortex is the true POC as this is the region receiving the greatest number of afferents from the OB (and it receives projections from the cortex and subcortex) (Brunjes *et al.*, 2005). Given these connections, it could realistically be considered an association cortex. It is larger in monkeys than it is in humans and occupies a central position in the nominal POC (Mackay-Sim and Royet, 2006). In monkeys, it is located on the posterior orbital surface, caudal to the anterior olfactory region, lateral to the olfactory tract and continues to the temporal cortex. In humans, it is located in two regions. In both, it has reciprocal connections with the OBs. There are also projections sent to the OB by the AON/region, the entorhinal cortex, amygdala, LOT, raphe nuclei, locus coeruleus and hypothalamus (Mackay-Sim and Royet, 2006). The region is sometimes divided into anterior and posterior sectors, comprising three layers – the superficial plexiform layer I, layer II (densely packed with pyramidal cells) and layer III (containing cell bodies and fibres). The first layer is well organized and its superficial layer 1a receives afferents from the OB and its 1b layer receives afferents from the PFC, entorhinal cortex and AON/region (Brunjes *et al.*, 2005). The third layer receives input from the amygdala, OFC and other parts of the PFC (Figures 3.2a, b and 3.3).

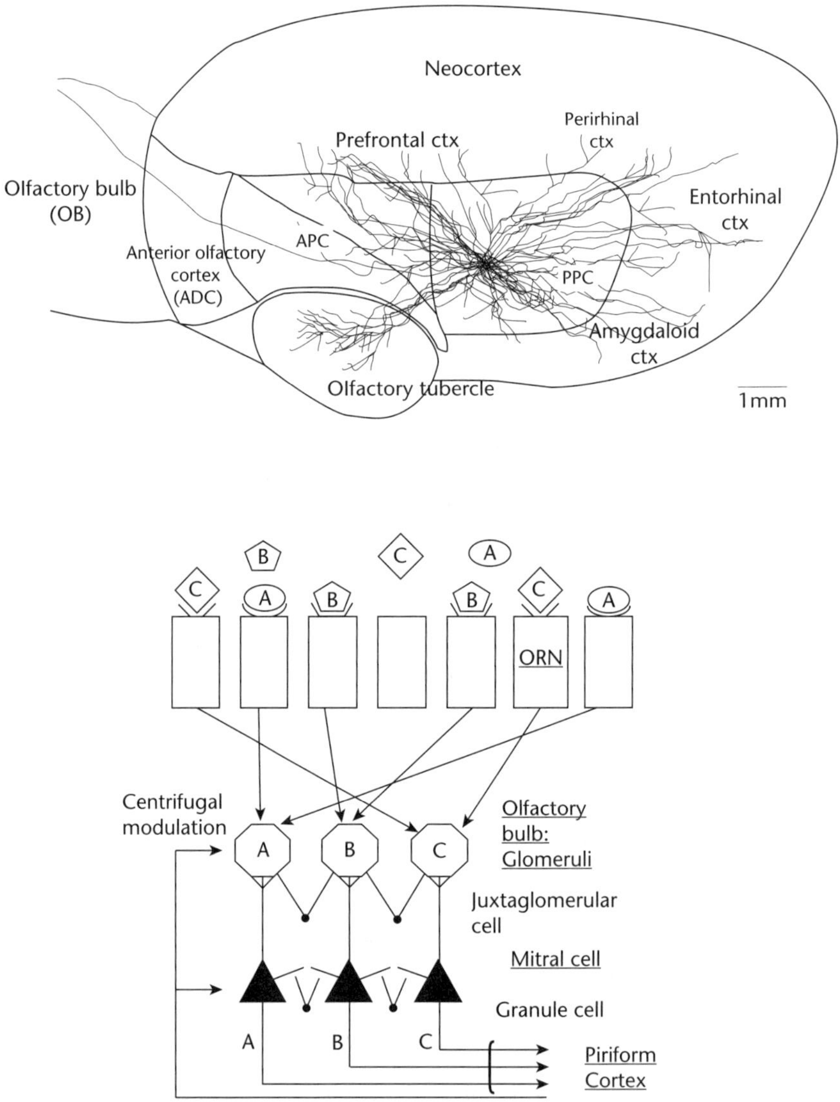

FIGURE 3.2 (a) and (b) Two schematic representations of the molecular mechanism of olfactory perception and the link between the olfactory receptors and piriform cortex; (a) shows layer 11 pyramidal cells in the rat prefrontal cortex

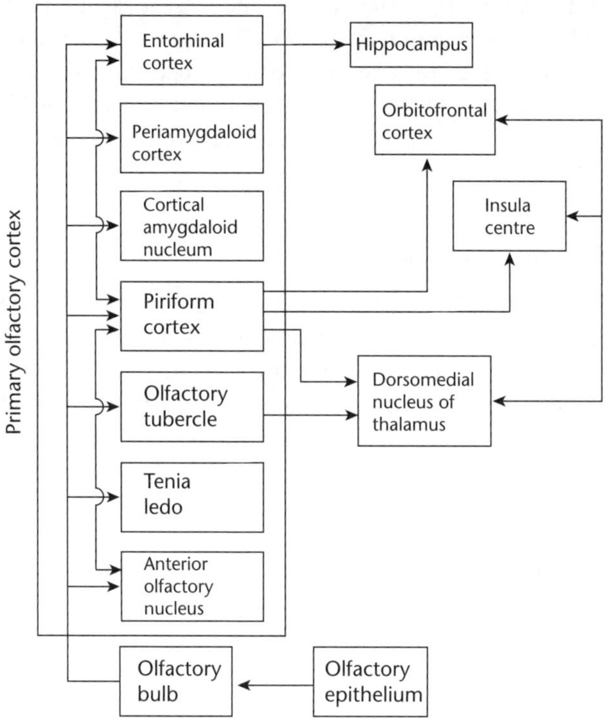

FIGURE 3.3 Schematic representation of the relationship between the peripheral and central olfactory structures and the sense's efferent connections

Advanced tracer methods have enabled a clearer delineation of the pathway from the OB to the POC. Thus, Sosulski *et al.* (2011) developed a tracer in mice that revealed a diffuse projection pathway to the piriform cortex but patches of projections to the amygdala from individual glomeruli that overlap with those sent to the POC. They suggest that as the projections to the amygdala are more specific and less diffuse this structure may generate the 'innate' response to odour given that the projections to the piriform cortex are random. Zou *et al.* (2005) examined how olfactory signals were transformed between bulb and cortex and how neuronal activation occurs in the anterior piriform cortex by exposing mice to various odorants. They found that each odorant was associated with activation in a small subset of neurons that were highly distributed. The greater the concentration, the greater the spatial distribution of activation and neuronal activation.

3.6 Anterior olfactory nucleus/cortex

The site that interconnects the OB and POC is the AON or cortex, which also connects homologous areas of the hemispheres. Its role is thought to be a messenger between the bulbs and the POC and between the left and right OBs via a small band of nerve fibres, the anterior commissure (Brujes *et al.*, 2005) but its precise role is unclear. It is currently regarded as a region of cortex, however, rather than a nucleus (it was first described as such in 1910 and as cortex in 1926) and it merges with the third layer of the piriform cortex. It projects mainly to the rostral PFC and area 1b.

3.7 Secondary olfactory cortex

The POC projects diffusely to what is loosely described as the SOC, which includes a combination of subcortical structures and cortical regions. The SOC comprises, at the subcortical level, the amygdala, the dorsomedial nucleus of the thalamus, ventral putamen, the hypothalamus, medial thalamus, the nucleus basalis of Meynert, the hippocampus, the septal region, the insula, the reticular activating system and, at the cortical level, the OFC. Nine of the twenty-two areas in the OFC receive projections from the POC without synapsing in the thalamus (Mackay-Sim and Royet, 2006).

The hippocampus is an especially interesting structure because of its role in memory consolidation and there are mutual interactions between various sensory cortices and the hippocampus (Rolls, 2008). Only two synapses are made before OB projections reach the dentate gyrus of the hippocampus (Vanderwolf, 1992). The hippocampus has reciprocal connections via the entorhinal cortex and directly from CA1 to the granular layer (Gulyas *et al.*, 1998). The frequency of neuronal oscillation, especially oscillations in the theta frequency, in the hippocampus may also be related to olfactory discrimination (Kay and Sherman, 2006) and sniffing (Macrides *et al.*, 1982). Increased beta frequency has been observed in the OB and PFC when rodents performed an odour discrimination task (Martin *et al.*, 2004). Increased beta in the OB and dorsal and ventral hippocampus has also been reported during the acquisition of odour discrimination in a go/no go task in rats (Martin *et al.*, 2007). With each new odour set discriminated, coherence between frequencies in the OB and hippocampus was reset.

There is some evidence to connect cortical thickness with olfactory ability (Frasnelli *et al.*, 2010). Grey matter thickness correlated with Sniffin' Sticks performance in forty-six individuals, especially in the right medial OFC and insula. Correlations were also found between performance and the thickness of the central sulcus bilaterally, and some sex differences were noted in the right regions.

3.8 The thalamus

The thalamus is of particular interest to neuroscientists, olfactionists and any other person interested in sensory processing in the brain because, unlike every other sensory system, olfaction does not appear to use this structure as a sensory relay station. The olfactory pathways have been thought to pass through the thalamus with synapses being made directly with the olfactory cortex from the OB, with perhaps a few fibres passing through the mediodorsal thalamic nucleus (Ongur and Price, 2000). Compare this with vision which projects retinal information to the lateral geniculate nucleus (LGN) of the thalamus and then to the cortex. Lesions to the thalamus do not produce anosmia nor do they impair olfactory discrimination (Price *et al.*, 1991b). Thalamic involvement in sensation reflects an 'anatomical bottleneck' (Kay and Sherman, 2006) – around a million relay cells are to be found in the LGN of the macaque monkey whereas around 150 million more than that exist in the primary visual cortex where higher order processing occurs. If the thalamus is not involved in olfactory sensation then how, to paraphrase Shepherd, does olfaction do it?

A study of single cell recording by Murakami *et al.* (2005) might suggest one explanation. They found that when rats were exposed to odour, the olfactory cortex showed large spikes and discharges during periods when cortical EEG was in a fast-wave state rather than a slow-wave state. This suggests that electrical flow through the olfactory apparatus was gated. This spiking was found in the anterior piriform cortex and olfactory tubercle but not the OB. When Murakami *et al.* measured activity intracellularly, the membrane potential alternated between depolarization and hyperpolarization during the slow-wave state; in the fast-wave state there was a change from hyperpolarization to depolarization. Thus, oscillations in the cells were synchronized with oscillations in the cortical EEG.

That said, recent studies from clinical neuropsychology suggest that the thalamus may not be as redundant as neurophysiology suggests. Animal work, for example, has shown that odour detection (but not discrimination) is impaired by thalamic lesions, as is complex learning (Eichenbaum *et al.*, 1980; Kawagoe *et al.*, 2007). A very early stimulation study found that the dorsomedial nucleus of the thalamus receives projections from the OB and sends projections to the olfactory cortex (Jackson and Benjamin, 1974). Human patient studies are rare and not particularly well controlled. A study of a patient with bilateral dorsomedial thalamic lesions found odour abnormalities (Rousseaux *et al.*, 1996); Asai *et al.* (2008) also claimed to have observed temporary odour (and taste) disturbances in a patient with left dorsomedial and right ventral posterior and ventral lateral nuclei lesions caused by ischaemia. The patient experienced nausea when smelling a food. Taste and smell disturbances disappeared two weeks following treatment but, as the authors acknowledge, no objective olfactory testing took place and the observations were anecdotal.

A better-controlled study of olfactory performance and localization of olfactory perception in the mouth, found that the four patients with medial dorsal thalamic nuclei lesions studied showed impaired attention to odour (Tham *et al.*, 2011). Sela *et al.* (2009) studied seventeen patients with unilateral lesions to the thalamus and found that detection was unimpaired but identification was significantly impaired. Right-sided lesions (only) were associated with reductions in the pleasantness ratings for pleasant odours, perhaps indicating a role for the structure in hedonic perception; an analogous task using auditory stimuli produced no such asymmetries. The objective olfactory impairments were mirrored in sniffing behaviour – in healthy controls, sniffing was controlled and appropriate to the odour sniffed; in the patients, this modulation was absent.

However, in a bold proposition, Kay and Sherman (2006) explain away the redundancy of the thalamus by suggesting instead a model in which the OB undertakes the work that the thalamus does in other senses. For example, they suggest that the thalamus' role in receiving inputs from the cortex via layer 6 and the brain stem, which affects the flow of information throughout a system in a way that depends on the attention directed to the stimulus, may be one shared by the OB. It would be olfaction's 'anatomical bottleneck', providing the final control over information before it is distributed to the rest of the cortex. Like the thalamus, they suggest, this bottleneck final stage is under the control of feedback received from the cortex and other subcortical structures. The thalamus shows a degree of specificity in the types of processing it undertakes (the LGN and the ventral posterolateral nucleus are innervated by different cells); it and the OB might process sensory signals that share a common feature. These signals are attenuated/amplified by sensory axons and/or other mechanisms in the glomerulus. A second stage of processing would be inhibitory, involving neurons that receive no direct sensory input (granule cells and reticular nucleus neurons in the thalamus). The neurotransmitter system in both is likely to be GABAergic. Kay and Sherman (2006) argue that there is no evolutionary reason why the two structures should not undertake the same task (in the sense that one structure performs a function conventionally undertaken by another); one did not evolve before the other. They suggest that they might have developed independently to solve a similar problem: sensory processing.

3.9 Lateralization (external) in olfaction

The issue of whether the left or right nostril is the superior naris for olfactory perception has been the subject of a number of investigations since the 1900s. This has important implications for the understanding of the neuroanatomy of olfaction because it is thought that the olfactory pathway is ipsilateral (Zatorre and Jones-Gotman, 1991). Therefore, if one nostril is superior to the other then this indicates a degree of hemispheric dominance for olfactory processing. Toulouse and Vaschide (1900), for example found a left nostril advantage for detecting camphor in 87 per cent of their sample and a right nostril advantage for the detection of ammonia in women – thresholds were lower for this nostril. The picture, however, is mixed. Studies indicate greater sensitivity for the right nostril (Youngentob *et al.*, 1982; Cain and Gent, 1991) or show no difference (Schneider and Wolf, 1955; Koelega, 1979; Betchen and Doty, 1988; Zatorre and Jones-Gotman, 1991; Bromley and Doty, 1995; Shimomura and Motokizawa, 1995; Brand *et al.*, 1999).

Findings of right nostril asymmetry include greater intensity sensitivity in right-handed women (Pendse, 1987), an advantage for odour discrimination but no threshold difference for phenyl ethyl alcohol (PEA) (Zatorre and Jones-Gotman,

1990), better discrimination (Cain and Gent, 1991) but poor naming (in patients with callosotomy) (Gordon and Sperry, 1969), better discrimination for unfamiliar odours, with symmetrical performance for familiar odours (Savic and Berglund, 2000), and a perception of neutral odours as more pleasant (Herz *et al.*, 1999), although disgusting odours are perceived as disgusting in both nostrils (Gordon and Sperry, 1969). The left nostril has been reported to be superior for odour naming (Gordon, 1974; Herz *et al.*, 1999). Gordon and Sperry's (1969) study is famous for being one of the first systematic demonstrations of ipsilaterality because they were able to study olfaction in a singular group of patients – those whose corpus callosum had been completely or partly sectioned. All five of their participants could name an odour presented to the left; none could when presented to the right. The reasoning behind this finding was that the active left hemisphere could receive information from the left nostril and thus demonstrated that the pathway was ipsilateral; the right nostril odours, on the other hand, could not be named (if the pathways were crossed, they would have been). Gordon (1974) later argued that the pathway between the right nostril receptors and the right hemisphere were inefficient. Gordon and Sperry also found that when disgusting odours were presented to both nostrils, stimulation was perceived as disgusting but only unpleasant odours presented to the left nostril could be named thus indicating that the pathways allowing the labelling and our cognitive construction of odour were different to that which allowed the hedonic response to odour. Herz *et al.* (1999) also noted a left nostril advantage for naming odours.

Bromley and Doty (1995) examined odour memory in men and women who were asked to remember, in a recognition paradigm, odours that had been presented to the left or right nostril or both, on two occassions. While no nostril difference emerged, men were better at remembering when the presentation was birhinal and they were better at the second testing session than the first. Millot and Brand (2000) extended this study further by videotaping male and female right-handers as they smelled various odours. The aim was to observe whether participants would show a bias to one nostril or the other when sniffing. They found that men used their right nostril more than they did their left, whatever the odour. On a broader level – whether the hemispheres are preferentially involved in olfactory perception – studies tend to indicate priming of the right hemisphere in healthy individuals (Olsson and Fridén, 2001). Zucco and Tressoldi (1988), for example, found that participants who smelled odours and then had to indicate whether a picture or a word flashed to the left or right visual field responded more quickly when the stimuli were presented to the right hemiphere. (In visual field studies, visual/verbal material delivered in the left visual field is processed more quickly in the right hemisphere and material delivered in the right visual field is processed more quickly in the left hemisphere.)

A complicating factor in nasal asymmetry, however, is handedness. Handedness is a curious psychological variable because of its relationship with the language areas of the brain. Thus, almost all right-handers, and most left-handers, will express left-hemisphere speech, but left-handers are significantly more likely than their dextral participants to show right-hemisphere or bilateral hemispheric dominance for speech. Youngentob *et al.* (1982) found that right-handers showed greater sensitivity in the right nostril and left-handers showed greater sensitivity in the left. This appears to have been an aberration, however, as most studies suggest a left nostril advantage for right-handers (Toulouse and Vaschide, 1900). Hummel *et al.* (1998) found that n-butanol threshold was not affected by handedness but that left-handers were better at discriminating between odours presented to the left nostril whereas the reverse obtained for right-handers. Gilbert *et al.* (1999), on the other hand, found greater left–right asymmetry in right-handers when participants made judgements about odours. Finally, Mohr *et al.* (2001) examined the relationship between nostril asymmetry and magical ideation – hallucination-like experiences and a belief that events can be caused by paranormal agents. They found higher thresholds in both nostrils in those scoring highly for magical ideation. In men, the higher the magical belief, the poorer the detection in the left nostril (but not discrimination).

3.10 Lateralization (cortical) in olfaction

If olfactory processing is ipsilateral, it follows that damage to olfactory areas in the brain should result in impaired olfactory ability in the ipsilateral nostril and preserved ability in the contralateral nostril. The evidence is mixed and early studies of brain injury were more concerned with investigating the effect of the hemisphere damaged on olfactory ability in general, rather than ability in each nostril. These studies, however, almost consistently show olfactory impairment following right (usually temporal) lobe lesions, although detection is relatively preserved (Eichenbaum *et al.*, 1983; Eskenazi *et al.*, 1983, 1986). Thus, Abraham and Mathai (1983) reported an impairment in a complex odour-matching task in patients with right, but not left, temporal lobectomy; Rausch *et al.* (1977) found impaired delayed odour matching following the same procedure; Jones-Gotman and Zatorre (1988b) found greater odour matching and recognition impairment in patients with right temporal lobe lesions (but the number of those with unilateral lesions was not stated); and Carroll *et al.* (1993) found an impairment in the retention of nameable odours in patients with right temporal lobe epilepsy (see Chapter 5). Carroll *et al.* concluded that 'structures within the right temporal lobe play a specific role in the short-term retention of odours'. Interestingly, their choice of odours included some that were not particularly identifiable – herbs and spices. 'The majority of subjects', they reasoned, 'were young men who might not be expected to cook with herbs and spices'. Their deflating assessment of the men's culinary ability aside, the right-sided group were particularly impaired at retaining these odours. Bilateral medial temporal lobe damage was found to be associated with an inability to distinguish between smells (Eichenbaum *et al.*, 1983).

In terms of ipsilateral and contralateral effects, Eskenazi *et al.* (1986) found that unilateral temporal lobe lesions produced identification deficits in the ipsilateral nostril (but no impairment in detection) while Zatorre and Jones-Gotman (1991) found discrimination, but not detection, deficits in the ipsilateral nostril in patients with temporal lobe excisions. Zatorre and Jones-Gotman also found an interesting series of interactions between the side of the frontal lobe injured and performance. Right frontal lobe injury was associated with discrimination impairment in the right nostril, but left frontal injury produced deficits in the contralateral nostril. Patients with right frontotemporal lesions produced the worst scores of any group, with both nostrils impaired. Also, right frontal lesions that spared the OFC produced discrimination deficits in the ipsilateral nostril but lesions not sparing the OFC produced impairment in both nostrils. (It is worth noting that Zatorre *et al.*'s (1992) PET study, described in Chapter 4, found OFC activation to scent only on the right side but found bilateral activation in the piriform area.) Tanabe *et al.* (1975a, b) found that bilateral lesions to the lateral posterior OFC produced odour discrimination deficits, but lesions to adjacent areas did not. This finding was confirmed by Potter and Butters (1980) who reported severe odour quality discrimination impairment following frontal lobe lesions, but no effect on detection. The task was performed in the nostril ipsilateral to the lesion although no left–right comparisons were undertaken. Bilateral OFC damage has also been associated with impaired identification ability (Jones-Gotman and Zatorre, 1988a), but not with detection impairment. Identification deficits have also been reported in patients with temporal lobe resection (Eskenazi *et al.*, 1983), but not in patients with temporal lobe epilepsy (Rausch *et al.*, 1977).

3.11 Airflow and nasal patency

Perhaps the single most important factor in nostril asymmetry is one that is easy to overlook. It used to be assumed that both nostrils behaved similarly – the only difference was that one sent signals to one hemisphere, the other, to the other. In terms of functional integrity, they were considered to be equivalent so that any impairment in one nostril was explained by its associations with the ipsilateral hemisphere. But this is not true. And it is not true because of a phenomenon called nasal patency.

The nasal cycle is a curious phenomenon – it describes the process whereby the velocity of airflow alternates between nostrils throughout the day (Keuning, 1968). The 'on' period for one nostril's increased velocity varies between 1–8 hours (periodicity varies between 50 minutes to four hours in rabbits and rats; Bojsen-Møller and Fahrenkrug, 1971) and there are thought to be about two to three changes in airflow velocity during the day (Hasegawa and Kern, 1977; Soubeyrand, 1964; Mirza *et al.*, 1997). This is similar to, and consistent with, Kleitman's (1963) notion of the Basic Rest Activity Cycle which describes circadian shifts in behaviour and physiology (such as the REM/non-REM sleep periods where one alternates with the other in ninety-minute cycles). The cause of patency is thought to be differential input from the sympathetic and parasympathetic nervous systems. Thus, when sympathetic innervation of the nasal mucosa to one nostril is dominant, there is vasoconstriction and decongestion (a swelling in the nasal cavity); when parasympathetic innervation is enhanced in the other nostril, this causes congestion. A very simple way of demonstrating patency is to breathe out of your nostrils, one at a time, on a mirror or stainless steel surface. The area of condensation caused by the congested nostril should be smaller than the one with freer airflow. Klein *et al.* (1986), using this very technique, observed that verbal performance was better in participants who showed right nostril dominance although forced nostril breathing through the right nostril has not been associated with better verbal analogy performance (Jella and Shannahoff-Khalsa, 1993), indicating that the free inhalation of air is important rather than forced inhalation. In this study, however, forced left nostril breathing was associated with increased mental rotation performance. Forced left nostril breathing has also been associated with a more negative emotional state, higher anxiety scores and the telling of more negative stories generated by ambiguous pictures (Schiff and Rump, 1995).

At a cortical level, some theorists argue that this airflow asymmetry is related to hemispheric dominance. Werntz *et al.* (1983, 1987), for example, argued that the nostril that is dominant depends on which hemisphere is dominant (as expressed via EEG). There are two to three shifts in hemispheric dominance across the waking cycle and these are reflected in nasal airflow asymmetry. The relationship between hemispheric and nasal asymmetry, however, is poorly understood, even if there is a relationship.

However, more recent research has questioned whether this phenomenon is physiological – caused by structural differences rather than functional differences (airflow) (Sobel *et al.*, 1999a). Sobel *et al.* exploited a property of odorant molecules which is that they attach (sorb) to the mucosa at different rates. High-sorption mocules elicit weaker responses when velocity is low and greater when high whereas low-sorption molecules show the opposite pattern. The reason for this is thought to be because a high-sorption molecule sorbs to the mucosa before moving along it when airflow is low; when it is high, it covers more of the mucosa before it sorbs. A low-sorption odorant inhaled at high velcocity will move past the mucosa and not sorb. In Sobel *et al.*'s study, a mixture of L-carvone (a high-sorption odorant) and octane (a low-sorption odorant) was used. When the odorant was delivered with low airflow, more people said they detected octane; when delivered with high airflow, most detected L-carvone. When they studied eight participants whose dominant nostril had changed, the nostril with the higher airflow was better at perceiving high-sorption odorants and the less dominant was more sensitive to low-sorption odorants.

3.12 The trigeminus

In addition to the olfactory nerve, chemosensation is also mediated by another of the CNs, the fifth and the largest. The trigeminal nerve or trigeminus, like the olfactory nerve, is stimulated by odour molecules but its function is not olfactory.

Instead, it mediates the sensation of irritation, stinging, pain, itching or cooling that is evoked by an odour or any substance that is inhaled and stimulates its nerve endings; the anterior third appears to be particularly sensitive (Hummel *et al.*, 1996). At the beginning of the nineteenth century, it was regarded as the 'common chemical sense' to distinguish it from olfaction (Parker, 1912). Towards the middle of the twentieth century, experimental work demonstrated that the trigeminal nerve was activated by almost all of the odours that stimulated the first CN (Beidler and Tucker, 1956).

The nerve endings of the trigeminus are found on its ophthalmic and maxillary branches and these processes extend to the nasal mucosa and OE; the nasal cavity also receives projections from both branches of the trigeminus (Brand, 2006). Because it responds to the chemical nature of an odour or an inhalant – its irritation, temperature, humidity, and so on – the trigeminus's function is principally somatosensory and, in some parts of the face, nociceptive (the lachrymal reponse to peeling an onion and the response to ammonia is due to stimulation of the trigeminus, as described on page 37). These responses are more than likely mediated by the unmyelinated C fibres and A delta fibres that are implicated in pain sensation and that innervate the respiratory epithelium (Anton and Peppel, 1991). The C fibres are probably responsible for the burning sensation; the A fibres are responsible for the stinging sensation (Mackenzie *et al.*, 1975). There is some evidence that the nerve endings exhibit selectivity in that the acid-sensing ion channel (ASIC) family of receptors, vanilloid receptors and purin (P2X) receptors have been found on the trigeminal nerve endings and are activated by stimulants that provoke the trigeminus (caipsaicin, heat).

The pathway of the trigeminus has been fairly well mapped. Thus, the trigeminal sensory nucleus emerges from the spinal cord and midbrain and projections from the pain nerve fibres are received here and in other trigeminal nuclei (Anton and Peppel, 1991). There is also evidence of connections between the trigeminus and the amygdala, via the parabrachial complex (Bernard *et al.*, 1989) and that trigeminal stimulation activates the primary and secondary somatosensory cortex, as well as other cortical regions (described in the next chapter).

One of the more well-known trigeminal stimulants is caipsaicin, the burn- and heat-producing chemical in chillies. This irritates the nasal cavity and the response is probably mediated by the C fibre afferents; the precise mechanism is unknown but in activating the C fibres, caipsaicin also stimulates the release of substance P, a hormone that is generated by pain. This process may also alter or change the operation of the olfactory receptors in some way. There is evidence that the presence of a trigeminal stimulant can increase olfactory sensitivity (Jacquot *et al.*, 2004) and also, conversely, that the presence of an odour can increase trigeminal intensity. This has been demonstrated with vanillin and H_2S (Livermore *et al.*, 1992; Kobal and Hummel, 1998).

We tend to become desensitized to caipsaicin and mustard oil (allyl isothiocyanate) when the interstimlus interval is 3–4 minutes (Brand and Jacquot, 2002). Acetic acid desensitizes whatever the inter-stimulus interval (ISI) (Jacquot *et al.*,

2005). As might be expected, trigeminal stimulation interacts with olfactory stimulation to create a percept of what is inhaled/sensed. The location of this interaction is unclear but might be the mediodorsal nucleus of the thalamus. Unlike olfactory projections, trigeminal ones tend to be contralateral. As Chapter 1 noted, there are some odours that may not stimulate the trigeminus (and are, therefore, pure odours) and others that stimulate the trigeminus strongly (pyridine or n-butanol, for example). Consequently, people who are demonstrably anosmic do respond to trigeminal stimulation albeit at a lower degree of sensitivity (Hummel *et al.*, 2003). There is more on this in the OEP section of Chapter 4.

3.13 Vomeronasal organ

A final organ or structure that deserves a brief mention in this section is the vomeronasal organ (VMO). The reason it merits a mention is that it is a conundrum and may not even exist in humans. It does exist in other vertebrates – and in reptiles, amphibians and mammals – and may undertake a pheromonal function in those organisms, allowing the detetction of biologically important odours that generate a specific behavioural response (see Chapter 2). A tube located at the base of the nasal cavity, it is enclosed by cartilage or bone and was described by Ludvig Jacobson in 1813 who noted a nerve alongside the trigeminus that had a good blood supply and that led to the OB (which Jacobson reasoned carried chemosensory information to the bulbs). For this reason, it is also called Jacobson's organ. Up until the 1970s, the olfactory nerve and vomeronasal organ were thought to function in parallel and perform different functions (with the VNO responsible for the perception of pheromones, although this has been cast into doubt by the finding that the perception of androstadienone does not require the VNO (Knecht *et al.*, 2003)). Research in the 1990s suggested that it developed two pathways – one to the accessory OB and one to the amygdala and hippocampus (Halpern *et al.*, 1998; Martinez-Marcos and Halpern, 1999) – but whether we posses this organ in any functional sense is unclear (Martinez-Marcos, 2009; Mast and Samuelson, 2009). Some thinking argues that it is present prenatally and then degenerates after birth. Others have suggested that, if present, it is non-functional (Keverne, 1999; Wysocki and Preti, 2004); it contains no sensory neurons, mainly epithelial cells, and no olfactory protein markers (Trotier *et al.*, 2000).

3.14 Central mechanisms: the cortex

Shepherd (2005) concluded that the early stages of olfactory processing – the neurophysiology – are quite well understood. 'At the olfactory cortex, however,' he contends, 'we reach an impasse' (p. 166). Chapter 4 considers the involvement of the cortex in odour and taste processing at the psychophysiological level and from studies using PET, fMRI and EEG. This section considers the evidence from studies of human brain lesions and the effect of lesions on olfactory perception. It also considers whether lesion data point to further lateralization of olfactory function. The two cortical regions consistently implicated in neuropsychological studies of olfaction are the medial temporal lobe and the OFC.

3.15 The temporal lobes

The temporal lobes were originally included in the collection of brain regions described as the rhinencephalon. Studies of olfactory aura in epilepsy (specifically, medial temporal lobe epilepsy (MTLE)) reinforced this view (Chapter 5 considers epilepsy's association with olfaction) and Hughlings Jackson had reported the involvement of the temporal lobes in odour impairment in the 1890s. The region contains significant parts of the POC (AON, piriform cortex, nucleus of olfactory tract of the amygdala and entorhinal cortex) and has significant, dense connections with the OB. The POC is found at the inferior frontal-temporal junction, with the secondary cortical area falling nearer the caudal OFC, medial and subcallosal PFC and insula. The POC receives the greatest number of projections from the OB and it also projects to the secondary area, the thalamus, hypothalamus and ventral striatum (Price, 1990).

Whether the POC is an association area rather than a sensory area has been challenged (Haberly, 2001). For example, as the section on the thalamus discussed, the OB provides the first stage in the analysis of sensory information, not the cortex. For this reason, some have argued that the POC is a secondary olfactory area containing secondary olfactory regions and structures (Cleland and Linster, 2003).

What also distinguishes the POC from the association areas of other senses is that it is laminarly less complex. The association areas of other senses are comprised of six layers. The POC, being a collection of regions belonging to allocortex, has three. The olfactory cortex is granular or agranular and may or may not have a poorly developed layer IV (Carmichael and Price, 1994). As the POC contains so many cortical and subcortical structures it does have significant connections within and outwith its regions: the perirhinal cortex projects to the amygdala and subiculum, the entorhinal cortex projects to the hippocampus. There is a direct connection between the amygdala and the hippocampus (Carmichael *et al.*, 1994).

These connections have been determined by tracing or lesion work in animals. In humans, lesions exclusively affecting the temporal lobes are not particularly common – other regions are usually damaged – but the disorder of MTLE provides a clinical model of cortical olfactory involvement as its (surgical) treatment can include anterior lobectomy or selective lesions to the amygdala and hippocampus. Both of these interventions remove the medial temporal cortex but one removes more than the other (Djordjevic, 2006), thus corticoamygdalohippocampectomy will involve sectioning 4–5cm of the cortex, plus portions of the amygdala and hippocampus. The second technique leaves the cortex virtually spared. An examination of the amount of olfactory cortex spared after surgery found that in twenty-one patients, around 35 per cent of the priform cortex remained bilaterally, the entorhinal cortex was removed as was the amygdala and there was partial sectioning of the hippocampus (Dade *et al.*, 2002). Dade *et al.*'s PET study asked participants to remember a list of six odours over a four-day training period. In a recognition memory paradigm, memory for the odours was associated with activation in the piriform cortex, bilaterally, suggesting to the authors that this region is involved in memory not simply olfactory perception (there was no significant increase during encoding). An increase in the piriform cortex was also found in another study of odour recognition, as well as perception of a single odour (Savic *et al.*, 2000).

The consequences of such removal in MTLE are not universally negative. For example, sensitivity and detection thresholds tend to be preserved but disrimination of odours in the nostril ipsilateral to the surgery, or when injury is bilateral, is impaired (Zatorre and Jones-Gotman, 1991). Unilateral lesions have been associated with problems in odour identification (Eskenazi *et al.*, 1983; Carroll *et al.*, 1993). Of the problems reported by patient HM, one of the least remarked is his inability to discriminate between odours or identify them (Eichenbaum *et al.*, 1983). Odour memory is also impaired after unilateral surgery (Abraham and Mathai, 1983). Of course, this presupposes that olfactory function was intact in these patients prior to surgery but the evidence for this assumption is uncertain (e.g. Jones-Gotman *et al.*, 1997). Ciumas *et al.* (2008) used PET to examine birhinal perception of familiar and unfamiliar odours (that were equally pleasant and intense) in twenty patients who suffered from either left- or right-sided MTLE. A control group showed activation in the amygdala, piriform cortex, anterior insula and cingulate cortex. Familiar odours also activated the right parahippocampal gyrus and left BA44, BA45 and BA47. The epileptic participants showed no amygdala, piriform cortex or insula activation; patients with a left-sided locus showed no activation in the left BA44, BA45 and BA47 to the familiar odours.

Activation in the entorhinal cortex continues to be little observed in the literature, but probably because it receives few olfactory inputs and, therefore, is not intricately involved in perception. However, these studies, using PET and fMRI, are considered in more detail in their own section in Chapter 4 so a discussion of these findings, including the first inconsistency, is postponed until then.

3.16 The orbitofrontal cortex and the insula

The two secondary olfactory areas of significance are the OFC and the insula. One of the earliest studies to demonstrate the importance of the OFC were those of Elsberg and Stewart (1938) and Elsberg and Spotnitz (1942) who developed olfactory tests to detect tumours in 1,000 patients and found that olfactory sensitivity was reduced in patients with 'lesions in or around the frontal lobes'. Experimental lesioning research with dogs in the same period found that the removal of the frontal lobe – but not the temporal or parietal lobe – resulted in a delay in learning of olfactory-conditioned reflexes and/or a disruption in the ability to discriminate between positive and negaive conditioned odours (other senses were unaffected) (Allen, 1940, 1943).

Of course, the OFC is not a unitary area and has several cytoarchitectonically distinct subdivisions. Thus, BA12 can be further subdivided into four regions, BA13 into three, BA11 into medial and lateral regions and so on (Carmichael and Price, 1994). Price *et al.* (1991a) posit that nine areas of the OFC receive input from the POC, including the anterior/posterior piriform cortex, AON, entorhinal cortex and periamygdaloid cortex. More systematic experimental work found that potentials could be recorded in the various pathways of the olfactory system in aware, unanaesthetized monkeys (Tanabe *et al.*, 1974, 1975a, 1975b). Field potentials in the lateral posterior OFC were recorded with stimulation of the OB and piriform cortex. This region Tanabe *et al.* termed the lateral prefrontal OFC (LPOFC) and included parts of BA12, BA13 and the frontal operculum. Tanabe *et al.* (1975a, 1975b) found that of the forty-four neurons they recorded from in the LPOFC, 50 per cent responded to at least one of eight odours; four was the maximum number of odours they responded to. Using tracer technology, Potter and Nauta (1979) were able to identify two pathways to the LPOFC – a thalamocortical route via the piriform cortex and a cortico-cortical pathway via the piriform and entorhinal cortices. A further, secondary projection area was identified in the OFC with a pathway between BA13 and the central/posterior OFC mediated by the thalamus (Yarita *et al.*, 1980). Pathways were also identified between the OFC and the anteroventral insula (Mesulam and Mufson, 1982a, 1982b).

Carmichael and Price (1994) concluded that the areas to which the POC projects in the OFC include five areas of the granular layer of the insula, areas 13a, 13b and 14c in the medial OFC, and the inferior medial BA25. The densest connections are to the Iam and Iapm regions of the insula (which projects deep into the cortex, layers I to VI, and which are bidirectional), followed by BA13a. The granular insula and posterior OFC are connected to the mediodorsal thalamus which receives input from the POC. Reciprocal pathways exist between the anterior medial OFC and anterior medial POC and between the LPOFC and POC. The further away the OFC structures from the POC, the less visible the connections. It is also noteworthy that there is a great deal of – but not complete – cytoarchitectonic and topographic convergence between the area described as the OFC in humans and that identified as the OFC in non-human primates, which are the subjects of the majority of the studies reviewed in this section (Chiavaras and Petrides, 2000; Ongur *et al.*, 2003).

In humans, lesions to the PFC, caused either by tumour, haemorrhage or other insult, have been associated with identification impairments, deficits in odour quality discrimination and odour memory problems (Jones-Gotman and Zatorre, 1988b, 1993; Zatorre and Jones-Gotman, 1991). Jones-Gotman and Zatorre

(1988a) reported that lesions to the OFC only produced identification impairment (problems were observed with temporal lobe excisions but these were not as great). Potter and Butters (1980) found that lesions to the PFC led to an inability to judge the similarity of odours but the ability to judge the similarity of hues was unaffected. Gottfried and Zald (2005) reported that in thirteen neuroimaging studies that (i) measured odour detection only, (ii) included no aversive odours and (iii) required participants to enage in no cognitive task, seven showed bilateral activation in the OFC, five found right-sided activation and one showed left-sided activation. They note that the right hemisphere was dominant during odour stimulation (but that the left was involved in two-thirds of studies). They localized odour perception in these studies to the bilateral medial orbital sulcus. The posterior, agranular part of the orbital surface shows little activation to odour in neuroimaging studies (Gottfried *et al.*, 2002; Gottfried and Dolan, 2003).

The agranular insula and the posterior OFC show activation if the olfactory task involves some decision-making involving the odour's pleasantness or when odour interacts with another sense (de Araujo *et al.*, 2003c). The lack of caudal activation in imaging studies may be explained by the possibility that fMRI is not sensitive to measuring activation here, the output/activation is too weak, the signal is too weak or the region may be too small. Gottfried and Zald (2005) also suggest that fMRI might be insensitive to the millisecond-level temporal changes in the olfactory system (note that fMRI's temporal resolution is weaker than its spatial resolution). There may be a short period of activation followed by a rapid return to baseline (Poellinger *et al.*, 2001).

The insula, as well as the OFC generally, are two of the regions that Zald and Pardo (2000) conclude are consistently activated in neuroimaging studies of olfactory perception (and this is taken up more comprehensively in Chapter 4). The insula is a polymodal region, involved in a number of functions incuding, consistently, taste perception, as well as ingestion-related behaviours, vomiting, diarrhoea, swallowing and pain perception. Lesions reduce people's ability to respond emotionally to threatening visual and auditory stimuli and the region is implicated in people's ability to recognize disgust in faces and voices (Calder *et al.*, 2000; Kipps *et al.*, 2007). Wilder Penfield's pioneering stmulation studies found that stimulation of the insula causes gastrointestinal sensations such as nausea, chewing and swallowing (Penfield and Faulk, 1955), findings that have been replicated and extended to include aversion and the sensation of unpleasant taste (Krolak-Salmon *et al.*, 2003). It is also involved in language and auditory processing, as indicated in a number of neuroimaging studies, and appears to be highly active during states of anxiety and drug addiction (Jones *et al.*, 2010). It is located at the lateral sulcus, covered by the frontal, temporal and parietal opercula and receives input from the auditory cortex and visual association areas. Lesions to the region result in disgust recognition impairment (Jones *et al.*, 2010) and activation is found here when people view faces expressing disgust and when they inhale disgusting odours (Wicker *et al.*, 2003). However, increases in activation in the left insula have been reported to pleasant odours (Fulbright *et al.*, 1998) and in the right to disgusting ones (Heining *et al.*, 2003), as well as bilaterally to unpleasant (not necessarily disgusting) odours (Royet *et al.*, 2001). Wicker *et al.*'s (2003) study also found that the disgust stimuli stimulated the left side.

Evidence suggests that the insula's involvement is broader than responding to smell and taste, as the studies on disgust cited earlier indicate, in that it is involved in the emotional evaluation of stimuli or in reflecting feeling states and bodily sensation (Damasio *et al.*, 2000; Jones *et al.*, 2010). Craig (2009) suggests that its general role is to interpret information about our bodily states, sensory states and cognitive states and create a unitary feeling that reflects the interaction of all three and specifically reflects the 'emotional now', 'the emotional self' or the 'global emotional moment'. This is returned to in the final chapter.

3.17 An anatomy of taste

Once a food or any substance is inserted into the mouth and placed on the tongue (ingested), it makes contact, and combines, with saliva. Saliva covers the tongue and the taste receptors and is secreted principally by three types of (tubular) salivary glands (parotid, submandibular and sublingual), with other minor glands supplying other parts of the mouth (Bradley and Beidler, 2003). The material – whether it is water, chilli, yoghurt or sand – activates receptors located on the tongue. It is at these receptors that taste transduction occurs and signals are sent to the cortex. Receptors for taste are located on the surface of the tongue and are found in one of thousands of papillae. These taste buds allow us to discriminate between any of the few basic tastes (Harper *et al.*, 1966), regardless of where they are located on the tongue. Papillae come in three major forms. The fungiform papillae (so-called because of their mushroom-like shape) appear as pinkish spots and contain one or more taste buds. They are located at the front two-thirds of the tongue. Circumvallate (wall-like) papillae are found at the back of the tongue and foliate (leaf-like) papillae are found at the sides. There are about 50–150 receptor cells in each papillae.

When our food, or a tastant, contacts the tongue, it stimulates the taste receptor by entering a taste pore. It is the interaction with the receptor here that causes an action potential to be fired, leading to impulses travelling along two major nerves (discussed in more detail below) – the chorda tympani, if stimulation occurs in the anterior tongue, or the glossopharyngeal nerve. Unlike smell, the chemical composition of a tastant is largely irrelevant. That is, even though saccharin and sucrose have different molecular structures, the sensation they always produce is sweetness.

The tongue is innervated by three CNs: the facial nerve (VII), the glossopharyngeal (IX) and the vagus nerves (X). These nerves innervate different regions and are thought to be mixed types – that is, a combination of motor and sensory nerves. The facial nerve has an important role in supplying innveration to muscles in the face, the stapedius muscle of the middle ear, the sublingual/mandibular salivary glands, the lachrymal glands and nasal mucosa.

It and the chorda tympani and superior petrosal nerves innervate the front (anterior) two-thirds of the tongue and the soft palate. CNIX, via its lingual tonsillar branch, innervates the posterior third of the tongue and the circumvallate papillae and foliate taste buds. The pharyngeal branch of this nerve innervates the taste buds in the nasopharynx. Finally, CNX, via its superior laryngeal branch, innervates the pharynx and larynx and because of the location of the branch, it is thought that these areas are not closely involved in taste discrimination but respond to pH levels in ingestants (Bradley, 2000). Injury to one of these nerves can affect taste perception, as Chapter 5 describes in more detail, but because these nerves innervate taste bud 'fields', the destruction of one field or group of fields may go unnoticed (Bartoshuk, 1989). Bartoshuk *et al.* (1996) also reported that when the chorda tympani was anaesthetized or injured, function in the glossopharyngeal nerve was enhanced (as enhanced bitter sensation at the back of the tongue).

The facial nerve pathways carry gustatory information from the papillae course via the lingual nerve and meet at the mandibular third molar where they form the chorda tympani. This nerve then makes its way through the skull via the stylomastoid foramen and to the tympanic cavity of the middle ear. Synapses are formed with the geniculate ganglion at a bend in the facial nerve and this region then forms the nervus intermedius. This traverses the internal auditory meatus and meets CNVIII. After reaching the cerebellopontine angle, it then makes its way to the pons and onto the wider brainstem.

The small, well-protected glossopharyngeal nerve innervates motor muscles in the face (the stylopharngeus), the parotid gland, the eustachian tube and the baroreceptors of the carotid sinus, the latter being sensory in nature and governs the slowing of the heart. It, therefore, performs a sensory and motor function. Its role in innervating the parotid gland is important because this gland allows salivary secretion and the stimulation of saliva produced by a seasoned tastant is permitted by the action of this nerve.

The vagus nerve generally performs a number of disparate roles – it is responsible for vocal cord sensation, our sensation of sound inside the head and the visceral sensations produced by the larynx and the gut, and mediates the severest form of gagging reflex (vomiting) and swallowing, coughing and sucking reflexes. It innervates taste buds found in the epiglottis and oesophagus. As you can see, this nerve is essential for the functioning of food flavour perception – allowing as it does swallowing and sucking and an awareness of the auditory elements of these actions (such as the crunch of a crisp) – and the rejection of food.

Other nerves involved in taste also include the trigeminus, as this responds to the heat and temperature of tastants (such as spicy food, as well as hot drinks). This nerve and some facial nerve fibres travel in parallel for a short way. A rare phenomenon, taste-initiated pain, is caused by injury to another (non-cranial) nerve, the auriculotemporal nerve. This pain is localized in the face (De, 1990).

The CNs synapse in the nucleus terminalis of the medulla. They descend and overlap here and form a gustatory nucleus. From here, projections are sent to the primary and secondary taste cortices. Input from the peripheral taste apparatus is thought to enter the nucleus tractus solitarius (nucleus of the solitary tract, NST) and diverges – these two pathways become united in the insula. These two areas – the primary taste cortex (PTC) and the NST – are thought to have reciprocal connections (Katz *et al.*, 2002). The region of the NST considered to represent sensation is thought to be in the rostral part; the part that responds to the affective quality of taste, more intermediately (Sewards, 2004). Projections also interact with other nuclei that mediate chewing and licking, facial expression and preabsorptive insulin release (Smith and Shipley, 1992). Somatosensory projections and those from the gut are present with NST and OTC taste neurons. The NTS projections ascend with the central tegmental tract and synapse with the parvicellular division of the ventroposteriormedial thalamic nucleus or taste nucleus (Beckstead *et al.*, 1980). There is evidence from rodents that different parts of the thalamus mediate different responses to taste. Thus, sensory responses are mediated by the dorsolateral aspect and hedonic responses by the ventromedial part. Unlike olfaction, therefore, gustation does send projections to the thalamus before it makes contact with cortical areas.

These cortical areas include the PTA, which includes the insula, the perisylvian regions and the inferior pre and postcentral gyrus, and the secondary taste cortex found in the caudomedial OFC and next to the primary cortex. In addition to their involvement in taste, these areas are also implicated in behaviours such as touch and temperature and functions such as disgust perception and the nature of the specificity of cells in the taste areas. See Figures 3.4, 3.5 and 3.6.

3.18 Sensing different tastes

3.18.1 Bitter

The list of tastants that elicit bitter tastes is large but bitter substances tend to belong to certain classes. These are hydrolyzed proteins, rancid fats, alkaloids and poisons. As this list suggests, bitterness is not normally a sign of pleasure but a signal of harm and danger and all of these substances are highly bitter (Rouseff, 1990), toxically so at high levels (Drewnowski and Gomez-Carenos, 2000). We tend to reject them unless we have built up a tolerance (for very, very strong coffee, for example) but tolerance for tastants such as quinine is not common unless, as Chapter 2 showed, people lack the gene for tasting bitter substances. Young children and pregnant women tend to avoid bitter substances (Drewnowski and Gomez-Carenos, 2000). Fortunately, we are very adept at detecting bitterness and are able to detect quinine at twenty-five micromolecules per litre; bitter plants are less concentrated than this (Drewnowski, 2003). We can detect sucrose only at much higher concentrations – in the region of thousands of micromolecules.

Bitterness is the one taste whose genetic coding is relatively well understood (Caicedo and Roper, 2001), principally because of evidence from PROP tasters

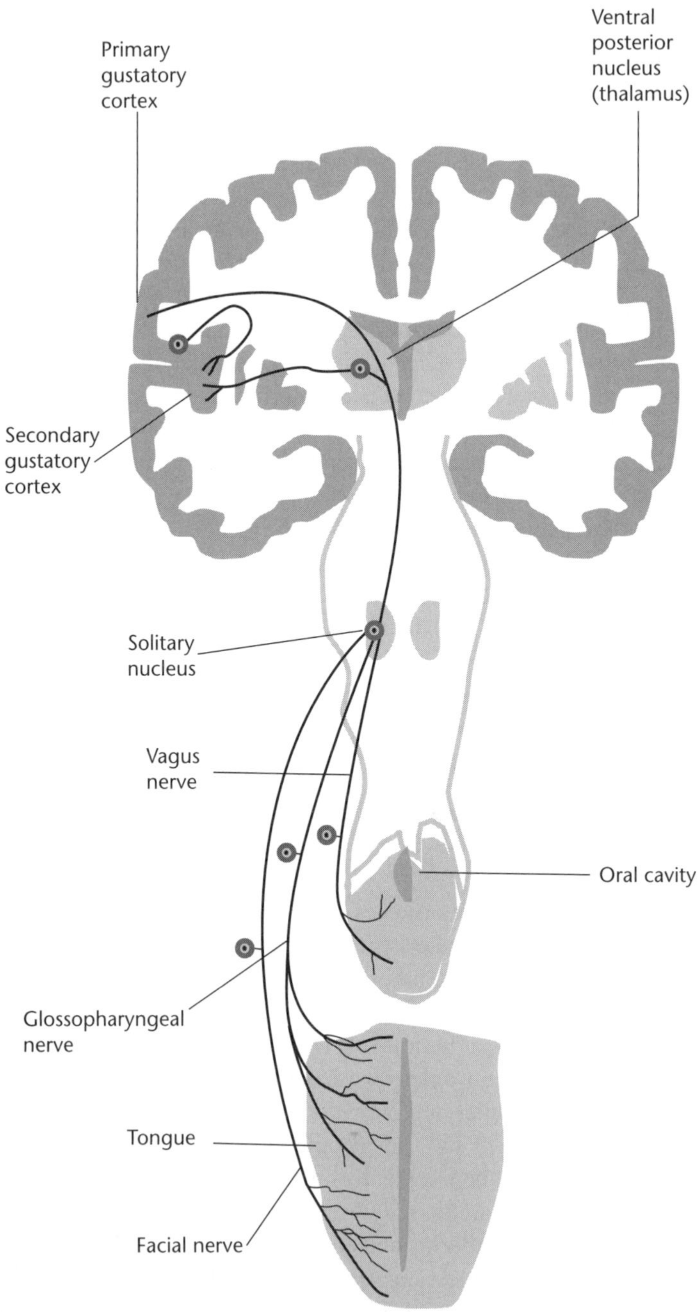

FIGURE 3.4 The major taste pathways

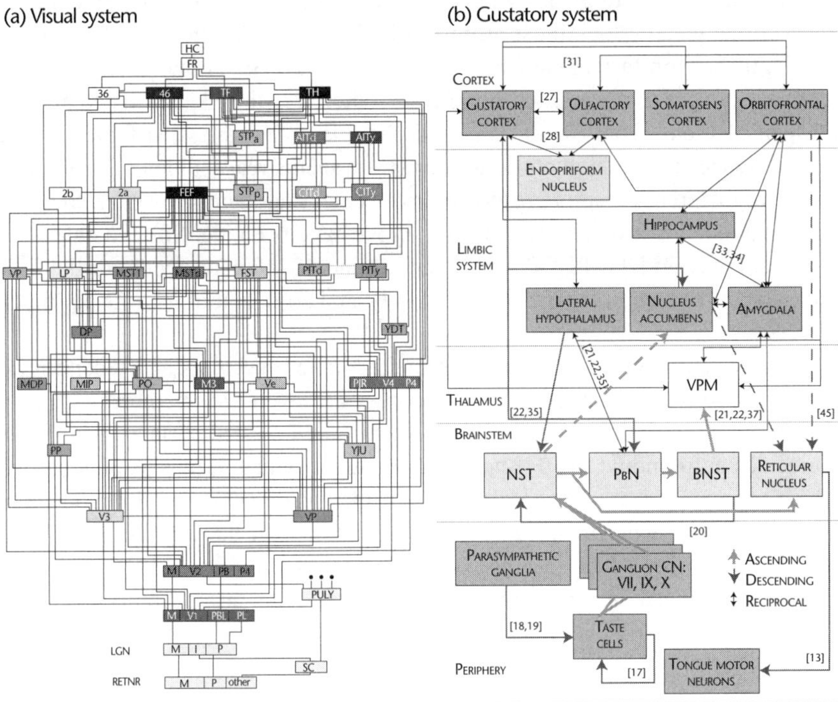

FIGURE 3.5 A comparison of the structures and regions involved in, and the interactions between these structures and regions, the visual and gustatory system

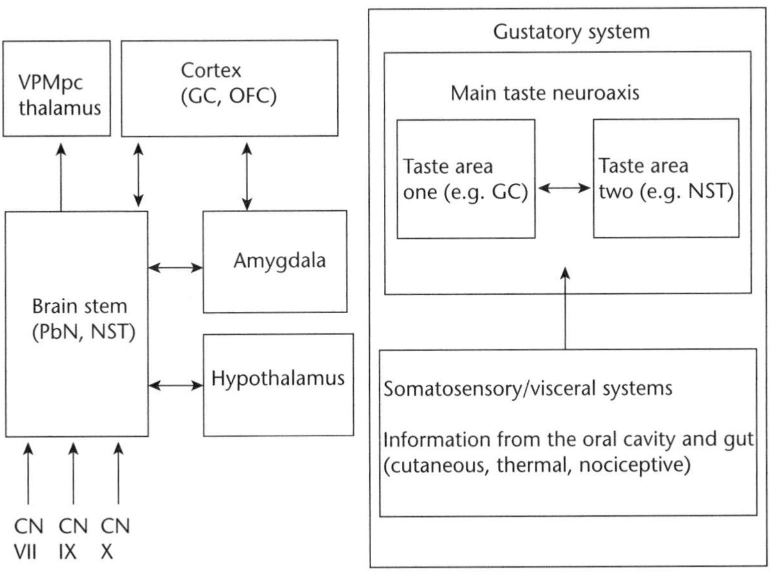

FIGURE 3.6 A schematic representation of the brain regions involved in gustation

and non-tasters. Tasting ability has been linked to having two recessive alleles and various genetic loci, most promisingly 5p15 and chromosome 7 (Guo and Reed, 2001). Non-tasters have two dominant alleles. Given the variety of unrelated substances that can elicit bitterness – and Drewnowski has listed at least thirteen of them, from esters, urea and phenols to Xanthies (caffeine) to sulfoxides (saccharin in high concentrations) – it seems reasonable to assume that there are different receptors responsible for detecting them. However, the pool may be small and around three receptor types have been identified to date – for quinine, urea and PTC/PROP. However, around 40–80 genes code for the T2R group of bitter receptors (Adler *et al.*, 2000) and so this pool is probably much larger.

Like sweet, bitter tastes act on G-protein coupled receptors and second messengers. G-protein receptors are also known as seven-transmembrane domain receptors and are a large family of receptors found on cell surfaces that detect molecules surrounding them. They provoke the production of phospholipase C which prompts the production of inositol triphosphate (IP3) and the production of calcium ions (which then leads to depolarization). The human T2R4 and T2R8 receptors have been found to respond to denatonium and the T2R16 to salicin. Thus it appears that there may be a large number of bitter taste reeptors that mediate responses to particular bitter tastants.

3.18.2 Sweet

All primates and almost all mammals can perceive sweetness and it is one of the tastes that can be detected early. This is not altogether surprising given that glucose is a vital source of energy and that sugar can be converted into glucose. Our ability to sense natural sugars such as sucrose depends on the activation of a pathway that suppresses potassium channels via cAMP (Tonosaki and Funakoshi, 1988). This mechanism differs from that responsible for the sensation of sweeteners (such as saccharin, aspartame, dulcin, sucralose, acesulfame-K and others), which activates phospholipase CB2. Each triggers the generation of enzyme-dependent second messengers but different ones are produced by sugars and sweeteners. For example, IP3 and diacylglycerol are stimulated by PCB2 (Bernhardt *et al.*, 1996). If the PCB2 gene is knocked out in rodents, however, this seems to reduce the neural and behavioural response to sucrose (Zhang *et al.*, 2003).

In 1999, two taste receptors for sweetness were identified – TAS1R1 and TAS1R2 (Hoon *et al.*, 1999) and these 'dimers' appear not to be activated by stimuli that are not sweet (Li *et al.*, 2002). A third receptor has been identified – TAS1R3 – which may mediate the response to saccharin (Matsunami *et al.*, 2000; Bachmanov *et al.*, 2001; Montmayeur *et al.*, 2001) but it appears that our sensitivity to aspartame may depend on the R2 variant of this rather than R3. These proteins are coded by the genes, TAS1R2 and 3. If these are deleted in rodents, preference for sweet tastes declines (Damak *et al.*, 2003). If the R1 gene is knocked out, however, rodents prefer sweeter tastes. The precise mechanism by which information from the taste bud passes to the peripheral nervous system processes is unknown.

As the concentration of sucrose increases, so does the neural response to it in the chorda tympani, NST and parabrachial nerve (McCaughey, 2008). It is the most effective sugar in doing so, followed by fructose, glucose and maltose. Various substances, however, can inhibit the sensation of sweetness, including lactisole, gurmarin and extracts of the Ziziphus Jujube plant (Yamada and Imoto, 1987) and these bind to various R receptors. The precise mechanism of inhibitory action, however, is unknown.

The pathways for bitter and sweet signals appear to be separate as they travel from the receptor cell to the cortex, but there is currently little evidence that the brain responds selectively to them (Sugita and Shiba, 2005). Neurons in the NST respond to sucrose in a similar way that they do to NaCl, HCl and quinine (McCaughey, 2007). Some sugar-oriented neurons have been reported in the chorda tympani and parabrachial nerve but these also tend to respond to umami (McCaughey, 2007).

Temporally, receptor reponses are slower to sweet tastes than the others. Whereas salty tastes tent to elicit responses in the NST at around 600msecs following lingual administration, this is 800msecs for sweet tastants, even when sucrose is placed at the front of the tongue (McCaughey *et al.*, 1997). It also takes people longer to identify sugar than salt – 854msecs versus 229msecs (Halpern, 1985). One possible explanation for this could be sweet's reliance on G-protein mediated transduction, rather than the direct pathway of ions that allows salt transduction (Heck *et al.*, 1984).

Sucrose is a strong reinforcer and one of the reasons for this is that it gives pleasure. Rats prefer tasting sweet tastants rather than brain stimulation of reward pathways (Conover and Shizgal, 1994). Sugars and sweeteners have been found to increase the amount of B-endorphin in the cerebrospinal fluid of rats (Yamamoto *et al.*, 2000) and, as Chapter 2 showed, it can provide some relief from pain and distress. There is an association between sweet sensation and activity in the nucleus accumbens (NA), a region that is involved in addiction and reward. Sugar appears to stimulate cholinergic receptors in the ventral tegmental area which project to the shell of the NA (Norgren *et al.*, 2006) and the preference for sucrose is correlated with preference for ethanol (Kampov-Polevoy *et al.*, 1999). Both appear to produce neural changes that are seen during drug abuse and withdrawal (Spangler *et al.*, 2004). The pathways for signalling sugar sensation are found in the parabrachial nerve and in the PTC. In a complex, but methodical, way, projections are sent by the parabrachial nucleus (PBN) to the thalamus and the ventral area (including the amygdala and hypothalamus which respond to sugar). Ventral projections are then projected to the ventral tegmental area which, in turn, projects to the NA (Norgren, 1976). The taste cortex projects to the PFC and amygdala and has recipiral connections with the hypothalamus (Yamamoto, 2006).

3.18.3 Salt

When sodium ions (Na+) enter an ion channel in the taste pore, we sense the taste of salt. Salt increases the production of saliva, the secretion of gastric acid, gastric motility and the secretion of pancreatic enzymes – in short, it prepares the body for absorption and digestion (Mattes, 1997). The process results in depolarization, and the entry of calcium into the cell, followed by repolarization and the opening of potassium channels and the exit of potassium. One mechanism by which salt is detected is thought to involve the amiloride-sensitive epithelial sodium channel ENaC (Rawson and Li, 2007). This may allow a pathway for sodium current into the taste receptor as amiloride blocks this passage and even administering it on the tongue may result in reports of reduced salt sensation (Schiffman *et al.*, 1983; but Desor and Finn (1989) report negative findings).

Salt is something that we routinely add to savoury food to add flavour. It can also be added to chocolate to enhance its flavour, giving it a tangy sweetness. According to various studies (e.g. Greenfield *et al.*, 1984), 20 per cent of people salt their food without tasting it. Around 5–10 per cent of our intake is under dietary control – i.e. what we consciously ingest by adding it to what we eat (Mattes and Donnelly, 1991). As the section on the development of taste preferences in Chapter 1 noted, we can show increased preference for saltier food when salt is added in small amounts. We can detect salt by the fourth month of life (Beauchamp *et al.*, 1986). It can take between 8–12 weeks for infants and adults to become accepting of additional salt in a diet, but acceptance can occur as early as two weeks (Bertino *et al.*, 1982; Mattes, 1990). Curiously, our sensitivity to salt is not correlated with our preference for it (Lauer *et al.*, 1976).

3.18.4 Sour

Most sour tastants tend to be acids (and these are sour because they contain hydrogen ions). Sour tastants trigger event potentials in a dose-dependent manner, that is, the sourer the tastant, the greater the firing. Potassium channels are blocked and positive ions are allowed to enter which leads to depolarization and the release of neurotransmitters. The mechanism for the sensation of sourness is poorly understood (and poorly studied). One suggestion is that the degenerin or ENaC family of ion channels is involved (Rawson and Li, 2007).

3.18.5 Umami

The fifth taste is one that responds to glutamate and is generated by amino acids that bind to G-protein coupled receptors, activating second messengers, like most tastants. T1R1 and T1R3 may dimerize to allow them to function as umami receptors on the tongue (Li *et al.*, 2002; Nelson *et al.*, 2002). Another receptor protein

– mGluR4 – has been identified in rats (Chaudhari *et al.*, 2000). Umami shares with sucrose the T1R3 receptor subunit (sweet involves the R2 type and umami the R1) and the two tastes may be linked. For example, the sweetness-inhibitor, lactisole, also inhibits the sensation of umami, but at concentrations that are sixteen times stronger than that needed to inhibit sweetness (Galindo-Cuspinera and Breslin, 2006). It leaves other tastes unaffected.

What is currently unknown is the precise mechanism that allows the tastes to be distinguished at the receptor level. Is there one cell with many different receptor types that responds to multiple tastes (Caicedo *et al.*, 2002) or does one taste receptor cell detect one taste modality?

3.19 Swallowing

After taste and chewing comes swallowing, a biomechanical activity that is under the control of the brainstem and parts of the cortex – some people after a stroke have difficulty swallowing (Daniels and Foundas, 1997). Swallowing serves the function of propelling the chewed food and saliva (bolus) to the back of the buccal cavity (the pharynx) which, by muscle contraction, delivers the bolus to the open oesophagus. The tongue is pressed to the buccal palate, front first, then back. This, coupled with the act of pharyngeal contraction and pressure, clears the cavity of any material. Different tastes provoke different degrees of reaction. Thus sour tastes tend to cause contraction of submental muscles and contractions are greater to this taste than salty, sweet or bitter. Pressure from the tongue to the palate increases when a person tastes 10ml of cold sour tastant than the way they taste water (Pelletier and Dhanaraj, 2006). Wahab *et al.* (2011) compared the biomechanical responses of sixteen people to high and low concentrations of lemon odour or taste. They found that both stimulants created an immediate biomechanical change – there was increased pressure of the tongue to palate, especially at the anterior end. Pressure to the second pharyngeal sensor decreased. The effects correlated with the transit time of the tastant and submental contraction. Repetitive transcranial magnetic stimulation research (rTMS) suggests that several cortical areas are involved in swallowing, including the lateral and premotor area, frontal operculum and insula (Martin *et al.*, 1999). Martin *et al.* (2001b) used fMRI to measure three types of swallowing in fourteen women. The participants either swallowed naturally/automatically or voluntarily when water was presented to the tongue. Both types of swallowing activated the lateral precentral gyrus, the lateral postcentral gyrus and right insula. Automatic swallowing selectively activated the superior temporal gyrus, middle inferior frontal gyri and frontal operculum. Voluntary swallowing was associated with caudal anterior cingulate cortex (ACC) activation.

In a comparison between swallowing water or tasting or smelling a lemon concentrate, Wahab *et al.*'s (2011) rTMS study found that motor-EPs were observed for tasting but also for swallowing, an effect that appeared to persist for ninety minutes.

3.20 Central mechanisms of taste: the insula and other regions

In Chapter 4, the evidence for taste-specific activation in brain areas, measured via neuroimaging, is described and discussed. Before these techniques were available, however, our understanding of the role of the brain in gustation, when not relying on experimental ablation work in non-human subjects, was based on our observation of the gustatory changes that followed brain lesions/injury/trauma and the effect of disorders (such as epilepsy) .

The first reliable, documented case of taste loss following brain injury was reported by Ferrier (1876) whose patient eventually recovered the ability to taste, but not smell. The patient could distinguish between bitter and sweet and beef and mutton. Ferrier attributed the initial loss to injury of the lower 'temporosphenoidal lobe', and it is this general area – the temporal lobe – that has been the focus of subsequent research and observations. However, there are other, more mesencephalic regions that, when damaged, produce taste impairment. For example, injury to the midbrain and pons can also result in post-traumatic ageusia (Johnson, 1996; Kojima and Hirano, 1999). Small *et al.* (1997b) reported that temporal lobectomy, involving the anterior temporal lobe, was associated with impairments in taste quality recognition and subsequent lesion research has found that the symptoms may depend on the side injured. Thus, right lesions lead to taste perception and recognition impairment whereas right injury is associated with impaired taste perception ipsilaterally and recognition impairments bilaterally (Pritchard *et al.*, 1999; Cereda *et al.*, 2002), suggesting that the left side may receive input from both sides of the tongue. Lesions here also cause impairment in previously conditioned aversions to taste (Ramirez-Amaya *et al.*, 1998) and people's ability to describe smells using taste adjectives (e.g. sweet, savoury, etc.) is disrupted by lesions to the left side (Stevenson *et al.*, 2008). People with epileptogenic insular tumours have reported gustatory auras (sensations not generated by sensory stimulation) before their seizures (Hausser-Hauw and Bancaud, 1987), although this is not common.

The earlier section described how the insula is implicated in olfactory perception and the perception of pleasant and unpleasant odours in particular. It also noted how the insula is involved in a number of behaviours and functions, including smell and taste. Its role in taste perception, however, appears to be one of its most important.

The insula (or insular cortex or Island of Reil) is one of the major regions of the primary cortical taste area or PTC, which also includes various of the opercula – Rolandic, frontal and parietal (Small *et al.*, 1999; Barry *et al.*, 2001). It is divided into anterior and posterior regions by the insular sulcus and the posterior, granular portion receives greatest input from the thalamus, and parietal, occipital and temporal association cortices. The anterior portion has reciprocal connections to various regions including the ACC, ventromedial OFC, the amygdala and ventral striatum (Jones *et al.*, 2010). Stimulating the left or the right tongue with a tastant activates the right side of the insula (Barry *et al.*, 2001) but both sides of the insula are thought to be involved in taste. In monkeys, the PTA is thought to include the insula, the frontal operculum, and the base of the pre and postcentral gyrus (the Rolandic operculum). Evidence from human studies suggests considerable overlap. In macaques, less than 10 per cent of the neurons in the PTC (which receives projections from the gustatory thalamus) are taste neurons; a similar proportion exists in proximal areas (Scott *et al.*, 1991).

Because of the location of the PTC it is difficult to measure activation there, via neuroimaging, accurately – it is located deep inside the Sylvian fissure – and the sense itself does not lend itself to precise timing of neural processes associated with a tastant once it has been delivered on the tongue. Taste also involves a somatosensory component, as a tastant will express a tactile quality, and so the response of the PTC to taste and touch may be combined. As noted in Chapter 2, tasteless water also stimulates the PTC so an aqueous solution that activates this area may not do so because of its taste but because of its liquid, textural, somatosensory properties.

There is also thought to be a secondary taste cortex whose constituents and boundaries are less well delineated but include the OFC and the anteromedial temporal lobe (AMTL). Taste memory may implicate the NA specifically, in addition to other areas typically involved in memory. The AMTL contains the amygdala and there is evidence that neurons here respond to taste, but not necessarily the various taste qualities but the hedonic information carried by these tastes, i.e. whether they are pleasant or not (Henkin *et al.*, 1977; Scott *et al.*, 1993; Yan and Scott, 1996; Small *et al.*, 1997b; Nishijo *et al.*, 1998; Zald *et al.*, 1998). The PTC projects to the amygdala and the medial temporal lobe in monkeys (Turner *et al.*, 1980; Amaral and Price, 1984). Aversive and pleasant tastes produce different levels of activation in the OFC in humans (O'Doherty *et al.*, 2001) and aversive tastants have also been associated with cingulate gyrus activation (Zald *et al.*, 1998).

Adolphs *et al.* (2005) reported an unusual case study: patient B, who is 72 years old and had suffered herpes simplex in 1975. The disease caused degeneration of the amygdala, hippocampus, temporal cortex, bilateral anterior insula, posterior and medial OFC and ACC. He was unable to taste (ageusic) or smell (anosmic), failed to recognize the food he ate and ate indiscriminately. Although he preferred sucrose over saline, he failed to recognize either. Thus, while perception was impaired, (intact) pathways existed that must mediate preference. One candidate suggested by the authors is that from the NST to the hypothalamus and thalamus via the PBN. They suggest that the patient's behaviour supports the distinction between a ventral taste pathway, involving the amygdala and hypothalamus which mediates basic discrimination, and another, dorsal pathway involving the insula that mediates more complex taste perception. B did not know why he preferred sucrose, only that he preferred it over saline.

Historically, cortical models of taste perception have been of two types: labelled line and across-fibre pattern. The labelled line model, the simpler of the two, argues that taste pathways mediate specific taste qualities (bitter, sweet, etc.) (Mueller *et al.*, 2005). A variant of this model argues that axons may code for different degrees of palatability – e.g. whether the taste is aversive or acceptable – rather than taste quality (Marella *et al.*, 2006). The across-fibre model makes no assumptions about dedicated pathways but instead argues that taste recognition relies on the pattern of activation across neurons/receptive fields. Ethanol, for example, produces across-neuron activation in a similar way to sweet tastes in the NST (Lemon *et al.*, 2004) and, as the section below notes, there is a correlation between ethanol and sweet preference. Although these models may be too simplistic, they are currently the only two models to have been developed systematically.

3.21 Lateralization of taste

The evidence regarding whether the gustatory pathway is crossed or uncrossed tends to the conclusion that it is uncrossed, i.e. information from the left side of the tongue is projected to the left hemisphere. However, there is evidence for some crossing. The evidence in support of both views has been provided by studies of callosal patients. For example, Aglioti *et al.* (2000) examined whether tastes could be named or identified by pointing to a name in three groups: a control group, a patient with the corpus callosum absent and a man who had undergone a complete section of the callosum. Frequently, there are no behavioural differences between those with callosal agenesis and healthy individuals. Aglioti *et al.* (2000) found no difference between the controls and the agenesis group, but the patient with the section showed left hemitongue advantage. He was more accurate and faster at naming a taste when it was administered to the left side of the tongue. Another study examined gustatory discrimination ability in two patients with corpus callosum resection – one complete, the other sparing the genu and rostrum, two important parts of the corpus callosum (Aglioti *et al.*, 2001). They pointed to words or images of objects corresponding to the tastants placed on the left or right of the tongue. There was an advantage for the left side, although not significantly so, and responses were generally above average. The left-sided superiority – with its association with the language-dominant left hemisphere – suggests to the authors that the 'gustatory pathways from tongue to cortex are bilaterally distributed with an ipsilateral dominance'.

Pritchard *et al.* (1999) examined quality and intensity ratings of tastants administered on the left and right side of the tongue by six patients with unilateral insula damage, three with injury outside this region and eleven controls. When the lesion affected the right insula, there were ipsilateral recognition and intensity impairments; when the left insula was affected, there was an ipsilateral impairment in intensity rating and a bilateral deficit in taste recognition. Thus, the authors conclude that information from both sides of the tongue goes through the left insula to enable gustatory recognition. Small *et al.* (2001b) reported that individuals with right anterior temporal lobe lesions rated an aversive taste (bitter) as more bitter than did controls. There was no group difference for sweet taste. This suggests the involvement of the right hemiphere, and the AMTL specifically, is greater for the pereption of unpleasant or, at least, very strong tastes. Small *et al.* (1997b) had also reported that patients who had undergone right anterior temporalobectomy showed increased recognition thresholds for citric acid but normal detection ability. In a PET study with citric acid, bilateral or contralateral activation was found in the OFC and more activation was found in the right AMTL and the right OFC. This suggests that taste recognition requires the AMTL or regions within it.

At the subcortical level, damaged regions in the brainstem produce either ipsilateral or bilateral impairment. Thus, lesions to the tractus solitarius and its nucleus produce ipsilateral deficits whereas lesions to the pontine-tegmentum produce bilateral ageusia. Bilateral damage to the thalamus leads to ageusia but unilateral lesions above the brainstem produce no loss of taste ability. Onoda and Ikeda (1999) reviewed fifteen cases of unilateral taste impairment and concluded that injury to the pons, thalamus and midbrain resulted in unilateral impairment. They argue that the gustatory pathway ascends ipsilaterally from the medulla, then crosses over at the pons and synapses with the contralateral thalamus. However, EP studies of taste find that unilateral taste stimulation can lead to contralateral or bilateral EPs (Kida, 1974).

3.22 Taste aversions and taste memory

Our preference for food and our preference for taste depends on a trace, template or schema for that taste having been encoded and then being capable of retrieval. In short, preference depends on memory encoding and retrieval and the clearest illustration of this dependence can be seen in the development of taste aversion. This is of vital benefit to our/the animal's survival although there are examples where the aversion is incidental and of no survival consequence (e.g. nausea to food during treatment for illness). There is evidence that our memory for safe tastes activates different receptors to memory for aversive tastes, specifically muscarinic and n-methyl d-aspartate (NMDA) receptors (Guitérez *et al.*, 2003). Regionally, the NA has been implicated in this process as it is also involved in addiction, reward, feeding and motivation (Ramirez-Lugo *et al.*, 2007). One study has found forty-seven neurons in the area that respond to sugar and fifty-seven that do not respond to quinine (Roitman *et al.*, 2005). Areas of the ventral pallidum are responsive to sucrose but not sodium.

The NA is found in the forebrain and is part of the ventral striatum. It is divided into three areas – the core and shell in the back two-thirds and a third, rostral area. The shell and the core project to different regions – the shell to the medial ventral pallidum, lateral hypothalamus, PBN and the substantia nigra and the core to the doroslateral ventral pallidus and substantia nigra. Neurons in the NA respond to novel foods (Lee *et al.*, 1998) but foods in general can increase the amount of dopamine in the NA shell and core (Bassareo *et al.*, 2002). Dopamine has been found to increase in the NA to saccharin, before it is paired with an aversive stimulus. When paired, the dopamine reduces (Mark *et al.*, 1991). A similar phenomenon has been reported with olfactory (rather than gustatory) stimuli (Louilot and Besson, 2000). When a dopamine antagonist is present in the shell, conditioned taste aversion is impaired (Fenu *et al.*, 2001). Glutamate is another neurotransmitter implicated in taste conditioning in the NA. It is released during aversive conditioning and its presence in the NA may be necessary before aversive taste memory can be formed (Ramirez-Lugo *et al.*, 2007). The learning of safe associations may depend on the activation of cholinergic muscarinic receptors in the core of the NA (Ramirez-Lugo *et al.*, 2006).

Thus, the picture so far of the role of the brain, its function and its anatomy, as well as the role of the sensory apparatus in our ability to smell and taste is relatively good. More recently, however, the meat and drink of neuropsychology and neuroscience – studies of brain injury in humans, experimental ablation and tracing in animals – has been complemented by a small explosion of research using neuroimaging which itself followed on from some very productive work in the 1980s and 1990s on the development of techniques to record electrical potentials to odour stimuli. These fields are reviewed in Chapter 4.

4

PSYCHOPHYSIOLOGICAL AND NEUROIMAGING STUDIES OF SMELL AND TASTE

4.1 Psychophysiology

At the beginning of the twentieth century, one of electrophysiology's pioneers, Lord Adrian, had an ambition: 'the electrophysiologist,' he wrote, 'looking at a series of these records, could identify the particular smell that caused each one' (Adrian, 1953). This vaulting ambition, based on electrical recording from the OB of rabbits, frogs and hedgehogs, has not been realized. There is currently no olfactory signature to be found in the electrical potentials produced by the brain but there are certainly changes in the brain's script that are produced by odour.

Adrian had observed that electrical potentials could be recorded from the surface of the OB and that this potential increased with the intensity of odour. Similar experiments in humans have also reported increased potentials recorded from the olfactory mucosa. Thus, Osterhammel *et al.* (1969) found that a flow of coffee-saturated air passed over the olfactory mucosa elicited positive potentials in two patients and that these potentials increased as the flow increased. Kobal (1981) similarly recorded positive potentials from the mucosa to the odours of hydrogen sulfide (H_2S), amyl acetate and eugenol. OEP research has attempted to, and succeeded in, finding similar intensity-related changes in potentials recorded from the scalp (see below). Adrian's legacy is the human OEP described in section 4.4. First, however, is a discussion of the forerunner of the ERP, the EEG, and its role in olfactory perception.

4.2 Electroencephalography (EEG) and olfaction

The EEG is an electrical signal generated by the post-synaptic dendrites of millions of brain cells, usually pyramidal cells. The signal itself is very small (not more than 100 microvolts) and, therefore, must be amplified many times before it is fed through to a line polygraph (or monitor) and seen as the EEG wave. In humans, EEG is recorded distally from the scalp via a standardized electrode array (whereas in animals it can be routinely recorded intra- and extra-cellularly from a small number of neurons, using very fine recording filaments). The EEG signal is conventionally divided into four classical wavebands or frequencies: delta (1–4Hz, i.e. the wave appears 1–4 times a second), theta (5–7Hz), alpha (8–12Hz) and beta (13–22Hz); and others (e.g. gamma). Nobody quite knows what these frequencies represent, although the appearance of more or less of some of them is characteristic of particular behaviours (opening and closing eyes, for example).

Since Adrian (1942) first placed electrodes in the OBs of hedgehogs, there has been a modest contribution from EEG to our understanding of the role of the brain in olfactory perception. Moncrieff's (1962) early human study using rudimentary EEG recording techniques found that almost all of the odours presented (essential oils, synthetic oils and unpleasant-smelling chemicals) produced a reduction in alpha activity (measured via eight, bilaterally placed electrodes) in five participants (ylang-ylang had no observable effect on the EEG). Similar desynchronization had been noted two years earlier (Bushteva *et al.*, 1960), although electrode sites were not specified.

Subsequent EEG investigations have produced a mixed bag of findings and this is probably attributable to the number and kind of odours used, the recording and analysis techniques employed, the period of EEG selected for analysis, and participant selection. One study, for example, has reported differences in the amount of alpha activity over right and left hemispheres (the specific direction of the differences is not reported) and greater theta activity using fine fragrances as stimuli and four electrodes as recording measures (Lorig and Schwartz, 1988). In a separate experiment the same group used three different odours and four electrodes, and reported decreases in activity were found in the theta frequency. Kline *et al.* (2000a), using the odours of valerian, vanilla and water, found relative left-sided frontal increases in 58 women. Others, using eight different odours and nineteen electrodes, have reported no alteration in alpha but selective increases in theta (Klemm *et al.*, 1992). Lorig and Schwartz (1988) found greater theta to fragrance in the left hemisphere whereas theta was reduced to the other odours especially in right frontal and left posterior areas. Specific inter-odour differences were also found, with spiced apple producing less theta than did lavender. In Klemm *et al.* (1992)'s study, birch tar, jasmine, lavender and lemon were associated with increases in theta. Thus, although studies now employ much more sophisticated EEG technologies and presentation techniques, there continues to be a persistent variability amongst these findings.

Most studies recruit adult participants in conditions of optimum relaxation (eyes closed, seated) and employ exposure periods lasting from 6–7 seconds per odour (Moncrieff, 1962) to two minutes (Klemm *et al.*, 1992). The time consideration in recording responses to odour is important. If habituation to an odour occurs too quickly (the odour is not suprathreshold), then it is possible that the brain does not register such 'subliminal' stimulation. Two studies, however, have demonstrated differences in the amount of alpha (Lorig *et al.*, 1991) and theta (Lorig *et al.*, 1990) as a result of exposure to undetected galaxolide, and low vs. medium concentrations of spiced apple and lavender oil, respectively. In another study, greater EEG speeding was reported to peppermint odour in sleeping subjects (Badia *et al.*, 1990) although the procedure employed by the experimenters whereby the participants were woken up if they did not stir, told that the odour was present and then requested to resume their sleep raises some doubt as to its ecological validity. Kline *et al.* (2000b) also found that alpha activity decreased to concentrations of isoamyl acetate that were less than 0.001ppm. Owen and Patterson (2002) reported increased left-sided theta and alpha and increased right-sided delta to 'dislike' responses to low concentrations of the fruity damascenone, but results were not statistically significant. Finally, Harada *et al.* (1996) report decreased frontal delta band coherence – a computation of two EEG signals recorded from different sites that measure the relationship between the signals in terms of their frequency – during inhalation of methyl-cyclopentenolone and skatole and an increase in alpha 1 and 2 in the temporal region.

The difficulty arises as to whether the brain, after sensing the odour for the first time, then recovers from its initial arousal following stimulation and continues behaving as chaotically as before. A form of cerebral habituation may be in evidence if an odour is presented for too long a time. Associated with this difficulty, however, is the amount of EEG that is necessary for any statistical analysis to show reliable effects. Authors are fairly flexible concerning the minimum amount of artifact-free data that is required in order to make the EEG sample 'representative' but a minimum of sixty seconds would appear to be advisable. With systems employing algorithms that may correct for eye movement, the necessity of recording a given set of frames is only governed by the need to collect a reasonable sample of data. When no such algorithms are available, however, the amount of data has to be collected carefully, bearing in mind that contaminated records will be discarded and will therefore reduce the amount of useable and analysable data. Thus Van Toller *et al.* (1993) use samples of a maximum of ten seconds with no apparent eye corrections. It is imaginable that a considerable amount of this data is corrupted which leaves the total number of used frames unknown and the interpretation of their results questionable.

Martin (1998), in two experiments, found reliable and distinct effects of food odour, one in particular, on human EEG theta activity. Food odour is usefully employed for two good reasons: it is relatively familiar and it is identifiable, in contrast to complex olfactory stimuli such as perfumes or certain non-food odours which generate less accurate verbal labels (Lawless, 1991). In one experiment, EEG response to the 'synthetic' odours of chocolate, spearmint, almond, strawberry, vegetable, garlic and onion and cumin or no odour was recorded from nineteen electrodes in all EEG frequencies (delta, theta, alpha, beta 1 and beta 2). Exposure to the odour of chocolate was associated with significant reductions in theta when compared with the odours of almond and cumin. Exposure to the odour of spearmint was associated with a significant reduction in EEG theta when compared with the no-odour control. No significant effects were observed in other frequencies. In a second experiment, EEG response to the odours of real foods (chocolate, baked beans, rotting pork) and two controls (no odour and hot water) was recorded as in experiment one. The odour of chocolate was associated with significantly less theta activity than was any other stimulus. It was hypothesized that the alterations in theta reflect shifts in attention or cognitive load during olfactory perception, with a reduction in theta indicating a reduced level of attention.

In a study of professional perfumers and naïve participants, general EEG was more apparent in frontal regions when smelling the odours of jasmine, lemon, lavender, ylang-ylang, basil and skatol, whereas temporal, parietal and frontal areas showed activation in the general population and perfume sellers (Min *et al.*, 2003). Sanders *et al.* (2002), in a re-analysis of previously published data, noted the baseline EEG of participants and reported that those with right-sided baseline activation shifted to the left during perception of rosemary and lavender whereas left-sided participants shifted to the right. The significance of this is unknown.

4.3 Inhalation and EEG

The method of inhalation has also been reported to affect the pattern of the EEG. Nasal inhalation produces greater anterior/posterior diversity in beta than does mouth inhalation and appears to reduce left-hemisphere alpha (Lorig *et al.*, 1988). Lehmann and Knauss (1976) reported increased desynchrony as a result of inhalation of room-air while others have demonstrated a correlation between increased EEG amplitude in one hemisphere to the contralateral nostril stimulated during inhalation of room-air (Werntz *et al.*, 1983), as noted in Chapter 3. The respiratory mode of the subjects in EEG experiments is, therefore, important.

4.4 Olfactory evoked potentials (OEPs)

EPs elicited by somatosensory, visual and auditory stimuli have been successfully recorded for decades. Olfaction, however, had originally been a more awkward sense because of the difficulty in delivering odour to the sense organs reliably, controllably and in a measurable way that allowed easy detection. While the delivery of a tone or a flash of light is easily manipulated – and can be assumed to be perceived – the delivery of an invisible scent is more problematic. Consequently, olfactometry has been developed (described in Chapter 1) which allows for the computer-controlled flow of an odorant via tubes placed in, or at the entrance of, the nares at regular intervals/pulses and at a standardized velocity and temperature.

Using this technique, OEP research in humans has helped answer nine fundamental questions about the relationship between the brain and olfaction:

1. Does a flow of odour and a burst or pulse of air affect the OEP differently?
2. What components of the OEP are elicited by scents?
3. Do different scents evoke different OEPs?
4. Do odours of different intensities affect different components of the OEP?
5. Do olfactory odours differ from trigeminal ones?
6. Do the OEPs of men and women differ?
7. Does age affect the OEP?
8. Does the ISI affect the OEP?
9. Is the OEP affected by disease and illness?

The first human OEPs were recorded by Finkenzeller (1966) and Allison and Goff (1967) from the scalp of awake humans. In Allison and Goff's experiment, an olfactometer delivered air at a flow rate of 15.1 litres per minute through a flask that contained an odour – amyl acetate, benzene, propanol or acetic acid. Two seconds before the odour was presented, a burst of white noise was administered to cue the participant. Every six seconds, a pulse of odour was delivered after a pulse of clean air. The pulses were perceived as moderate- or strong-smelling odours, some stinging. The odours produced a positive-going peak at around 500–600msecs, which was sometimes preceded by a negative-going wave at 300msecs and followed by a negative wave at 600–700msecs, thus providing an early and (now well replicated) answer to Question 2. It is important to note here that this is where olfaction introduces one of its conventional confusions. Normally, the N1, P2, P3 and so on in other sensory modalities reflect peaks/troughs seen roughly after 100, 200 and 300msecs after stimulus onset. These undulations or components are labelled (usually) N100, P200, N200, P300 (P3a, P3b), N400, etc., depending on polarity and time of onset, i.e. the direction of their peaks (N = a negative peak, usually meaning downward-pointing; P = a positive peak, upward-pointing) and their time of appearance (in msecs) following stimulus presentation. The polarity refers to the wave's amplitude; the time of onset refers to latency. In olfaction, these labels are used metaphorically and as shorthand to refer to the first appearance of such waves. Thus, the N1 may appear at 300–400msecs in the olfactory modality because this is when the first peak appears. Similarly, the P3 occurs at approximately 700msecs. The onset time reflects the time taken for the signal to be processed by the cortex following transduction and analysis in the OE and OB.

The largest OEPs recorded by Allison and Goff, for example, were to odorants such as benzine and the smallest to acetic acid. The peak latency for amyl acetate was around 10–50msecs shorter than that for propanol in three-quarters of participants. The response was greatest in the vertex – a finding that has been replicated repeatedly since and which is why any OEP study always records from the Cz (the vertex) and invariably the Pz and Fz electrodes. In fact, Cz is recommended as essential for OEP recording (Evans *et al.*, 1993).

Some had suggested that such effects were due to stimulation of the trigeminus, rather than the olfactory nerve. To determine whether the effect was or was not due to stimulation of the intranasal free trigeminal nerve endings, Smith *et al.* (1971) attempted to record potentials to amyl acetate from three patients who had lost trigeminal function, and from three healthy individuals. In the healthy individuals, they simulated olfactory loss by anaesthetizing the olfactory area with cocaine flakes. The odours were delivered by a stream of air along a glass tube fitted to one nostril; air was delivered to the other nostril.

Smith *et al.* failed to observe EPs to odours in these patients suggesting that the OEP is 'elicited by stimulation of nasal trigeminal afferents and not to stimulation of olfactory receptors'. They also suggested that the OEP may be olfactory in origin but the trigeminus was needed to maintain mucosa autonomic function or the excitability of the OB; they dismissed this possibility. They concluded that it was 'questionable whether any substance is a purely olfactory stimulus'.

Later experiments, described below, have applied more sophisticated methods to the recording of OEP, including using odours considered to be pure olfactory odours. The purest test of the trigeminal hypothesis would be to compare the responses to pure odours and trigeminal odours on people with anosmia. This is what Kobal (1982) did, finding no observable OEP in anosmic patients exposed to vanillin and phenyl ethyl alcohol, two of the only demonstrably olfactory, non-trigeminal stimuli. Trigeminal stimulation, however, (via CO_2 administration) did produce a response (and larger amplitudes and shorter latencies than those elicited by odour in normosmic individuals). Hummel and Kobal (1992) compared OEPs elicited by vanillin (a pure odour), acetaldehyde, H_2S and ammonia in twelve 23–34 year olds from seven electrodes. The two olfactory stimuli showed maximal amplitudes at parieto-central sites; the trigeminal stimulants at the vertex (both findings are also reported by Kobal *et al.* (1992), who noted that latencies were shorter and amplitudes smaller when CO_2, H_2S and menthol were delivered to the left nostril and vanillin to the right nostril). Responses to the trigeminal odours were largest in the hemisphere contralateral to the nostril stimulated but there was no asymmetry for the olfactory stimuli. For H_2S and ammonia, amplitudes were higher and latencies were shorter than for the odours. These were also rated as more intense. This partly answers Question 5. The laterality effect for the trigeminal odours has also received some support from Rombaux *et al.* (2008) who noted higher right hemisphere amplitude in frontal sites regardless of the nostril stimulated. It is interesting that the right frontal region shows relative increased activation during experience of negative emotion in the visual modality (Davidson *et al.*, 2004).

These studies suggest that there may be two types of potential elicited by stimulants that are inhaled – those that stimulate the trigeminus and those that stimulate the olfactory nerve. Kobal and Hummel (1988) have suggested using the terms chemosomatosensory EPs (CSSEPs) and OEPs for these phenomena, respectively, an innovation that is sometimes adopted in the literature, but not always.

4.5 Methodological considerations: olfactometry

Chapter 1 described two methods of olfactometry used in EP research: the 'pulse method' and the 'flow method'. In the former, a puff of air of varying duration is directed towards the olfactory mucosa. Its concentration can be controlled and its effect is rapid – within 20msecs after stimulus onset. A problem with the pulse method, however, is that it also produces potentials elicited by non-odorous stimulation, usually at the time the odour is presented. That is, the EPs are the result of artefact, not olfactory stimulation.

The 'flow method' circumvents this problem by delivering a constant, gaseous flow over the mucosa. When required, an odorant can replace the gaseous air without affecting the volume or flow rate of the (now odorized) air, a little like a train being switched from one track to another (Kobal and Hummel, 1991). Plattig and Kobal (1979) found that no meaningful EPs were produced to non-odorous air using the flow method and EPs were produced to the odour of eucalyptol. The odour produced reliable positive and negative changes in amplitude and changes in EP latency. When the pulse method was used, EPs were also generated by non-odorous air suggesting that the potentials were produced by the somatosensory effect of the air. There were also component differences in response to the odour. Using the pulse method, N1 amplitude was smaller and the latency was 20msecs shorter during odour exposure, thus answering Question 1 in the list above. Extra-olfactory stimulation is an important artifact to remove in OEP studies. Eye-blinking, for example, commonly accompanies the perception of odours of high concentration and this can contaminate the EP by either producing larger components or masking others. The activity associated with blinking (the electronystogram – a positive potential at the vertex) is thus controlled in most modern experiments (Evans *et al.*, 1993).

The use of the vertex is important because OEPs are not always recordable from all areas of the scalp. Early studies and descriptions (Plattig and Kobal, 1979) found that no measurable potentials were found in some sites. Greatest amplitudes – in this experiment, to eucalyptol – were found at the vertex with the largest early components being recorded at the vertex; the second largest amplitude was recorded next to the vertex and contralateral to the stimulated nostril – in this study, the right (electrode C3). The second component's amplitude was highest at Fz (and F4). The more occipital the electrode location, the less recordable the OEP. This study showed that OEPs could be recorded by a small number of (largely central) electrodes and that the pattern of response was specific and non-random.

To examine the reliability of OEPs, Thesen and Murphy (2002) tested responses to amyl acetate on two occasions, separated by four weeks, at four electrode locations in twenty 26 year olds and ten 76 year olds. They found that reliability was higher for latency than amplitude and best for P2 latency at Cz and Pz (at 450–700msecs after stimulus onset) and P3 amplitude and latency at Cz and Pz (approximately 700msecs after stimulus onset). Thesen and Murphy call these potentials olfactory ERPs, a term that is usually used interchangeably with OEPs.

The early research described above stimulated three decades of work into the OEP and its determinants and, after having established a valid and reliable method of recording OEPs, studies have addressed, amongst others, issues such as whether odour concentration affects the OEP (e.g. Lorig *et al.*, 1993), whether young participants differ from old (Evans *et al.*, 1995), whether odour valence modulates the OEP (Masago *et al.*, 2001), whether men and women differ (Lundstrom and Hummel, 2006; Stuck *et al.*, 2006), whether early experience with an odour affects OEPs (Poncelet *et al.*, 2010), whether exposure to the left and right nostril produces different effects (Olofsson *et al.*, 2006), whether retronasal and orthonasal administration exerts different effects on the OEP (Rombaux *et al.*, 2007) and whether olfactory testing in ERP research can co-occur with recording of responses to other sensory modalities (Nordin *et al.*, 2011). This is discussed below.

4.6 Other methodological issues

Methodological issues such as odour intensity, ISI, passive/active smelling and effect of concentration have all been investigated using ERP methodology (Questions 4 and 8). Thus, concentrations of 16,000rpm of toluene produces significant amplitudes at Fz, Cz and Pz, but 1,600 and 1,800rpm do not (Prah and Benignus, 1992). Lorig *et al.* (1993) reported a concentration-dependent, P300-like component at 320–520msecs when fifteen participants were asked to press a key whenever they smelled an odour (concentrations of n-butyl alcohol), and responses were recorded from nine electrode locations. Mouth-breathing affects the OEP differently to nose-breathing (Lorig *et al.*, 1996). Participants in this experiment inhaled 2 per cent or 4 per cent of butanol in one nostril when they inspired or smelled the same but through the nose. Breathing affected the N1, and P2–P2 amplitude was higher during passive breathing.

In terms of comparisons with other senses, the P2 and the P3 can be similarly elicited by an odour (pyridine), a 1,000Hz tone and a white circle (Olofsson *et al.*, 2008) and an olfactory oddball – where the participant decides on whether the concentration of the odour is high or low and where there is always three times as many of one concentration than the other – can be recorded with different concentrations of citral (Pause *et al.*, 1995). There is also evidence that different sensory modalities can be recorded in the same testing session without any detrimental effect on wave integrity (Nordin *et al.*, in press). That is, although P3 amplitude was larger for auditory and visual stimuli and the P2 was smaller to visual stimuli, there were no significant differences between this protocol and an exclusively single-stimulus protocol.

The ISI is an important consideration when recording OEPs because habituation to odour can occur rapidly (as Chapter 1 reviewed). Kobal (1981) found that reducing the ISI from 52 to 32 seconds reduced amplitude by around 15 per cent. When reduced to 12 seconds, the P2 was only seen in 25 per cent of recordings thus suggesting that people were not able to detect any discernible difference between the first and second presentation (i.e. the odourless air in-between presentations). In a more sophisticated experiment, Wang *et al.* (2002) delivered amyl acetate to six young men and women in pulses of 35, 50, 75, 100 and 2,000msecs and with an ISI of 2.5, 5, 10 or 60 seconds. At all concentrations, amplitude decreased by 47 per cent with a shorter ISI. When the ISI was 2.5 seconds, people were unable to discriminate the strength of an odour. Similarly, Kassab *et al.* (2009) found that amplitudes to phenyl ethyl alcohol and H_2S decreased with decreasing ISIs (30, 20 and 10 seconds). Trigeminal stimuli, however, appear to be not as affected – the largest amplitude to CO_2 is seen when the ISI is 10 seconds.

The time course of OEPs as an odour is processed at various stages throughout the olfactory system has been difficult to study in humans due to the relative inaccessibility of the olfactory system and problems with odour delivery. However, using an ERP technique called source localization analysis, Lascano *et al.* (2010) were able to determine, using H_2S as a stimulant, that specific stages occurred between 200 and 1,000msecs. Activation begun in the hemisphere ipsilateral to the nostril stimulated, in the mesial and lateral temporal cortex – the amygdala, parahippocampal gyrus, the superior temporal gyrus and the insula. Following this stage, activity would move to homologous contralateral structures, then frontal structures.

4.7 Individual differences

4.7.1 Age

As Chapter 2 demonstrated, one of the most powerful individual differences that can affect olfactory perception is age. The older the individual, the greater the olfactory impairment. OEP research has, not surprisingly, reflected this phenomenon (Question 7). Thus, Murphy *et al.* (1994) administered a forced choice detection threshold test (described in Chapter 1) using amyl acetate to seven 53–84 year olds and seven 20–35 year olds and recorded OEPs from Fz, Cz and Pz. They found decreased N1 and P2 amplitude in the older group. This amplitude shortening negatively correlated with odour threshold so the higher the threshold the smaller the OEP. An experiment in which the same odour was administered to thirty-three 18–83 year olds in an odour identification test found increased latency in the older participants to the odour, suggesting that they took longer to identify it (Evans *et al.*, 1995). The authors calculated that the latency increased 2.5msecs for every year of age. The peak between the N1 and P2 decreased in the older men. The youngest women in Stuck *et al.*'s (2006) study of H_2S and CO_2 perception showed the largest amplitudes and the shortest latencies, thereby demonstrating a sex difference favouring young women (see section 4.7.2 below).

One particularly noteworthy study examined the effect of early experience with an odour on OEPs (Poncelet *et al.*, 2010). They examined OEPs to rose and mint in a group of European-French and Algerian-French participants. The mint and the participants were chosen because mint is experienced early in life in the Algerians and they are more exposed to it. When asked to talk about objects or events involving mint, they generate more discourse than to other odours (and than do non-Algerians). Poncelet *et al.* found that P2 latency was longer to mint than rose in the Algerian-French participants, suggesting to the authors that this component is 'sensitive to experience, development and age'.

In a study of the effect of attention and age on OEP appearance, Morgan and Murphy (2010) administered amyl acetate for 200msecs with an ISI of 30 seconds to ten young (23.9 years), ten middle-aged (50.6 years) and older (70.2 years) participants. Participants were also exposed to an image, thereby allowing a cross-modality comparison. In a passive condition, participants were told to ignore the stimuli; in the active, attention-salient condition, they were asked to press a button when they could detect an odour. The P3 latency increased with age to the odour and the image. The responses to the image in the attention conditions were similar across all ages, but latency was longer in the passive condition for the older participants (thus indicating a reduction in spontaneous processing speed with age).

4.7.2 Sex

Another important individual difference is sex and OEP studies have shown some differences in men and women (Question 6). Olofsson and Nordin (2004) reported more identifiable N1 and P2 amplitudes in women to three concentrations of pyridine (and these women gave the highest-intensity rating to the strongest concentration). Their P2 and P3 amplitudes were larger at all electrode sites and latency was shorter at Cz. When thirty-four men and women between 21–30 years old were exposed to peppermint oil (a bimodal odour – both trigeminal and olfactory), no sex difference was found in terms of sensitivity or hedonic rating of the odour, but larger amplitudes and longer latencies were seen in the left hemisphere in women (the right hemisphere showed larger amplitudes in men) (Lundstrom and Hummel, 2006). Savic *et al.* (2002) also found increased activation in left-hemisphere structures in women smelling acetone (another trigeminal stimulus) in their neuroimaging study.

4.8 OEPs and valence

Comparisons between olfactory and trigeminal stimulants aside (where the trigeminal stimulants are always perceived as the less pleasant), OEP investigations of the effect of the hedonic quality of odour are mixed. Masago *et al.* (2001) examined the effect of orange and eugenol in seven men using an oddball paradigm and found that latency was unaffected by odour but P3 amplitude was larger to orange when the orange odour was the rare/infrequent stimulus. Unpleasantness was associated with increased amplitude in Pause and Krauel's (2000) study. In a study of the sex-steroid androstenone, Lundstrom *et al.* (2006b) found that labelling this scent as 'body odour' or 'not body odour' had different effects on amplitude and latency. The body odour label led to larger amplitudes of late positive OEPs, a finding that is consistent – in terms of cognitive mediation affecting the brain's response to what is the same odour – with some neuroimaging research that has found different types of brain activation depending on how the odour is labelled. There was no difference in intensity ratings in Lundstrom *et al.*'s study but the less pleasant the participant found the odour, the greater the P3 amplitude.

4.9 OEPs and lateralization

Cerebral lateralization in the olfactory system is described in more detail below, but some ERP studies have shown that one hemisphere shows larger amplitudes than the other. Whether the nostril stimulated has consistent effects on (1) latency and amplitude and (2) asymmetry is unclear. A study of twent-eight 18–30 year olds administered amyl acetate monorhinally found that amplitudes in the left hemisphere were larger but that the nostril stimulated had no effect on amplitude. N1–P2 amplitude was larger for left-nostril stimulation and larger in the left hemisphere with stimulation of the same nostril (Olofsson *et al.*, 2006). This, as the section on the trigeminus showed above, is in contrast to the right hemisphere dominance for trigeminal stimulation.

4.10 Olfactory disorders

As the early OEP studies showed, olfactory disorders such as anosmia have clear and consistent effects on electrophysiology and these effects continue to be demonstrated (i.e. no OEPs are recorded in people who are anosmic; Rombaux *et al.*, 2007). The nature and extent of this disorder are described in Chapter 5. Other, neurodegenerative conditions also affect the OEP and some of the most important are described and also discussed in Chapter 5. But one or two studies, not discussed there, are worth noting. For example, the OEPs of patients with nasal polyposis are correlated with their olfactory test performance – the worse the performance, the later and smaller their OEPs (Rombaux *et al.*, 2006). Olfactory loss tends not to affect trigeminal perception – unless the trigeminus is damaged – but in this experiment, OEPs were smaller to CO_2 in these patients compared to those suffering post-traumatic olfactory loss and post-infection olfactory loss. Alcoholism is another disorder that affects olfactory performance, specifically identification, and OEPs (Maurage *et al.*, 2011). In this study, ten alcohol-dependent patients, with a mean age of 50 years, were exposed to the stimulants benzyl carbinol (a floral odour) and CO_2. Delayed N1–P2 latency was observed as was reduced P3 amplitude. Olfactory test performance correlated with OEP appearance, but the response to the trigeminal stimulus was not affected.

4.11 Psychological effects on the OEP

As with research examining the effect of psychological variables on auditory, visual and somatosensory EPs, investigation has also explored psychological effects on the OEP, although they are much less common. In one study, helplessness was induced by giving participants insoluble tasks and providing them with false (negative) feedback (Laudien *et al.*, 2006). Participants were exposed to two test odorants. P2 and P3 amplitudes were reduced during exposure to both odours. The latency for N1, P2 and P3 was longer when participants received the negative feedback. The same research group also examined the way in which context could affect the OEP (Laudien *et al.*, 2008). In their experiment, participants were told that the odour of *isobornyl acetate* was healthy, potentially hazardous or a common test odorant. If the participants believed the odour was harmful, N1 and P2 latencies were longer; if they believed it was healthy, they were shorter.

If participants expect to experience an unpleasant stimulus, aspects of the OEP can change (Bulsing *et al.*, 2010). For example, after having been exposed to an irritation-free odour (phenyl ethyl alcohol or H_2S), participants were told that they may or may not experience an irritating stimulus, CO_2 (depending on a visual cue presented before the odour). When expecting an irritating stimulant, N1 and P3 amplitude to H_2S increased and N1 latency decreased (N1 amplitude increased to PEA). The more intensely the odour was perceived, the faster the processing.

OEPs can also change when a person smells an odour while tasting (Welge-Lussen *et al.*, 2009). Sixteen men and sixteen women were exposed to a food-like odour (vanillin) or a non-food odour (PEA) while tasting sweet or sour stimulants. When taste and smell were congruent (vanillin-sweet) and smelling was retronasal, P2 latency was shorter than when the stimuli were incongruent. However, when smelling was orthonasal, P2 latencies were shorter in the incongruent condition (vanillin-sour). Intensity rating of the odours was lower after retronasal smelling. The study suggests that congruent stimuli in related senses can enhance the processing of odour when smelling is retronasal. Processing is also faster with incongruent stimuli during orthonasal smelling because the odour induces a feeling of conflict in the context of the taste.

4.12 Magnetoencephalography (MEG) and olfaction

Neurons generate magnetic as well as electrical currents and these magnetic fields can be measured from the surface of the head via a superconducting quantum interference device (SQUID). The machine detects the activity of magnetic fields from a large number of neurons because the magnetic fields generated by single neurons are very weak. These changes in magnetic fields can be recorded from the suface of the scalp. The subsequent recording is called the magnetoencephalography (MEG). Unlike EEG, MEG can be used to localize sources of activity fairly well and these sources can be plotted on a three-dimensional image of the participant's head. There are a few studies that have exploited the versatility of MEG to investigate olfactory processes.

The first MEG study in olfaction examined whole-scalp neuromagnetic signals elicited by PEA, H_2S and vanillin (Kettenmann *et al.*, 1996). This found that responses peaked at around 700msecs and were found in overlapping areas bilaterally in the superior temporal sulci. In a follow-up study, the magnetic changes that mirrored the typical OEP components from 200–700msecs were studied (Kettenmann *et al.*, 1997). Five men and five women between 27–40 years old were exposed to the odours of H_2S and vanillin (via olfactometry) for 200msecs with an ISI of 40 seconds. Between 200 and 800msecs, activity was found to both odours in regions between the superior temporal plane, the peri insular cortex, anterior-central insula and the superior temporal sulcus. An asymmetry in response was found by Hamada and Yamaguch (2001). In this study, two participants sniffed air, lemon, lavender and soy sauce. Peak latency occurred at 350msecs and there were odour-specific responses in the right hemisphere, especially in the right frontal cortex.

Recent MEG studies have employed an increasingly sophisticated methodological approach. Thus, Walla *et al.* (2003a) required twenty participants to engage in the shallow or deep encoding of words during exposure to PEA or no odour. In the shallow task, participants made judgements about the order of letters in the words; in the deep task, they decided whether the words described animate or inanimate objects. At 250–450msecs and 650–1,000msecs, activity was changed by the odour and by depth of processing. When the odour accompanied the word during encoding, recognition performance declined, a finding that is unusual in the literature (which tends to support a facilitative effect of odour on recall/retrieval). In a related experiment, the same group explored the effect of odour on the encoding of faces (Walla *et al.*, 2003b). Some faces were associated with an odour and participants were asked to recognize the faces they were exposed to in a later session. Recognition was poorer for the faces associated with an odour. At 200–300msecs, the faces encoded without odour present, more electromagnetic activity was seen. At 600–900msecs (when conscious processing of odour occurs, as the OEP studies above suggest), activity was reduced.

But findings are inconsistent. Walla *et al.* (2002) found increased activity to odour stimulation, and in those participants who consciously perceived the odour of PEA, when visually presented words were encoded. Walla and Deecke (2010) also asked ten young men and women to judge the intensity and hedonic quality of pictures of babies, flowers, erotic images, fear-inducing images and disgust-evoking images that had been paired with no odour, or high or low concentrations of two odours (PEA and H_2S). The positive evaluation of flowers increased during exposure to both odours; the negative evaluations of the disgusting images were also enhanced by odour exposure. The only decrease was observed for the baby condition in the highly unpleasant odour condition. At 300msecs, MEG activity was observed to all pictures during odour exposure; at 700msecs, this activity was only seen for the flower, fear and disgust pictures, suggesting that the interaction between odour and emotional responses to visual images is quick and may occur below the level of consciousness.

4.13 Olfaction and neuroimaging

Since the 1980s various neuroimaging methodologies have been employed to understand the neurophysiological processes and brain regions involved in human olfactory perception. Early studies using PET have been complemented by fMRI studies, those using MEG and, recently, near-infrared spectroscopy (NIRS). These studies have generally found activation in areas of the primary and secondary cortex in basic odour perception and the recruitment of other areas depending on the nature of the task undertaken during imaging – discriminating between odours, making judgements about the hedonic quality of odour, allocating attention to odour and perceiving odours under different physiological conditions (e.g. hungry vs. sated). To date, over 70 studies have been published using neuroimaging to study olfaction and trigeminal response. As an indicator of how specific the field has now become, one recent study has examined the effect of good and bad olive oil aroma on brain activation (Garcia-González *et al.*, 2011).

4.14 Neuroimaging and odour perception

The first proper neuroimaging study of human response to odour was published by Zatorre *et al.* (1992). In their PET study, they examined blood flow responses to the odours of Shalimar, butter extract, patchouli, kirsch, citronelle, lavender, lemon oil, maple extract and an odourless cotton wand in eleven young adults. They found activation at the junction of the temporal and inferior frontal lobes, an area equivalent to the piriform cortex, and in a third area that, as Chapter 3 showed, receives projections from the piriform cortex: the right OFC (BA11) (Figure 4.1).

It is probably accurate to conclude that subsequent studies have invariably found activation in the OFC and, often but not always, the piriform cortex, with other areas also showing activation depending on the odour and the nature of the task. For example, for each study that finds piriform cortex activation during odour perception (Zatorre *et al.*, 1992; Savic *et al.*, 2000, 2002; Sobel *et al.*, 2000; Bengtsson *et al.*, 2001; Poellinger *et al.*, 2001; Gottfried *et al.*, 2002; Cerf-Ducastel and Murphy, 2003; Kareken *et al.*, 2004; Porter *et al.*, 2005; Garcia-González *et al.*, 2011), there is an almost equal number showing no activation (Fulbright *et al.*, 1988; Zald and Pardo, 1997, 2000; Royet *et al.*, 1999, 2001; Yousem *et al.*, 1999b; Dade *et al.*, 2001; Plailly *et al.*, 2007a; Katata *et al.*, 2009;). There is also a tendency to find right-sided activation (e.g. Yousem *et al.*, 1997) although this is by no means consistent (e.g. Zald and Pardo, 1997) (Figure 4.2).

As Chapter 1 showed, sniffing affects the degree of olfactory perception that occurs – we take several sniffs of a low concentration odour but reduce sniffing when concentration is high. The effort reflects the degree of information we obtain about our olfactory environment. Breathing rate also affects brain activation (Colebach *et al.*, 1991), specifically in the primary and secondary motor

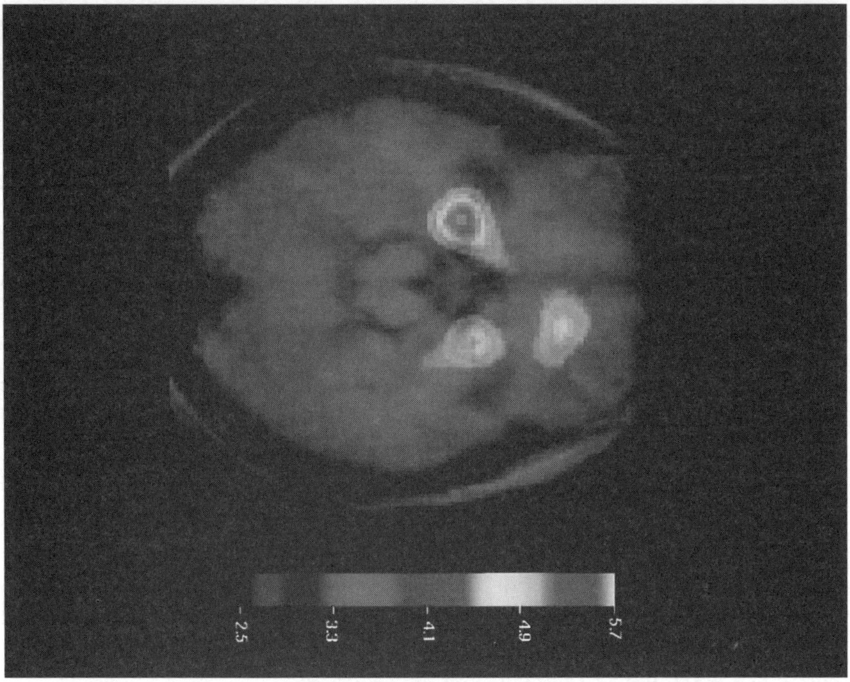

FIGURE 4.1 An image from the first neuroimaging (PET) study of olfactory perception

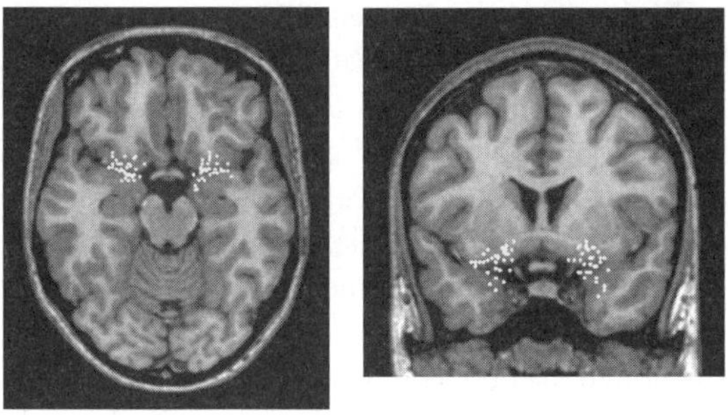

FIGURE 4.2 Temporal lobe/piriform cortex activation reported in olfactory neuroimaging studies (N=23)

cortices and the cerebellum (Gozal *et al*., 1995). Sobel *et al*. (1998) asked nine men and eight women beween 20–39 years of age to behave in two ways as fMRI recorded brain activation. In one condition, participants smelled the high and low concentration odour of vanillin and propionic acid and were asked to take a sniff and indicate whether an odour was present; in the second, they were asked to sniff for 800msecs, every 40 seconds when there may have been an odour or non-odorized air present. Both tasks activated the posterior lateral hemispheres of the cerebellum but activation was stronger to the highest concentration odour. Sniffing odourless air resulted in activation in the anterior part of the cerebellum. Sobel *et al.* suggested that cerebellar involvement provides a rapid feedback mechanism about odour concentration which then modifies output motor behaviour (i.e. our degree of sniffing) but conceded that the pathway undertaking this was unclear.

Deep brain stimulation (DBS) of the cerebellum appears to disrupt specific olfactory abilities (Kronenburger *et al.*, 2010) and there is evidence that atrophy of the cerebellum is associated with olfactory impairment (Abele *et al.*, 2003; Connelly *et al.*, 2003). Kronenburger *et al.* examined the effect of DBS treatment on 21 unilateral and bilateral, non-medicated patients with essential tremor. None of the patients had reported any olfactory problem but testing with Sniffin' Sticks found impaired odour threshold and slightly impaired discrimination.

Further up the cortex, Sobel *et al.* (1998) found that sniffing decanoic acid and vanillin was associated with activation in the piriform cortex and medial and temporal orbitofrontal gyri. The perception of a smell alone, with no active sniffing, activated the lateral/anterior OFC. They concluded that in humans, this is where the POC lay – the temporal and frontal cortex. They also hypothesized that impairment in olfactory perception in PD may be an impairment in sniffing rather than olfactory ability (a point taken up in Chapter 5).

In a study of sniffing without odour administration, eight men and three women were asked to undertake a single, brisk sniff every 20 seconds (this, the authors suggest, allows 60 per cent of maximum sniff pressure) (Koritnik *et al.*, 2009). They found that an extensive network of areas was activated, including the bilateral primary motor cortex (PMC), lateral premotor cortex, secondary motor cortex, anterior cingulate, basal ganglia, thalamus, upper pons, cerebellar vermis, PFC, entorhinal cortex and parahippocampal gyrus. Thus, sniffing activates those regions of the brain involved in motor execution and planning, not necessarily those involved in olfaction. Similar results had been reported by Kareken *et al.* (2004) who argued that the baseline comparison in neuroimaging studies is odourless sniffing and thus could produce activation. They used PET to examine activation to the odours of peach, lime, clove, amyl acetate, coconut, pine and orange or to the perception of a low flow of odorized air or non-odorized air. Sniffing the odourless air did not activate the piriform cortex but did activate the right OFC. When retronasal and orthonasal breathing were explicitly compared using fMRI, retronasal breathing was associated with activation in the mouth area of the central sulcus and greater activation than was orthonasal breathing in the perigenual, cingulate and medial OFC to chocolate odour, but not lavender butanol or farnesol (Small *et al.*, 2005b). Activation was greater during orthonasal breathing in the insula/operculum, thalamus, hippocampus, amygdala and caudolateral OFC.

Individuals can fairly rapidly habituate to odour, as Chapter 1 showed, and this is reflected in brain activation. When an odour (PEA) was presented to six men and four women between 24 and 44 years of age, for nine seconds or for 60 seconds, the POC was activated during short exposure and the piriform cortex, entorhinal cortex, amygdala, hippocampus and anterior insula activation increased (but only for a short while) during the longer duration (Poellinger *et al.*, 2001). After the increase, there was a return to baseline. OFC activation increases were maintained throughout and increases in the mediodorsal nucleus of the thalamus and caudate nucleus declined and returned to baseline after 20 seconds. Habituation is also thought to be responsible for the increase in blood flow to the secondary, rather than primary, cortex in individuals receiving intravenous odour – thiamine propyldisulfide and thiamine tetrahydrofurfuryl disulfide monohydrochloride which create a weak odour sensation (Miyanari *et al.*, 2007). Their fMRI study found that both stimuli activated the olfactory gyrus in the left hemisphere and that the first odorant activated the left subthalamic nuclei, precentral gyrus and right insula. The second activated the right superior frontal gyrus.

One view of how the neurophysiology of olfaction works is that the greater the amount of olfactory information received, the greater the number or neurons activated during its processing (Zou and Buck, 2005). Therefore, if a person perceives two odours in one stimulus, this should recruit a greater number of neurons or regions of the cortex, or recruit different regions. Boyle *et al.* (2009) created five different binary mixtures of odorant from pyridine and citral and administered them to twelve men and women in their 20s. The mixtures were associated with increased activation in the left cingulate cortex and right parietal and superior frontal cortex. The mixtures also activated the anterior OFC, but the single odour did not. Activation in the lateral OFC increased with the increasing impurity of the mixture, especially mixtures made of half of each odorant, perhaps indicating that this region plays a role in comparing and contrasting incoming stimuli.

4.15 Neuroimaging and valence/hedonic response

The first systematic study of neuroimaging and odour pleasantness exposed twelve 19–49-year-old women (only) to a mixture of sulfide gases (what the authors called a 'sulfide cocktail' comprising dimethylsulfide, methanethiol, ethanethiol and four malodours from the UPSIT) and so was an investigation of response to aversive stimulation exclusively (Zald and Pardo, 1997). It found increased activation bilaterally in the amygdala and in the left OFC to the cocktail. The more moderate malodours from the UPSIT activated the OFC, but not the amygdala. Activation in the left amygdala correlated with pleasantness scores (the higher the aversiveness rating, the greater the activation). The participants also altered their breathing during exposure to the sulfide and their muscle tension increased. This study clearly highlighted a role for the amygdala in the perception of an aversive stimulus and, given its role in mediating our response to threat, fear-inducing stimuli and other unpleasant stimuli, the amygdala's involvement in the perception of the sulfides is predictable. Studies since have also found activation in this structure to odours differing in valence, but not always. Fulbright *et al.*

(1998) exposed five men and eight women to beads saturated with clementine or isovaleric acid or to clear air in an fMRI study. The odour of clementine was associated with activation in BA8, left BA32, BA46/9 and the insula. Isovaleric acid activated all of these areas, apart from BA8. A comparison of activation found that activation to clementine was greater in the left insula and to isovaleric acid in BA6 (intriguingly, no OFC activation was noted either).

Royet *et al.* (2000) investigated whether there were any sensory-specific activational differences to pleasant and unpleasant stimuli. They asked twelve 20–30-year-old men to evaluate a range of pleasant, unpleasant and neutral odours, sounds and sights. Examples of the odours included, respectively, mint/rose, butyric acid and pepper. Examples of the different noises and sights included a flowing river/birdsong, a woman crying/explosion and wind; and a lake/picnic, road accident victim and a squirrel. Participants had to indicate whether they found the stimulus pleasant or unpleasant (in the neutral condition, they were asked to press a button at random). They found that all emotional stimuli – regardless of modality – activated the same core area/region. Activation in the OFC, temporal pole and superior frontal gyrus in the left was greater when people smelled, heard and saw emotional stimuli. Olfactory and visual stimuli also activated the hypothalamus and subcallosal gyrus. The only area showing odour-specific effects was the amygdala bilaterally. Rolls *et al.* (2003a) found activation in the OFC for pleasant odours only in eleven participants, using fMRI. They also found activation in the middle ACC to pleasant and unpleasant odours and that anterior activation correlated with the pleasantness score. Rolls *et al.* reported no activation in the amygdala but Anderson *et al.* (2003) and Royet *et al.* (2003) did. Royet *et al.* also measured electrodermal response while twenty-eight participants smelled pleasant and unpleasant odours. They found piriform cortex activation and that the amygdala and insula showed greater activation during perception of unpleasant odours (the more extreme the odours were rated, they greater the participants' EDR). The unpleasant odours also activated the left ventral insula in right-handers and the right in left-handers.

Anderson *et al.*, however, suggested that amygdala effects were intensity – rather than valence – dependent. They administered high and low concentrations of citral and valeric acid to eight young men and eight women. Activation in the amygdala was associated with odour intensity, not valence; OFC activation was associated with pleasantness. In Rolls *et al.*'s study, it was piriform cortex not amygdalar activation that correlated with intensity, indicating that these early studies were inconsistent. To examine whether the amygdala's role was mediating intensity, rather than pleasantness, Winston *et al.* (2004) administered high and low concentrations of citral, anisole, valeric acid and 2-heptanol (a neutral odour) to twelve women and six men aged between 22 and 29 years of age. They found an interesting pattern of results in that the amygdala was more responsive to high-intensity pleasant and unpleasant odours (suggesting that it mediates intensity primarily) but not to high-intensity neutral odour (which muddies the water a little).

To disentangle the contribtion of the cortex to odour mixtures, Grabenhorst *et al.* (2007) used fMRI to examine responses to mixtures of odorants that included jasmine and indole (unpleasant) in fourteen men and women in their 20s. Activation in the medial OFC correlated with odour pleasantness. The mixture produced activation that was similar to jasmine, not indole. When the mixure was more unplesasant than pleasant, activation was observed in the ACC and the middle OFC.

Of course, the preceding studies all assume that the odours are pleasant or unpleasant. Katata *et al.* (2009) asked participants to describe the quality of, and rate the intensity and pleasantness of, phenyl ethyl alcohol and undecatalactone directly and used these assessments rather than the a priori assessment of the experimenters to analyse brain activation. They found that the odours activated many of the areas already described – the middle OFC bilaterally, the left lateral OFC, the right insula, and the middle and anterior cingulate gyri bilaterally. In those who described the odour as unpleasant, there was activation in the left middle OFC and right lateral OFC; in those who described it as pleasant, activation occurred in the right ACC. The ACC appears to be a significant region for the perception of unpleasant stimuli and it is one of a network of brain regions involved in nociception. One case study of a 49-year-old man with neuropathic pain in the elbow and wrist and who reported that some odours increased his pain, found that unpleasant odour (pyridine, presented for 3–4 seconds) increased activation in areas involved in pain – the ACC, thalamus, amygdala and insula (Villemure *et al.*, 2006).

To examine whether context affected participants' responses to affective odours, Grabenhorst and Rolls (2009) asked seven men and seven women in their 20s to rate the pleasantness of the second of two odours presented, separated by six seconds, as fMRI recorded brain activation. The odours were citral, vanillin, hexanoic acid and isovaleric acid. The preceding odour would be rated as pleasant or less pleasant. They found that the anterolateral OFC tracked the relative subjective pleasantness and the anterior insula subjective unpleasantness. Absolute pleasantness was tracked by the medial/middle OFC.

A further context or priming effect was observed by Seubert *et al.* (2010). Seubert *et al.* had previously reported that participants were better at identifying disgust – but not happy or neutral expressions – in people's faces if they had been primed with an odour (pleasasant or unpleasant). Their second study employed fMRI to identify the neural correlates of this phenomenon. All odours activated the insula, superior temporal gyrus and entorhinal cortex. Again, disgust expressions were better identified when primed by an odour. Reaction time to this, the emotion that people tend to be slowest at identifying, was reduced to that seen for happy faces when the decision was preceded by an odour. Hypoactivation during this was found in the fusiform gyrus (the human face area), the middle frontal and middle cingulate cortex. Activation in the anterior insula was affected by the unpleasant odour. When images were conditioned to neutral odours (so that pleasant vanillin, neutral PEA and unpleasant 4-methyl pentanoic acid delivered via an olfactometer were paired with neutrally expressed faces of men and women), all odours were associated with activation in the posterior piriform cortex and pleasant and unpleasant odours were associated with the anterior piriform cortex (Gottfried *et al.*, 2002). The amygdala was activated during all odours but was strongest to the unpleasant odour.

Finally, one study has examined whether anxious sweat produced different brain activation to non-anxious sweat (there is more on biologically significant odour in section 4.17). Researchers took sweat from forty-nine donors before an exam or after a 30-minute bout of cycling (Prehn-Kristensen *et al.*, 2009). The odours were delivered, via olfactometer, to twenty-eight men and women. They were not intense odours – fewer than half of the presentations were detected. Odours from the anxiety condition activated the fusiform gyrus, insula, precuneus, corpus callosum, dorsomedial PFC, thalamus and cerebellum. The greatest activation was found in the insula suggesting to the authors that this area decodes emotional information of a social nature from the body and face.

4.16 Neuroimaging and trigeminal stimulation

Neuroimaging studies of trigeminal stimulants have shown selective activation in a range of areas, particularly the cerebellum and superior temporal gyrus with other specific regions and structures activated depending on the study. There is considerable overlap with olfactory areas. Hummel *et al.* (2005a) were the first to study systematically the effect of intranasal stimulation on brain activation using fMRI, comparing the effects of the olfactory odorants, PEA and H_2S, with the trigeminal stimulant CO_2 in seven women and twelve men in their 30s. Both types of odour activated the ventral insula, middle frontal gyrus and supplementary motor areas, particularly on the right. The CO_2 specifically activated the midbrain areas, dorsolateral prefrontal cortex (dlPFC), the frontal operculum, the superior temporal gyrus, the medial frontal gyrus and the anterior cingulate nucleus. The SII activation to odours and trigeminal stimulants is interesting given the much earlier MEG study of Huttenen *et al.* (1986) that found (in four participants) activation in this area to CO_2 (and at around 350–400msecs).

In a meta-analysis of neuroimaging studies of trigeminal stimulation, specifically stimulation by CO_2, Albrecht *et al.* (2010) concluded that consistent activation was found in: the brainstem, ventral lateral posterior thalamic nucleus, ACC, insula, precentral gyrus and primary somatosensory cortex (SI) and SII. Activation was also seen in the piriform cortex, OFC and association cortices. The latter suggests that trigeminal and olfactory stimulants activate similar, overlapping areas but that trigeminal stimulants activate two networks in the brain: those associated with nociception and those associated with chemosensory perception (Figure 4.3).

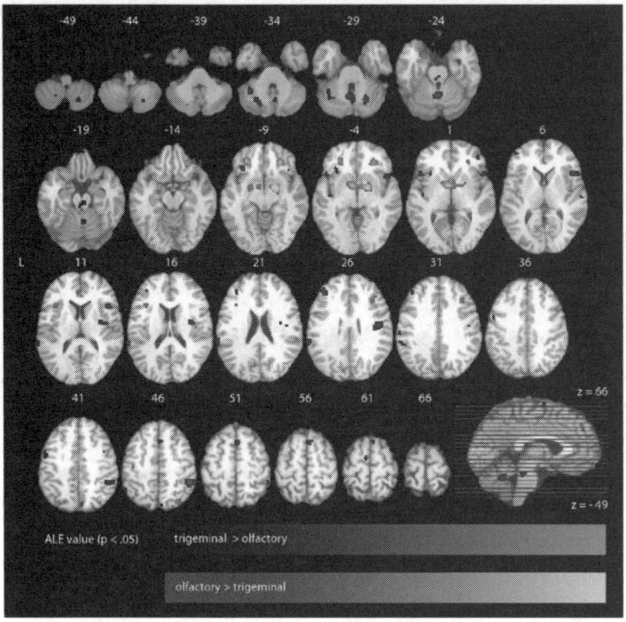

FIGURE 4.3 Areas of cortex activated by trigeminal stimulation in neuroimaging studies

4.17 Neuroimaging and 'biologically significant' odours

Studies suggest that a small collection of regions or structures is implicated in the neural processing of 'biologically significant' odours in humans. The first systematic study of the effect of semiochemistry on human brain activation was published by Savic *et al.* (2001). They exposed men and women to alpha 4, 16-androstene-3-one (AND) and an oestrogen-derived semiochemical, oestra-1,3,5 (10), 16-tetraen-3-ol (EST) using PET. They found that AND increased activation in the hypothalamus in women (specifically, in the preoptic and ventromedial nuclei) whereas EST increased hypothalamic activation in men (in the paraventricular and dorsomedial nuclei). The effect with EST in men had originally been reported by Sobel *et al.* (1999b) who found similar hypothalamic activation (and inferior frontal gyrus activation), even when the odour was not consciously detected. In this study, high and low concentrations were administered – activation was greater in the right side of these regions when the concentration was high (although not detected). In a study of heterosexual women exposed to androstenol and four ordinary odours (with air as the control), activation was found in the anterior hypothalamus in response to the semiochemical (Savic and Berglund, 2010). It was perceived as weaker than the other odours (which activated the typical odour areas – piriform cortex, amygdala, insula and ACC). In women, the brain also responded more quickly to AND than it did to control odours (Lundstrom *et al.*, 2006a). Jacob *et al.* (2001), however, found more widespread changes when alpha 4, 16-androstradien-3-one was swiped under the noses of ten 20–35-year-old women. Here, activation was seen in the hypothalamus, as well as the PFC, cingulate cortex, amygdala, visual cortex, thalamus, basal ganglia, premotor cortex and cerebellum.

As these semiochemicals derive from glands that also contribute to body odour, Lundstrom *et al.* (2008) examined the effect of body odour on brain activation in fifteen young (18–28 years old) heterosexual women, using PET. Body odour is regarded as an olfactory stimulus rather than a trigeminal one (Zernecke *et al.*, 2010).

Body odour in general activated the posterior cingulate cortex, occipital gyrus, angular gyrus and the ACC (there was no prefrontal or orbitofrontal activation that exceeded baseline whereas the control odour was associated with increases in this region and in the piriform cortex). A deactivation in anterior OFC was found. A stranger's odour activated the inferior frontal gyrus and the amygdala and was judged to be more intense and less pleasant smelling than other odours. The odour of a friend activated the central sulcus, occipital cortex and pre-somatosensory cortex, areas also known to be active during the recognition of a friend's voice or face (Shah *et al.*, 2001).

The electrophysiological effects of androstadienone and another component of sweat (2-methyl-3-mercapto-butanol or 2M3M, the chemical that contributes to body odour's repellent aroma) have also been demonstrated (Chopra *et al.*, 2008). Chopra *et al.* measured OEPs in prepubescent, pubescent and postpubescent boys and girls. The girls' thresholds for both odours were constant across the ages; pubescent boys, however, were less sensitive and had higher thresholds. Pubescent boys showed increased P1 and N1 latencies and an increase in the P2 to both odours, suggesting that it took them longer to evaluate these odours. P2 amplitude was larger to 2M3M in prepubescents, a finding the authors attribute to the novelty of this odour for prepubescents.

Curiously, homosexual men, like heterosexual women, show increased activation in the hypothalamus when smelling 4, 16-androstadien-3-one (Savic *et al.*, 2005). In Savic *et al.*'s study, common non-steroidal odours were found to activate similar areas in homosexual men and heterosexual men and women, suggesting that the response to the chemical can depend on sexual orientation more than sex. Maximal responses to AND were found in the preoptic area and anterior hypothalamus (common odours activated primarily the olfactory cortex regions). These data suggest that biologically significant odours recruit or activate different brain mechanisms and regions than do non-steroidal odours – the anterior hypothalamus is required for processing the pheromonal component of odour whereas the conventional olfactory cortex processes other odour. In a PET study of responses to AND and EST in lesbians, and heterosexual men and women, lesbians did not activate the preoptic hypothalamus when presented with AND, unlike the heterosexual women (Berglund *et al.*, 2006). In addition, their responses to EST was very similar to those of heterosexual men in the hypothalamus (specifically the dorsomedial and paraventricular regions). Their responses, therefore, were more like heterosexual men than women.

In a study in which the body odour preferences of homosexual and heterosexual men and women were measured, researchers found that heterosexual men and women and lesbians preferred the body odour of heterosexual men to gay men; gay men preferred the body odour of gay men (Martins *et al.*, 2005). None of the participants knew the sex or sexual orientation of the person who donated the body odour. All participants over 25, apart from gay men, preferred the body odour of lesbians to that of gay men. Gay men preferred the body odour of heterosexual women to men. The results suggest that attraction to body odour can be sex – and sexual orientation – specific.

A further extension to this work has been reported by Berglund *et al.* (2008). They examined responses to AND and EST in twelve non-homosexual male-to-female transsexuals and twelve male and female control participants, using PET. The male-to-female transsexuals and the female controls activated the hypothalamus in response to AND, and the amygdala and piriform cortex in response to EST. The male controls showed hypothalamic activation to EST. The data, suggest Berglund *et al.*, indicates 'a pattern of activation away from biological sex', which opens up a new avenue of investigation and a new interpretation of the significance of neural activation to steroidal odours in men and women.

4.18 Neuroimaging and imagining odour

As with mirror activation in other senses, imaging odour activates areas very similar to those activated during actual olfactory perception. The first study of its kind (fMRI), asked nine men and twelve women between 22–48 years old to imagine or smell the odour of amyl acetate (banana) or menthone (peppermint) (Levy *et al.*, 1999c). Levy *et al.* found that brain activation during imagery was less pronounced than during actual smelling but there was a significant overlap in the areas activated by both conditions: the bilateral frontal cortex, the entorhinal cortex, the bilateral temporal lobes and areas near the hippocampus and cingulate cortex. They also examined the responses of three patients with reduced olfactory sensitivity and observed that responses were greater during imagery than perception before these patients received treatment and that the pattern reversed after treatment. Levy *et al.*'s (1999a) earlier fMRI study of patients with poor olfactory sensitivity found a reduction in the posterior OFC, the cingulate cortex, piriform cortex, hippocampus and amygdala during perception. In Levy *et al.*'s (1999c) study, men in general appeared to show greater activation than women to the imagining and perception of banana. Another, later, systematic study compared brain activation in twelve good odour imagers who smelled or imagined phenyl ethyl alcohol, pine needle oil, lemon and strawberry (Djordjevic *et al.*, 2004b). During imagery, they found increased activation in the left primary cortical areas, including the piriform cortex, and left secondary areas, including the OFC. The left-sided activation is comparable to that seen during visual imagery (Sack *et al.*, 2005). There was bilateral rostral insula activation. Areas in which activation was seen in both conditions were the left and right rostral insula and left POC.

Bensafi *et al.* (2007) compared activation during the imagining of pleasant and unpleasant odours (ammonium sulfide and strawberry) in sixteen participants with an average age of 28. Consistent with Djordjevic *et al.*'s (2004b) findings, both activities activated the POC and the insula. Differences in odour type were reflected in different degrees of activation in the left piriform and anterior insula and the posterior insula during perception and imagery. They argue that the activation in the piriform cortex may reflect early attempts at categorization of odour, 'deciding very quickly whether an environmental stimulus is noxious or dangerous'. In participants thought to be expert at the analysis of odour and in imagining scent, Plailly *et al.* (2012) found that student and professional perfumers who imagined odour showed activation of the piriform cortex. The more expert the perfumer, however, the less the activation found in the posterior piriform cortex, the OFC and the hippocampus which suggests to the authors that there had occurred a 'reorganization of key olfactory and memory brain regions' with experience.

4.19 Cognitive variables: making decisions about, and remembering, odour

Almost all of the neuroimaging studies described so far have used odour in a fairly cognitively neutral way. An odour – of varying quality – is presented to men and women, young and old. The participant is a passive recipient of the stimulus, except in the few studies where they have to indicate degrees of perceived pleasantness and unpleasantness. There are, however, a handful of studies that have examined the neurophysiological basis of the interaction between odour perception and cognition.

One of the earliest studies, if not the earliest, asked fifteen individuals to indicate whether a large range of odorants was familiar, associated with something that could be eaten and whether the stimulus delivered via an olfactometer was an odour or air (Royet *et al.*, 1999). When participants were required to make judgements about familiarity, activation was observed in the right OFC, the subcallosal gyrus and the right and left interior frontal gyri and the anterior cingulate. When required to make judgements about edibility, activation was observed in the primary visual cortex and a decrease in activation in the OFC. The right-sided activation is consistent with that reported by Zatorre *et al.* (1992). A similar paradigm was adopted by Savic *et al.* (2000) who asked eighteen 22–27-year-old women to sniff single odours (eugenol, butanol, lavender oil, cedar oil), discriminate between odours based on their intensity, discriminate on the basis of their quality and recognize odours they had previously been exposed to. The perception of odour was associated with increased activation in the amygdala, piriform cortex, insula and cingulate cortex. When making judgements about intensity and quality, activation was observed in the left insula and right cerebellum, and when making judgements about quality, in the right caudate nucleus and subiculum. Recognition memory was associated with activation in the piriform cortex. A replication by the same group using an auditory task as a control also found increased right OFC activation during all tasks – detection, intensity judgement, hedonicity and familiarity (Royet *et al.*, 2001). The degree of activation was highest during familiarity judgements and lowest when participants had to judge whether an odour was present or not. Increased left OFC activation was seen during familiarity judgements and when people made judgements about pleasantness (and was highest in this second condition), suggesting to the authors that the region undertakes a degree of parallel processing. As before, judgements about edibility was associated with primary visual cortex activation.

When men were asked to make judgements about the familiarity of odour or to indicate whether they could detect it, activation was found in the right piriform cortex (the most strongly activated), hippocampus, left inferior frontal gyrus and amygdala when the first condition was subtracted from the second (Plailly *et al.*, 2005). Discrimination between two odours resulted in greater activation in the left anterior insula and frontopolar gyrus compared with the detection of a single odorant (Plailly *et al.*, 2007a). As discrimination involves holding information in the mind and storing it for future comparison, the task clearly includes a working memory component. It is of interest, therefore, that compared to the odourless control condition, the discrimination condition was associated with increases in left lateral OFC and middle frontal gyri.

In an extension to this paradigm, Rolls *et al.* (2009) examined whether the brain regions responsible for making judgements about intensity were different to those for assessing the affective quality of odour. Participants were exposed to two odours and asked to indicate which was more intense/pleasant. They also gave the second odour an absolute rating of intensity and pleasantness. When making judgements about pleasantness, activation was found in the dorsal medial areas and in the agranular insula. When making decisions, activation was higher in the medial PFC (BA10). When making judgements about intensity, activation in the dorsolateral PFC, ventral premotor and anterior insula was observed. When giving the absolute pleasantness ratings, activation was seen in the mid OFC. The study follows on from another Rolls *et al.* investigation in which participants were asked to remember and then rate the intensity of jasmine (Rolls *et al.*, 2008). Compared with individuals who were asked only to rate the odour, individuals who were asked to remember and rate for intensity showed greater activation in the medial OFC, inferior frontal gyrus and pregenual cingulate. The activation occurred in the period before the odour was actually presented (indicating that the areas were involved in preparing to experience the stimulus) and continued after the odour was removed. The results suggest that the areas required to make judgements about the emotional attributes of an odour may be different to those required to make judgements about its physical attributes.

A comparison between such judgements made about odour and about visual stimuli (to examine whether changes were olfaction-specific or modality-general) in eight Japanese men under 26 years of age found that visual and olfactory stimuli activated the left OFC and right piriform cortex and the occipital cortex bilaterally (Qureshy *et al.*, 2000). When naming odours, there was specific activation in the left cuneus, right anterior cingulate, left insula and cerebellum bilaterally. Remembering an unfamiliar odour was associated with activation in the cerebellum bilaterally and in the left cuneus. Selectively attending to an odour was associated with stronger activation between the mediodorsal thalamus and OFC, compared to attending to a tone (Plailly *et al.*, 2008).

Multisensory interaction has been explored in a very few imaging studies but the studies that do exist are very promising. One of the most remarkable exploited our famed inability to name odours well. De Araujo *et al.* (2005) asked twelve men between 23 and 35 years of age to smell a pleasant (alpha-ionone) and unpleasant odour (isovaleric acid) as fMRI measured brain activation. However, they told one group that the malodour was 'cheddar cheese' and the other that it was 'body odour'. When asked to rate the odours for pleasantness, the group that was told that the malodour was body odour judged it to be less pleasant than the group told it was cheese. The remarkable point about the study is that these judgements were matched by differential brain activation. Increased activation was observed in the ACC and medial OFC when the odour was called cheese and this activity correlated with pleasantness ratings. The odour, of course, was the same for both groups (Figure 4.4).

Gonzalez *et al.* (2006) examined whether the scent of a stimulus activated similar regions to its name. They used fMRI to measure twenty-three young women's brain activation as participants passively read odour-related words such as garlic, cinammon and jasmine, and neutral words (there was no direct comparison between scent and word, however). Increased activation to the odour words was observed in the piriform cortex bilaterally, and the amygdala. No activation was seen in the OFC. The result is very similar to that found with other modalities, such as studies showing that the generation of colour words activates areas similar to those activated during the perception of those colours (Martin *et al.*, 1995).

An examination of whether the senses share a network of brain regions involved in making judgements about familiarity compared fMRI responses to forty-eight familiar and unfamiliar odours and pieces of music (Plailly *et al.*, 2007b). Familiarity was based on the participants' own judgements, rather than a priori. Activation was seen in a network of areas that was common to both types of stimuli regarded as familiar – left superior/inferior frontal gyri, precuneus, angular gyrus, parahippocampal gyrus and hippocampus (cf. Royet *et al.*, 1999, 2001). Odours and music that were unfamiliar activated (to a much lesser extent) the right insula, which the authors suggest may reflect novelty detection.

A study, the topic of which is given more space in Chapter 6, that exposed individuals to either different types of sweet smell (food-related – chocolate/strawberry, or floral – lilac, roses) or a sweet taste found that each sense activated some overlapping regions (Veldhuizen *et al.*, 2010). The taste of sucrose activated the insula, which was also activated by the odours (as was the piriform cortex and OFC). But the response to the food odour in the insula correlated with its perceived

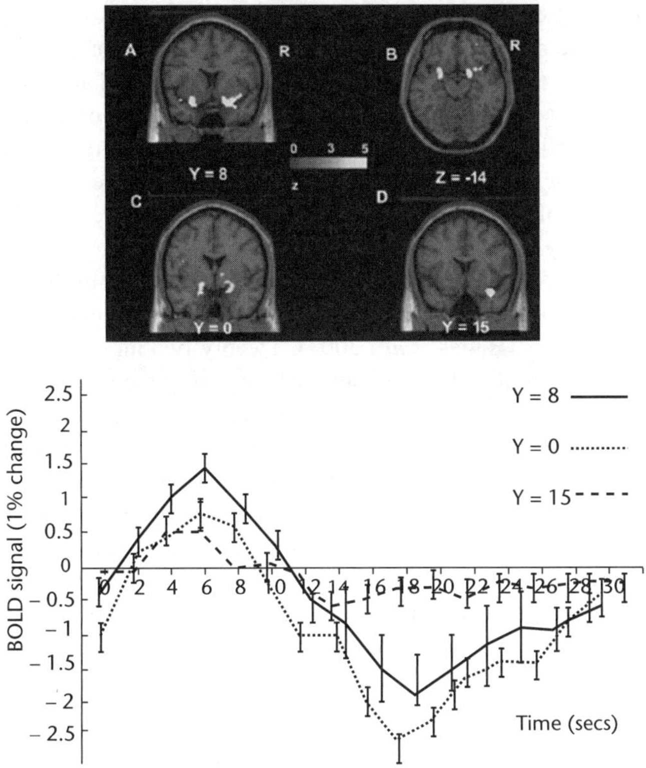

FIGURE 4.4 Scans from de Araujo *et al.*'s (2005) study in which areas associated with odour pleasantness are highlighted

sweetness. This, according to the authors, suggests that the insula acts as a form of multimodal area which processes the taste-like aspects of food odours.

Recognition memory for odour tends to localize activation in the piriform cortex (see Chapter 3). In a twist to the traditional memory paradigm, Herz *et al.* (2004) asked five young women to recall a memory in which their perfume was a key component. These perfumes (e.g, Royal Secret, Opium, White Musk, etc.) were then bought and presented to the participant who was asked to think about any memory these scents evoked. Memory evoked by odour was associated with increased activation in the amygdala during recall (the experimental odour also activated the cerebellum compared with a control odour), suggesting that the structure played a role in recalling personally meaningful (emotionally provocative) events. It also has a role in encoding and retrieval that does not involve affective stimuli.

In a more conventional, cross-modal study, Cerf-Ducastel and Murphy (2006) exposed 2–25 year olds to sixteen familiar odours (spices, herbs, fruits, etc.) before fMRI scanning and then later presented them with the names of those odours (or odours that had not been experienced). Participants were asked to press a button if they had experienced the odour corresponding to the label. In the first run of the experiment, increases in a variety of brain regions were seen to the odour name: these included the piriform cortex, amygdala, superior temporal gyrus, anterior cingulate, inferior frontal gyrus, OFC, superior medial frontal gyrus, lingual/fusiform gyrus and parahippocampal gyrus. By the third run of the experiment, activation had declined in the right hippocampus, frontal gyrus, lingual gyrus, parahippocampal gyrus and middle and frontal gyrus. The results suggest that there is a network of regions, focused on the mesiotemporal area, that is involved in episodic memory.

One study has examined the activational changes that occur during the performance of a standard working memory task while participants are exposed to unpleasant odours (Habel *et al.*, 2007). Twenty-two men, with an average age of 31 years, completed the n-back task requiring them to press a button when they saw an x on the screen or a letter that matched the last one seen, in the presence of unpleasant yeast odour (or air) as fMRI measured brain activation. Performance was worse, as expected, in the malodour group and reaction time was slower. The researchers divided participants into those who had been affected by the malodour and who had not. The unaffected group produced greater activation in the frontal, parietal and cerebellar cortices. The affected group showed activation in the temporal and medial frontal cortex. Poorer performance in the presence of malodour was associated with higher activation in the inferior frontal gyrus. Higher activation in the precuneus was correlated with less of a performance decrement.

4.20 Odour-specific reactions

This section has been separated, not because the studies here occupy a special place in the gallimaufrey of neuroimaging and odour research, but because they have examined the effect of individual types of odour explicitly, and this may be how the field will productively move forward (cf. Howard *et al.*, 2009). One of these, chocolate, is discussed in the section on taste. To date, studies have employed a huge variety of odours – administered almost willy-nilly – without necessarily examining the significance of each odour. Is one odour more visually evocative than another, is it more relaxing or arousing?

The scent of food has been used in previous studies (intermingled with other odorants), and food odour is a form of biological odour in that it conveys biological significance (such as triggering or exacerbating hunger). The first study to examine the effect of hunger on olfactory perception using fMRI took five hungry participants and exposed them to 334msecs bursts of vanillin and banana before and after being fed (they were fed bananas) (O'Doherty *et al.*, 2000). The OFC was activated in both conditions (hunger and satiety), as was the insula, part of the secondary motor cortex, premotor cortex and dorsal anterior cingulate. OFC activation to banana declined after eating but increased to vanillin suggesting that this area was responding to the reward-value of the stimulus (such as whether it delivered pleasure). A later study found that activation in the amygdala and mediodorsal thalamus to food odours predicted the delivery of an associated drink (peach, pineapple), but not the expectation of a tasteless drink (Small *et al.*, 2008). At a gustatory level, the left insula and operculum responded to the drinks and the right insula and left OFC responded to the relevant odour and drink. The pattern of results suggests that there may be separate but overlapping areas of the brain mediating the anticipation of food and the consummation of that anticipation.

Bragulat *et al.* (2010) used fMRI to examine the response to food and non-food odours by five obese and five lean individuals who had fasted for 24 hours. Most research in neuroimaging and obesity/eating has used visual stimuli to examine brain activation in eating (and to measure brain activation during hunger and satiety as explained in Chapter 6). Two odours of sweet foods and two of fatty foods, plus four non-food odours (e.g. Douglas fir) were administered. The food odours generally produced greater blood flow than did the non-food odours, particularly in the OFC, bilateral insula, operculum, anterior/posterior cingulate and ventral striatum. In the obese, activation was greater in the hippocampus and parahippocampal gyrus bilaterally. The lean showed more activation in the posterior insula. The authors note that the areas activated are similar to those activated by cues to addictive substances.

One such addictive substance is alcohol and it is well known that a small dose of alcohol can lead to a desire to drink. The odour of an unfamiliar alcoholic drink was associated with increased activation in the NA in participants not at risk of alcoholism (Kareken *et al.*, 2004). The same group also compared the responses of ten hazardous drinkers (who drank an average of 5.9 drinks a day) to the odours of alcohol, grass, leather, chocolate and grape (Bragulat *et al.*, 2008). Greater activation in the OFC, medial frontal lobe, retrosplenial areas and precuneus/posterior cingulate was found to the alcohol odour than to the grass and leather odours, particularly in the posterior cingulate gyrus.

Finally, Garcia-González *et al.* (2011) compared fMRI responses to three high-quality olive oils (Royal, Arbequina and Picual – green and fruity) and three others described as rancid and fatty in fourteen individuals. All odours were associated with activation in BA10 and BA11 bilaterally (see Figure 4.4) but pleasant odours were associated with activation in the inferior frontal gyrus, an area previously associated with recognition of familiar odours (Savic and Berglund, 2004). Unpleasant odours were associated with activation, strongly, in the inferior parietal lobe (BA40) bilaterally and to the left BA38 (temporal gyrus), BA24, BA32 and BA33.

4.21 Neuroimaging and taste

The first (PET) neuroimaging study of taste observed what a number of subsequent studies have found since. Tasting of a 0.18 per cent mixture of saline and water resulted in increased blood flow in the insular cortex, ACC,

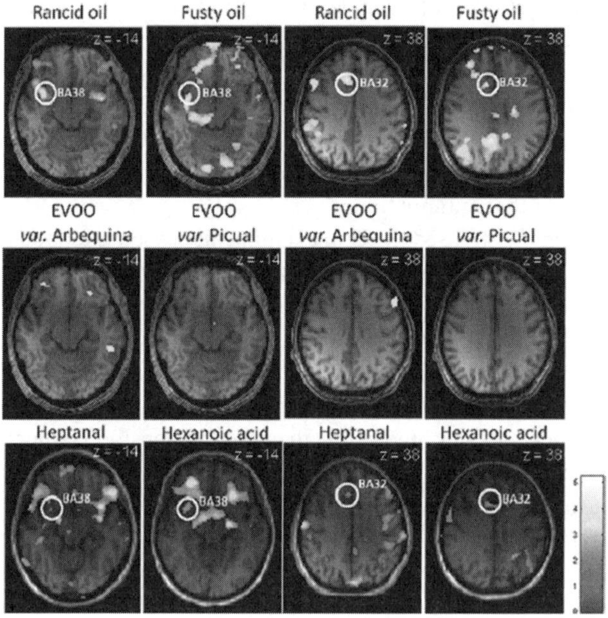

FIGURE 4.5 The field of olfactory neuroimaging has progressed so rapidly in the past 20 years that studies can now focus on specific qualities of odour, in this example good and poor olive oil

parahippocampal gyrus, thalamus, lingual gyrus, caudate nucleus and the temporal gyri (Kinomura *et al.*, 1994). The most consistently activated of these areas in taste research is the insula and subsequent studies have implicated, to a greater degree than others, the frontal operculum. These regions, together with the base of the pre- and postcentral operculum (Rolandic) and supra temporal plane, are considered to represent the PTA (Pribram and Bagshaw, 1954). MEG studies also implicate these regions, and the area between the frontal operculum and the insula, in gustatory processing (Kobayakawa *et al.*, 1996a, 1996b, 1999; Onoda *et al.*, 2005). These areas may not necessarily play a part in the hedonic response to taste, however – this may be the role of the caudolateral OFC (Rolls *et al.*, 1988, 1989; Critchley and Rolls, 1996; Small *et al.*, 1997a). The activation in the insula appears to increase over time and with repeated stimulation. Thus, five participants who were made familiar with specific tastes over ten weeks showed increases surrouding the Sylvian fissure, upper insula, frontal operculum, the foot of the precentral and postcentral gyri, and the ACC and caudomedial thalamus over this period (Faurion *et al.*, 1998). Small *et al.* (1997b) combined smell and taste and ensured that the stimuli matched in terms of quality or clashed. They found that activation was seen in the PTC and secondary

TC, as well as the POC but that this activation was stronger when the different stimuli were presented together. The clash produced activation in the amygdala and basal forebrain.

The inclusion of water as a control is important because, as the section on swallowing in Chapter 6 demonstrates, mouth movement and swallowing also activate brain areas. Several studies of macaques have shown that the PTC, esecially the anterior insula and frontal operculum, and secondary taste areas are activated when the animal imbibes water (Rolls *et al.*, 1990; Yaxley *et al.*, 1990). Reponses in the PTC do not appear to dissipate over time, but responses in the secondary area do (Rolls *et al.*, 1989). Extending the previous neuroimaging work to include sweet taste (sucrose) and comparing responses to water, Frey and Petrides (1999) found that taste activated the insula and operculum bilaterally. De Araujo *et al.* (2003b) compared brain activation when participants were either thirsty or sated. They found that water was associated with increased activation in the frontal operculum, anterior insula and caudolateral OFC and the activity in the first two areas was maintained when the participants were sated. Activation in the third area, however, declined when the participants were sated.

Comparing NaCL with aspartame, quinine hydrochloride (bitter), glycyrrhizic (liquorice), umami and D-threonine, Faurion *et al.* (1999) found that when the tongue was stimulated bilaterally in five left and five right-handers, activation was observed in the insula and perisylvian regions bilaterally. The study replicated, mostly, that of Cerf *et al.* (1998) who used fMRI to study similar tastants. They found activation in the insula and perisylvian areas, with bilateral activation in the superior insula. In terms of unilateral projections to the insula, Faurion *et al.* found ten such occurences in the left hemisphere, versus two in the right, in right-handers and twelve occurrences in the right hemisphere versus four in the left in left-handers, suggesting that handedness interacts with taste perception. Cerf *et al.* also observed an effect of handedness on activation in the lower insula and suggested that this might be the first stage of lateralization in the gustatory cortex.

A comparison of what the authors called unimodal or multimodal tastants, using fMRI, Kringelbach *et al.* (2004) also found activation in the PTC. Unimodal tastants were those such as glucose, salt and mineral water. Multimodal tastants were those such as tomato juice and chocolate which are gustatorily and cognitively more complex. Activation was observed in the anterior insula, frontal operculum, caudolateral OFC and dlPFC. A comparison between lemon juice and chocolate also found activation in the same areas – the OFC and insula (Smits *et al.*, 2007) – as have studies of caffeine and sucrose (Haase *et al.*, 2007), the latter also finding activation in the frontal and Rolandic operculum. A comparison between glucose and artificial saliva perception found activation in the PFC to glucose and also that activation in the medial and temporal cortices distinguished between the two stimuli (Frank *et al.*, 2003). Umami activates similar brain areas to glucose – the PTC and secondary taste cortex, especially the left OFC (de Araujo *et al.*, 2003a). When de Araujo *et al.* presented a mixture of two umami tastants, greater activation was found in the anterior OFC than when the tastants were presented singly.

The red berry of the miracle fruit (Richadella dulcifica) has an unusual property in that it reduces the sour taste of citric acid and adds sweet taste. The brain responds faster to citric acid than sucrose or citric acid after chewing miracle fruit but both stimulate the insula and frontal operculum (Yamamoto *et al.*, 2006). When the time course of activation to citric acid, sucrose, and citric acid after chewing miracle fruit is examined, activation is seen in the frontal operculum and insula but the response to citric acid after chewing the fruit is similar to that seen after tasting sucrose, i.e. 250msecs longer than to citric acid alone (Yamamoto *et al.*, 2006).

The brain, however, appears to be able to tell the difference between sucrose, a natural sugar, and sweeteners. As Chapter 3 showed, the cell types that mediate these two types of sweet are the T1R3 receptor and T1R2 receptor. A comparison of sucrose and sucralose (artificial) tastants found that both activated similar taste pathways but that sucrose elicited a stronger response in the anterior insula, frontal operculum, striatum and anterior cingulate (Frank *et al.*, 2008), suggesting to the authors that the brain has a mechanism for distinguishing between sweets that deliver caloric as opposed to non-caloric input.

The somatosensory qualities of taste also appear to stimulate the insula (Cerf-Ducastel *et al.*, 2001). Cerf-Ducastel *et al.*'s fMRI study presented twelve participants with six tastes – four were pure (e.g. NaCl, aspartame) and two were tastes that were also somatosensory stimulants for the tongue (HCl and aluminium potassium sulphate). All stimuli acivated the insula, Rolandic operculum, frontal gyrus and temporal operculum. The only difference between the two groups of stimuli was that the pure taste stimuli produced different patterns of activation in the superior and inferior parts. The somatosensory stimuli activated the left and right Rolandic operculum (no bilateral activation was seen to the pure tastes), suggesting that this region represents the tactile areas of oral structures. The authors also reported that the left inferior insula was activated with the left angular gyrus and suggest that this represents semantic processing.

In 1752, Swiss Professor of Mathematics Johann Sulzer observed a peculiar phenomenon in which pieces of lead and silver making contact with the tongue produced a taste sensation, specifically the taste of green vitriol. Since then, studies have shown that different directions of current can produce different taste sensations – anodal stimulation produces a metallic or sour sensation (or an 'electric' taste), whereas cathodal stimulation produces a bitter or sweet sensation. This sensation of 'electric tastes' depends on the concentration of fungiform papillae on the tongue. When electric taste stimuli were administered in eleven right-handers, activation occurred in the insula, superior temporal lobe, inferior frontal lobe and inferior postcentral gyrus, irrespective of which side of the tongue was stimulated (Barry *et al.*, 2001). In terms of lateralized regional responses, superior insula activation was largely right-based, with the central insula showing more bilateral activation. The premotor area (BA6, BA44) was more dominant in the right and the superior temporal cortex (BA22, BA42) in the left. The first area of activation of the bilateral anterior insula, the medial OFC, appears to occur at 70–80msecs after stimulus onset (Ohla *et al.*, 2010), with the more intense sensations activating the primary and secondary somatosensory cortices. Yamamoto *et al.* (2003) found that induced electrical taste activated the operculum and the insula bilaterally with a latency of 350msecs. When the stimuli were subthreshold there was a cortical response, but not strong; when stimuli were stronger, somatosensory cortex activation was also seen, a finding also reported by Ohla *et al.*

4.22 Neuroimaging and hedonic response to taste

While it is arguable that any taste evokes a hedonic response, some studies have specifically studied possible differences in brain activation to tastes varying in pleasantness. A number of these has studied aversive taste, as this is functionally important. The first study to make the comparison explicitly used PET to investigate blood flow activated by salinone (unpleasant), water and chocolate (3g solid) (Zald *et al.*, 1998). When salinone was compared with water, increases in the right amygdala, left anterior OFC, medial thalamus, pregenual area, dorsal anterior cingulate and right hippocampus were observed. Compared with chocolate, it activated the right amygdala, left OFC and pregenual cingulate cortex. Activation in the right caudolateral OFC has been associated with perception of pleasant tastes, regardless of intensity, and the left dorsal anterior insula/operculum to unpleasant tastes (of equal intensity to pleasant tastes) (Small *et al.*, 2003).

Given the role of the amygdala in responding to appetitive stimuli and to stimuli that signify threat or reward, some other studies have reported increased activation in this structure to aversive tastes (Zald *et al.*, 2002). Zald *et al.* compared PET activation during water, sucrose and quinine perception and found that bitter tastants activated the left amygdala when compared with the water condition and a non-gustatory eyes open condition. Further areas of the cortex also showed differences with the right posterior OFC and left inferior pole/anterior OFC activation distinguishing between pleasant and unpleasant tastes.

One food that reliably produces positive behavioural, cognitive and affective responses is chocolate. Consumption of *Theobroma cacao* has not only pleasure-giving and potentially cardio-protective properties (Ding *et al.*, 2006) but has also been found to increase positive mood and physical activity and to decrease negative mood (e.g. Karp *et al.*, 2006; Macht and Dettmer, 2006). Exposure to the aroma of chocolate is associated with looking longer at slides, recalling more words in a context-reinstatement memory task and rating paintings as artistically better (Schab, 1990; Herz and Cupchik, 1993). Human EEG theta activity, a frequency considered to be attention-related, is reduced during exposure to the odour of chocolate (Martin, 1998).

Small *et al.* (2001a) were one of the first to explore the effect of chocolate eaten to satiety, on brain activation using PET. They measured the pleasantness of pieces of chocolate eaten one chunk at a time in five women and four men and correlated the reward value of the food with brain activation. They found that as the reward value decreased (and pleasantness declined), so did blood flow bilaterally in the insula; conversely, there was an increase in the caudolateral OFC. When participants ate chocolate even though they were satiated, activation was found in the parahippocampal gyrus and caudolateral OFC; when eating the chocolate while still finding it pleasant, there was activation in the caudomedial OFC, insula, operculum, striatum and midbrain suggesting that two regions of the OFC contribute to the hedonic response to this food. They note that Tataranni *et al.* (1999) had also observed increases in the dlPFC with satiety and note that this may be an important cortical region for the termination of feeding. Only the cingulate cortex was active in both behavioural states and effects in the amygdala were observed. A similar experiment asked twelve men and twelve women to rate the pleasantness of chocolate milk before and after eating chocolate to satiety, as fMRI recorded brain activation (Smeets *et al.*, 2006). They found significant interactions between satiety and sex showing that in men, tasting while sated resulted in increased activation in the ventral striatum, insula, OFC and medial OFC and a decrease in the somatosensory area. In women, increased activation was observed in the precentral gyrus, superior temporal gyrus, putamen and decreased activation was found in the hippocampus and amygdala.

Studies of intensity find that intense tastes activate the cerebellum, pons, middle insula and amygdala, regardless of valence (Small *et al.*, 2003). P2 amplitude also increases with increased intensity and P1 and N1 latency become shorter (Hummel *et al.*, 2010b). Small *et al.* employed two concentrations of sucrose and quinine sulphate and also examined the participants' hedonic response to the stimulants. They found that tastes perceived as pleasant activated the anterior insula and operculum up to the OFC and right caudolateral OFC. Unpleasant tastes were shown to activate the left dorsal anterior insula and operculum. Kikuchi *et al.* (2005) exploited an acquired taste for the Japanese pickled plum, umeboshi. Umeboshi are a species of apricot that are unripe, dyed with papilla leaves and then pickled. They are extremely sour, so much so that there is 150 per cent more saliva flow to this than to lemon juice and 300 per cent more than to ascorbic acid (Kikuchi *et al.*, 2005). Kikuchi *et al.* found activation in the right insula, bilerateral operculum and bilateral OFC.

One final note: Recall in the section on odour, how the presence of an odour influenced people's reactions to faces expressing disgust. A similar phenomenon has been reported for taste. Participants exposed to disgusted, pleased or neutral faces while tasting either quinine or sucrose and whose empathy was measured showed increased inferior frontal operculum activation when watching disgusted and pleased faces if empathy scores were high (Jabbi *et al.*, 2007).

4.23 Fats

Fats are unusual in terms of taste because they contribute primarily somatosensory information such as texture – specifically, feelings of viscosity, smoothness, greasiness or creaminess. Behaviourally, they have reinforcing effects although they do not make people satiated in the same way that carbohydrates do (Bell *et al.*, 1998). Natural oils are primarily triglycerides – fats made up of one glycerol esterized to three fatty acids. Fatty acids at certain temperatures give off different tastes; below 10°C, they are perceived as sour and not as fatty. At high temperatures, they are virtually tasteless. Free fatty acids, such as saturated, monounsaturated and polysaturated fats are volatile and are used to enhance food flavours (they, with one or two other chemicals produced by the lipolysis of fat in milk, are what makes blue mould cheese, blue, mouldy and cheesy). Glycerol is not volatile, but its degeneration in deep-frying is what gives fried food a distinctive flavour.

Rolls *et al.* (1999) have maintained that neuronal response to fats is textural rather than due to some other flavour-related sensation. To explore this further, Verhagen *et al.* (2003) varied the viscosity of carboxymethylcellulose, linoleic acid, lauric acid and natural oils such as sunflower and coconut oils, presented to macaque monkeys. They found, as did Rolls *et al.*, that the OFC neurons responded to the fats and non-fats that had a similar texture. A separate population did respond to viscosity but not to fats with the same viscosity thus suggesting that two separate routes mediate viscosity perception – one fat-based, the other nonfat-based. A human fMRI study has found that high concentrations of fat emulsions were associated with increased activation in the anterior insula, frontal operculum, secondary somatosensory cortex, ACC and amygdala (Eldeghaidy *et al.*, 2011). Eldeghaidy *et al.* also found that the preference for fat correlated with activation in the primary and secondary somatosensory cortices, middle insula and anterior insula.

4.24 Memory and attention

When individuals are told that they are about to taste a bitter substance, they rate it as mildly bitter and not as bitter as those who are not primed. The priming is accompanied by commensurate activation in the brain – the milder the perception, the less the activation in the PTC (Nitschke *et al.*, 2006). The act of making decisions about tastes appears to activate specific areas. Thus, when asked to try and detect taste stimulus from a tasteless one, activation is found in the insula and operculum but not the OFC; when an unpredicted stimulus is tasted, the OFC is activated (Veldhuizen *et al.*, 2007). When asked to judge either the presence, intensity or pleasantness of sweet, sour, salty or tasteless stimulants, or asked to passively perceive them with no instruction to judge, the passive condition is associated with greater connectivity between the anterior dorsal insula and frontal operculum and amygdala (Bender *et al.*, 2009). The lateral OFC is activated during the hedonic evaluation task, suggesting that there exist 'parallel pathways encoding taste beyond the initial representation in the anterior insula' (Bender *et al.*, 2009: p. 335). The response also appears to be dictated by experience. A study of seven sommeliers and seven naïve participants found that the former group activated the left insula and OFC and bilateral dlPFC when tasting (Castriota-Scanderbeg *et al.*, 2005). In the naïve group, the PTC and amygdala were activated, suggesting that a degree of more specific, honed responding occurred in those trained to examine and analyse taste.

Taste memory is important for survival – unpleasant gustatory experiences, especially those that lead to illness or vomiting or feelings of nausea will not be repeated. In addition to the NA (highlighted earlier in the discussion on neuroanatomy), the region of the brain that also seems to play an important role in mediating this experience – remembering aversive and safe gustatory experiences – is the same as that responsible for taste perception, the insula. The effects appear to be modulated by NMDA receptors. When blocked in the insula, conditioned taste aversion and short- and long-term memory formation is not found (Ferreira *et al.*, 2002). The inhibition of protein synthesis in the insula also impairs conditioned taste aversion extinction but whether this synthesis is needed in all gustatory regions is open to debate. Early expression genes such as c-fos are associated with taste learning and c-fos is found in gustatory structures. In the amygdala and PBN, c-fos expression is necessary for conditioned taste aversion (Yasoshima *et al.*, 2006). One view is that the insula is necessary for novel taste memory and the PBN is necessary for the formation of taste memory but not retrieval (Nunez-Jaramillo *et al.*, 2010). Our ability to keep gustatory information in mind is likely to depend on the PFC. Goldman-Rakic (1996) suggested that different parts of the PFC processed different types of working memory material depending on their connections with other brain regions. The OFC may be the region that allows us to keep gustatory information in working memory. Two participants who completed a gustatory delayed match-to-sample task and were distracted during the delay were shown to demonstrate neuronal activation in the OFC and PTC (but not dlPFC and ventrolateral OFC) when they encoded identity across the delay (Lara *et al.*, 2009).

4.25 Taste imagery

Taste imagery is associated with activation in the frontal gyri (Kobayashi *et al.*, 2004) but also with activation in the insula (Levy *et al.*, 1999b). Levy *et al.*'s results were similar to those reported for the imagining of banana and peppermint odour, discussed in section 4.18. However, activation to taste imagery was more frequent and greater in the insula. Similarly, while extracellular recording from the anterior insula and frontal operculum shows activation here during the tasting of bitter, sweet, salty and sour stimuli (Scott and Plata-Salaman, 1999), perception and imagery of taste have been found to activate very similar areas but activation during the imagery task was greater in the left insula (Kobayashi *et al.*, 2004). Taste imagery in the last study also activated the middle superior frontal gyri but tastes did not. Thus, as for olfaction and other senses, imagery in taste activates areas that overlap with actual taste perception. In terms of the time course of imagery, one study has reported activation in the insula 401msecs after stimulus onset in participants when asked to recall taste prompted by words or pictures related to food and drink (Kobayashi *et al.*, 2011). Activation in the insula was seen 77.8 per cent of the time to words and 71.4 per cent of the time in the PFC prior to insula activation. To pictures, 66.7 per cent showed activation in the insula and 11 per cent showed activation in the PFC (Figure 4.6).

The conclusions we can reasonably draw from neuroimaging studies of smell and taste are that the functional neuroanatomy of taste is more consistently localizable than is the equivalent in smell. For olfaction, a network of structures is activated but not always consistently, not always in the same hemisphere but always dependent on the nature of the olfactory task taken and the type of odour used. Trigeminal stimulants and so-called biologically relevant odours activate different structures to more conventional inhaled stimuli – fruits, other foods, perfumes. The way in which these two senses might interact to produce flavour is discussed further in Chapter 6.

In Chapter 5, the contribution of clinical neuropsychology to the study of smell and taste is examined more explicitly by discussing disorders of the two systems and how neurodegenerative and psychological disorders might affect chemoreceptive performance.

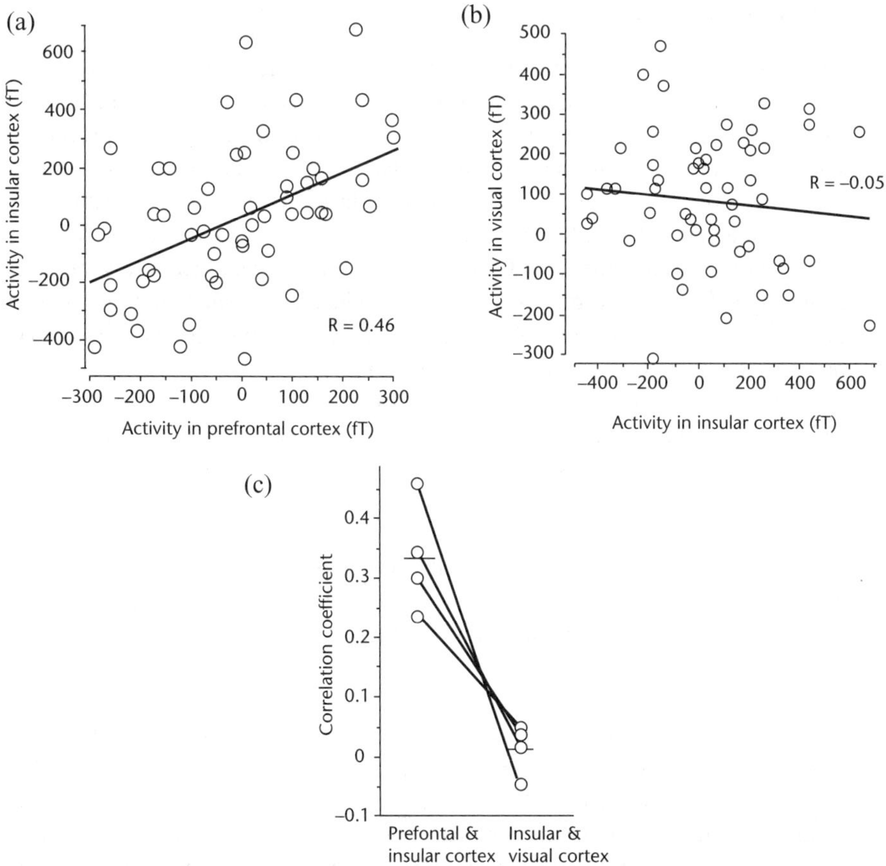

FIGURE 4.6 Correlations between insular cortex activation and prefrontal corex (a), the visual corex (b) and both combined (c) when people imagine tastes

5

DISORDERS OF SMELL AND TASTE, AND DISEASES ASSOCIATED WITH CHEMOSENSORY IMPAIRMENT

5.1 Introduction

Disorders of smell and taste can involve the complete absence of function, a partial absence of function, distorted function and heightened function. In some cases, function might be imagined – patients experience taste and smell phantoms, sensing stimuli that are not actually there (phantosmia/phantogeusia). Loss of perception for *specific* smells and tastes is thought to be rare – but probably only because the phenomenon has not been widely investigated and is not sufficently important to warrant study. As Chapters 2 and 4 discussed, there are some congenital specific smell dysfunctions (such as the inability to smell androstenone or the semen-scented l-pyrroline; Amoore *et al.*, 1975) and some people fail to detect bitter taste while others are highly sensitive to it.

The causes of smell and taste disorders are myriad and range from mechanical or pathological damage to peripheral processes and structures, dysfunctional neurons, and injury to CNs and cortical and subcortical areas. The dysfunction and lesions can be due to either a virus, head injury, exposure to toxins/pollution/solvents, disease, medication and a number of other pathological processes that affect these two systems specifically. Olfactory and gustatory deficits may also accompany – and may even be symptomatic of – neurodegenerative disease, endocrine disorder and psychiatric illness, including AD, PD, Kallmann's syndrome (KS), epilepsy and schizophrenia. The evidence for this is reviewed below.

The direct and constant exposure of the olfactory sense to the environment makes it particularly susceptible to viral infection and, therefore, dysfunction. For this reason – exposure coupled with susceptibility – smell disorders are probably much more prevalent and common than are taste disorders. In fact, upper respiratory infection accounts for two-thirds of partial or total smell loss, (Deems *et al.*, 1991). Olfactory loss is also reported after closed head injury: early prevalence estimates indicated this was around 31 per cent (Sumner, 1964). Later data suggested that the incidence following traumatic brain injury (TBI) was between 4–33 per cent (Callahan and Hinkebein, 1999).

Smell loss accounts for 84 per cent of chemosensory patient complaints and partial loss is reported more than is total loss (75 per cent compared with 23 per cent; Hoffman *et al.*, 2009). As Chapter 2 noted, these losses are much more common in the elderly (those over 65 years) and men are more likely to report losses in smell and report this earlier in life (Corwin *et al.*, 1995; Schubert *et al.*, 2009) (Figure 5.1).

However, estimates of smell and taste loss vary, and objective data differ from subjectively reported data. As Chapter 2 showed, people complaining of taste disorders rarely have a verifiable taste disorder but an olfactory one. The incidence of olfactory disorders is around 1.6 per cent. Extrapolated to a country the size of the US, this represents three million people, four million if taste disorders are included (Hoffman *et al.*, 2009). Across time, olfactory impairment has been found in 0.94 per cent of 18–24 year olds, and increases to 4.21 per cent in 85 year olds. Taste is less seriously affected – the equivalent figures are 0.7 and 1.7 per cent (Hoffman *et al.*, 2009).

Although objective data do exist and attempts are being made, especially in the US, to make chemoreceptive dysfunction testing more systematic, clinicians and physicians are not particularly well equipped to assess these dysfunctions. Neither do they usually have the time to undertake a comprehensive chemosensory evaluation. Physicians may undertake a basic ear, nose and throat examination, make

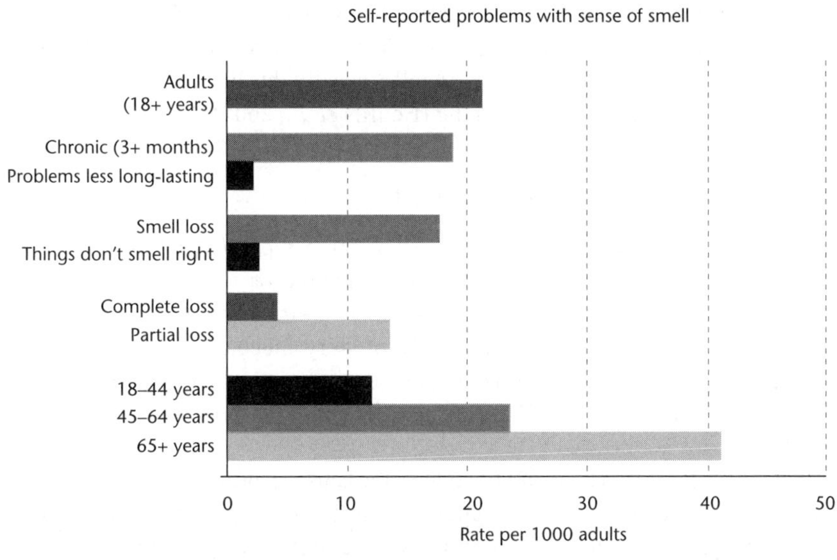

FIGURE 5.1 Some of the self-reported problems with sense of smell and some demographic features of the impairment

a note of dental hygiene and perform an oral examination (for taste) and take a patient history. Rarely and perhaps not surprisingly, would a validated smell and taste test be used. Patients might be referred to a specialist centre (of which there are several in the US). Furthermore, as Murphy *et al.* (2003) note, some patients, such as the old and the demented, may not even be aware of their loss and so may be oblivious to their actual impairment.

5.2 Disorders of smell

According to the American Medical Association's Guide to the Evaluation of Permanent impairment – a system of ascribing a pecentage to the quality of life that is impaired by a permanent ailment – bilateral loss of smell accounts for a 3 per cent impairment of the whole person. Small, but not negligible. Anosmia is a term used to describe a complete loss of smell. The most common disorder of smell, however, is partial anosmia and describes a condition where a person can detect some but not all olfactory stimuli. Other olfactory complaints include decreased sensitivity to odour (hyposmia, microsmia), distorted smell perception, usually unpleasant (dysosmia, cacosmia), over – or acute – sensitivity to odour (hyperosmia) and olfactory hallucinations (phantosmia). A rarely reported condition is olfactory agnosia where an individual cannot recognize an odour despite intact olfactory apparatus and normal language (and is, therefore, the olfactory equivalent of visual agnosia). The term presbyosmia is sometimes used to describe the general decline in olfactory function with age (Murphy *et al.*, 2003). As these disorders appear to be characterized by different classes of symptoms, Holbrook and Leopold (2006) have suggested that they be categorized as qualitative and quantitative olfactory disorders. Thus, quantitative disorders involve decreases in sensitivity or the absence of olfactory function; qualitative disorders involve distorted perception of actual smell – usually this distortion is unpleasant and described as rotten, foul or burning (Bonfils *et al.*, 2005).

5.3 Causes of olfactory impairment

Causes of olfactory disorders tend to be of three types (Mattes *et al.*, 1990; Murphy *et al.*, 2003): there may be (i) conductive or transport impairments because the nasal cavity/nasal sinus is obstructed by infection or polyps – masses filled with serous fluid and covered in respiratory mucosa; (ii) damage to the OE, resulting from viral infection or exposure to toxins; and (iii) injury to central structures such as the cortex. There are also congenital disorders that affect smell and taste.

If the disorder is transient and fluctuates, the cause is more likely to be obstructive rather than sensorineural in nature. A nasal endoscope can assess whether there is any pathology in the nasal cavity whereas a rhinoscope can detect pathology further along the system, in the olfactory meatus. At the level of central structures and peripheral structures close to the brain, neuroimaging (MRI) can be used to detect atrophy, inflammation or absence of the OBs, although this can be problematic because of the location and accessibility of these (small) structures. There are also conditions such as liver disease, diabetes, medication and genetic disorders such as KS (see section 5.5) which are also associated with smell impairment.

The most common cause of olfactory disorders is upper respiratory infection – cold, influenza, HIV – and this might also be accompanied by phantosmia. The precise mechanism of the dysfunction is unknown – one cause might be a failure of receptors to regenerate: the number of basal or immature cells, for example, have been found to be lower during infection (Loo *et al*., 1996). It is more likely to be caused by direct insult – the number of OE receptors declines, as does the number of cilia; the OE may be replaced by respiratory epithelium (Douek *et al*., 1975). Another view is that abnormal olfactory neurons are unable to form an image of an odorant.

A second cause of impairment is nasal or sinus damage. Here, the problem is normally one of conduction: airflow to the OE is obstructed in some way, usually by disease or inflammation (rhinosinusitis). There may also be changes to the olfactory mucosa (Jafek *et al*., 1987); less tissue is found in the epithelium of patients with nasal disease and the receptor arrangement may be more disorganized, or sensory receptors may be replaced by respiratory cells (Lee *et al*., 2000). Surgery or medication provides some relief from these problems, and although recovery is never full (Doty and Mishra, 2001), olfactory impairment can resolve itself over a period of months. Patients receiving surgery for hypertrophic adenoids were found to show 20–40 per cent reduction in airflow resistance (Crysdale *et al*., 1985). Other interventions such as middle turbinate medialization, polypectomy, uncinectomy and ethmoidectomy produce a relief in symptoms and better olfactory function.

Tumours, whether they are cranial or intranasal, can impair olfactory function. In fact, olfactory tests were once used to diagnose cribriform plate meningioma in the 1930s where increased thresholds were found on the side of the neoplasm that caused the pressure. The position of the olfactory apparatus, especially the OBs, makes them vulnerable to pressure from tumours that emerge from the dura mater, cribriform plate and prefrontal lobe. Bilateral and unilateral frontal glioma can result in anosmia, as can lesions produced by aneurysms (Graff-Radford *et al*., 1997). Lateral wall tumours produce impairment, and temporal lobe tumours (as Chapter 3 showed) are also associated with dysfunction, especially cacosmia. Rarely, tumours will affect receptor cells in the nasal vault – olfactory neuroblastoma and esthesioneuroblastomas (Castillo and Mukherji, 1996). These are malignant, aggressive and more common in young men.

TBI can result in olfactory impairment when damage occurs to the sinonasal tract, if there are fractures to the skull or head, or when the axons and tract and fila olfactoria (in summary, the olfactory nerve) are sheared. The loss of the latter is greater if impact is to the occipital rather than the frontal lobe

(Doty *et al.*, 1997; Collet *et al.*, 2009), presumably because the forward force of impact causes shearing of the olfactory nerve across the bony cribriform plate. The impact could also cause contusions to the OFC and anterior temporal lobe. Injury to the OB or tract, subfrontal area or temporal lobe can result in anosmia that is either severe or mild (Yousem *et al.*, 1996) and the more severe the bulb and tract loss, the greater the smell loss. Estimates vary: some studies report 4–7 per cent or 12.8 per cent incidence of anosmia after trauma (Haxel *et al.*, 2008); others report incidence as high as 50–60 per cent (Callahan and Hinkebein, 2002). Patients tend to report smell losses immediately after the trauma (Schecter and Henkin, 1974). A study of UPSIT performance in sixty-six patients with head trauma who were followed from one month to 13 years found that 36.6 per cent showed olfactory impairment, 18 per cent became more impaired and 45 per cent showed no change. Five per cent recovered olfactory function.

Varney and Bushnell (1998) suggested that olfactory deficits may be a marker for OFC impairment. Varney (1988) reported that 92 per cent of patients with anosmia reported behavioural problems; 64 per cent of patients with partial anosmia did. Callahan and Hinkebein (1999) found that 65 per cent of adults with TBI (N=44) showed anosmia and severe deficits in complex attention, new learning and memory and on the Wisconsin Card Sorting task (although the specific region damaged was not specified). In an explicit investigation of the association between olfactory impairment and frontal lobe function, Sigurdardottir *et al.* (2010) examined the olfactory performance, executive test performance and Iowa Gambling Task performance (thought to be sensitive to PFC injury) of 115 patients with mild, moderate or severe TBI over a period of one year. Anosmia was found in 10 per cent of severe injury cases and in 3 per cent of mild injury patients. There was no association between olfactory test performance and the gambling task but olfactory dysfunction was associated with verbal fluency impairments. Given that the olactory test was the B-SIT, the linguistic component of this may explain the association between the two tests (see discussion in Chapter 1).

5.4 Anosmia (not congenital)

The majority of cases of anosmia have been caused by some physical insult; there are some indvididuals who report not having been able to smell (or if they do, report only being able to smell one or two odours) since birth and these individuals are considered to have congenital anosmia (see below). Anosmia can be bilateral or unilateral (i.e. affecting one nostril or both) (Castillo and Mukherji, 1996).

Apart from the inconvenience caused by losing the sense of smell and the resulting lack of pleasure and enjoyment of life, a person's diet might also be affected, as noted in the section on ageing effects in Chapter 2. For example, some studies found a relationship between severe olfactory dysfunction and weight loss or complaints about food palatability (Markley *et al.*, 1985); other studies, however, find no change in diet or physical condition (Mattes-Kulig and Henkin, 1985). Clark and Dodge (1955) report the case of a 44-year-old woman who had complained of smell loss over twelve months. She was diagnosed with a tumour that destroyed the olfactory neurons. Two years before, she had reported being unable to taste chocolate or distinguish between spices. Pumpkin pie tasted of plain egg custard. She could not identify the odour of wintergreen, camphor, cloves, ginger and cinnamon but she could identify the four tastes. She identified nutmeg 'by the feel of it', rather than its flavour or taste.

Mattes *et al.* (1990) examined food acceptability and nutritional risk in 119 patients with chemosensory disorders and found that while food was considered less acceptable than it was by a control group, there was no significant change in diet (although those reporting dysosmia were more greatly affected than those with anosmia). According to the patients' self-report, a quarter put on weight, 42 per cent altered their spice use (it increased, presumably to enliven food perceived to be bland) and there was an increase in sugar and salt use. This demonstrates clearly the smell–taste confusions common to everyday life.

Recovery from partial anosmia is possible but it is never complete. Gudziol *et al.* (2010) examined thirty-five individuals with olfactory loss on one side of the nose who were retested 4.6 years later using Sniffin' Sticks. Compared to controls, olfactory performance was poorer in the unilateral group and there was also a risk of birhinal impairment within 4.5 years. Mueller and Hummel (2009) studied a 54-year-old man who had developed anosmia following a car accident which occurred when he was 38 years old. He reported being unable to perceive food flavour and developed his first olfactory impressions nine years later. After three years, his sense of smell improved and this was confirmed psychophysically but it is rare for recovery to occur after two years; recovery from trauma tends to occur between six and 12 months (Constanzo and Zasler, 1992).

Treatment varies. Steroid therapy is effective for conductive problems (and can distinguish between these and sensorineural problems that do not respond to steroids; Davidson and Murphy, 1997). Sometimes the septum has deviated in patients and this can be corrected by surgery. Currently, there is no successful treatment for congenital malformation of the OB and tract.

At the level of the cortex, it has been consistently reported that OEPs in patients with anosmia are barely recordable. For example, patients can detect trigeminal stimulants and produce CSERPs to these but show no electrophysiological response to olfactory odours, H_2S and vanillin. Cui and Evans (1997) studied ERPs to phenylethyl alcohol, isoamylacetate and chloracetyl phenone (a trigeminal stimulant) in nine (congenital) anosmic patients. OEPs were recordable in the control but not anosmic participants.

A reduction in OB volume has been associated with olfactory impairment (Mueller *et al.*, 2005). Reduced olfactory afferent input has been associated with reduced OB volume and the bulb has been reported to be smaller in patients with parosmia – patients who incorrectly perceive that an odour is present – than in patients with olfactory impairment but without parosmia (Abolmaali *et al.*,

2008). VBM is a technique that allows the investigation of volume differences across the entire cortex. Bitter *et al.* (2010a, 2010b) used VBM to examine cortical loss in seventeen anosmic patients and seventeen controls. They found atrophy in the ACC, the dorsolateral (and medial) PFC and NA (Figure 5.2).

This follows on from studies reporting volume loss in the right piriform cortex, right insula and right OFC (Cerf-Ducastel and Murphy, 2006) and in the superior temporal gyrus, superior frontal gyrus, inferior parietal lobule, hippocampus, parahippocampal gyrus and cerebellum (Dade *et al.*, 2001). A PET study by Varney *et al.* (2001) found hypometabolism in the medial PFC and OFC of individuals with partial anosmia (and an increase in the visual association cortex).

Volume loss in grey and white matter has also been reported in hyposmia and parosmia. The first reported white matter loss in hyposmia was by Bitter *et al.* (2010a) who noted grey and white matter volume loss in the ACC, OFC, cerebellum, frontal gyrus, precuneus, middle temporal gyrus and piriform cortex. White matter loss was observed in the insula, cerebellum and middle frontal gyrus – all of which, with one exception, were connected to the grey matter areas. Grey matter loss has also been observed in parosmia, particularly in the left anterior insula, right anterior insula, ACC, bilateral hippocampus and left medial OFC (Bitter *et al.*, 2011). There is also an association between thickness in some of these areas and olfactory function in healthy individuals. Grey matter thickness in the right medial OFC, right insula and areas around the central sulcus has been correlated with olfactory performance; the thicker the matter, the better the performance (Frasnelli *et al.*, 2010).

Patients with phantosmia and phantageusia experience imagined odours or tastes. These hallucinations tend not to be pleasant. Most are cacosmic and the phantoms are described as 'rotten'. In phantageusia, tastes may be described as

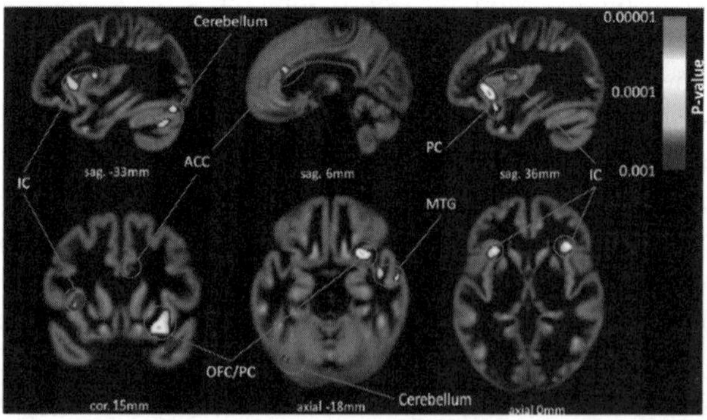

FIGURE 5.2 Areas of grey matter reduction in hyposmic patients

sour or metallic. Around 60 per cent of patients report this type of symptom; around half of that report cacogeusic or cacosmic symptoms and 10 per cent show mixed symptomatology. Phantosmia has been associated with activation in the OFC, frontal lobe and cingulate gyrus in two patients during phantom odour perception compared to actual odour perception (Henkin and Levy, 2001). Those perceiving taste phantoms activated the cingulate cortex. Phantosmic patients believed, erroneously, that they smell hallitosic (had foul breath) or that they emit bad odour through the nose (atmatosis).

5.5 Congenital anosmia

Some individuals are constitutionally incapable of smelling certain odours. These odours tend to belong to a particular chemical or psychological category which is why they tend to be unusual or slightly unpleasant. Examples include musk, trimethylamine, hydrogen cyanide, n-butyl mercaptan and isovaleric acid. As Chapter 1 showed, repeated testing with these odours can lead to an ability to detect (e.g. Doty and Ferguson-Segall, 1989; Wysocki *et al.*, 1989) which suggests that either the anosmia is really hyposmia or that repeated exposure activates the relevant but dormant receptor population. The studies on androstenone, for example, have led to the proposition that there may be three types of anosmic disorder: total anosmia, inducible anosmia and constitutionally sensitive anosmia (Murphy *et al.*, 2003).

Total congenital anosmia is rare and, when it does occur, is normally due to the endocrine disorder, KS. This autosomal, genetic disease expresses itself mainly in men as hypergonadtrophic hypogonadism and anosmia. The olfactory tract and bulbs are absent (Yousem *et al.*, 1993).

Congenital – or any total – anosmia can present a serious problem to health. Vowles *et al.* (1997), for example, report a case study of a five-year-old girl who presented at an ENT clinic because she, amongst other behaviours, drunk a glass of rancid milk and on another occasion failed to notice the smell of burning.

Structurally, individuals with congenital anosmia may show a failure to develop the OB or stalks (or have malformed structures – agenesis or dysgenesis) (Yousem *et al.*, 1996). Congenital, heritable anosmia is rare. It was first reported by Glaser (1918) who found a family of Russian Jews with an X-linked anosmic condition. Lygonis (1969) found evidence of total anosmia in twenty-seven members of four generations of families raised on the isolated community of Sandoy in the Faroe Islands. Singh *et al.* (1970) found six members of one family with cluster headaches who were also anosmic. Ghadami *et al.* (2004) report two unrelated Iranian families with isolated congenital anosmia. There was no neural abnormality in these families nor was there hypogonadal disorder. Nine members were affected in one family; three were affected in another. There was no difference between affected and unaffected OB volumes suggesting to the authors that the deficit is expressed at the level of the OE or cortex.

Does anosmia affect the abilty to taste? Some studies have suggested that anosmia is associated with a reduction in the ability to sense the four classical tastes (Hummel *et al.*, 2001). Other studies have reported no taste impairment: in 1,312 hyposmic cigarette smokers (Vennemann *et al.*, 2008) and four patients with KS (Hasan *et al.*, 2007). Stinton *et al.* (2010) undertook whole-mouth and regional taste testing of the left, right, anterior and posterior tongue in 581 patients who also completed the UPSIT and found that anosmia was not associated with taste alteration. Interestingly, when age, sex and aetiology were controlled for, an association did emerge.

5.6 Schizophrenia

Schizophrenia is the most common psychotic disorder and has several subtypes (undifferentiated, catatonic, paranoid and disorganized). It is characterized by two categories of symptoms: positive and negative. Positive symptoms include thought disorders, hallucinations and delusions. Negative symptoms of schizophrenia are known by the absence of normal behaviours: flattened emotional response, poverty of speech, lack of initiative and persistence, inability to experience pleasure and social withdrawal. Onset occurs in adolescence or early adulthood, and genetic factors account for 80 per cent of the disorder's appearance (Tandon *et al.*, 2008). Verbal fluency – the ability to name as many objects beginning with a particular letter or belonging to the same category – appears to be impaired in schizophrenic individuals (Gruzelier *et al.*, 1988), although the category version of this test appears to be better performed (Joyce *et al.*, 1996). There is a loss of total brain volume as well as reduced grey matter in the temporal lobe, PFC and thalamus (Keshavan *et al.*, 2008). PET studies of schizophrenia suggest that there is a decrease in dopamine receptors in the PFC (Okubo *et al.*, 1997) and a decrease of N-acetyl aspartate in the frontal and temporal lobe (Keshavan *et al.*, 2008). Executive function and episodic memory performance, in general, is poor (Reichenberg and Harvey, 2007).

To students of chemoreception, schizophrenia has been of great interest: Doty (2003) noted that there are more olfaction-related studies published on schizophrenia than on any other neurological disorder. Of particular interest is the possibility that olfactory impairment might indicate a degree of orbitofrontal function impairment (Purdon, 1998), as this impairment has been associated with frontal hypoactivation in schizophrenia (Wu *et al.*, 1993) (Figures 5.3a and 5.3b). And there are significant olfactory impairments that accompany schizophrenia. The first, pioneering study of the relationship between the illness and olfactory performance reported significant impairment in odour recognition memory in schizophrenia (Campbell and Gregson, 1972). At least eighteen studies using the UPSIT have reported identification deficits (Minor *et al.*, 2004), with schizophrenic participants scoring an average of 32.5 (Moberg *et al.*, 1997b). In Minor *et al.*'s study of fifty-four chronic unmedicated schizophrenic outpatients and 133 controls, thirteen specific items on the UPSIT distinguished the clinical group

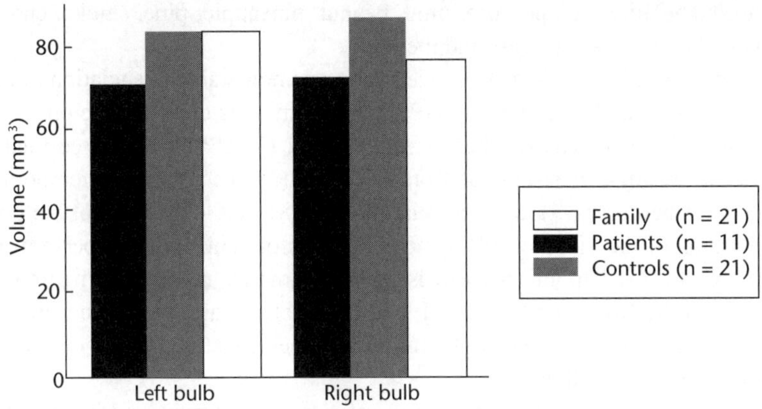

Patients vs. controls Left F(1.45) = 4.05, p = 0.050
Right F(1.45) = 10.04, p = 0.003

Patients vs. family members Left F(1.45) = 5.25, p = 0.027
Right F(1.45) = 0.99, p = 0.324

Family members vs. controls Left F (1.45) = 0.07, p = 0.793
Right F (1.45) = 6.08, p = 0.018

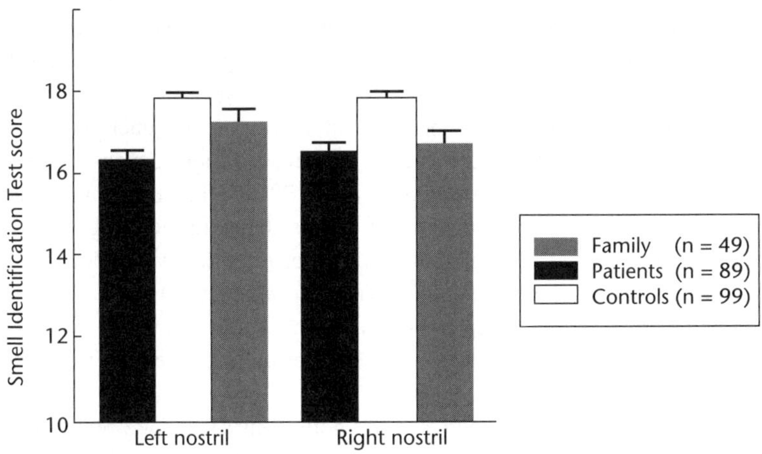

FIGURE 5.3 (a) Comparisons of olfactory bulb volumes in schizophrenic patients, families of patients and controls; (b) UPSIT scores for left and right nostrils in schizophrenic patients, unaffected families and healthy controls

from the healthy – grape, rose, mint, peanut, pineapple, pine, smoke, chocolate, cheese, banana, peach, grass and menthol.

Doty (2003) has identified seven other demonstrable associations between schizophrenia and olfaction: (1) UPSIT impairment is considerable but less than that seen in PD or AD (Moberg *et al.*, 1997a), (2) UPSIT scores correlate better with cognitive measures of frontal lobe function than medial temporal lobe function (memory), (3) these correlations apply to the negative, not the positive symptoms of the disorder, (4) some subgroups of schizophrenia perform worse than others, (5) hypometabolism is most commonly reported in neuroimaging studies in the frontal and temporal lobes, (6) deficits are seen sooner in patients with a family history of schizophrenia and (7) identification deficits correlate with the duration of the illness.

While Doty's summary is a good reflection of current understanding, it is worth noting that performance on most olfactory tests is impaired in schizophrenia (Moberg *et al.*, 1999). Threshold, for example, has been reported to be lower (Purdon and Flor-Henry, 2000), but it is improved especially in the left naris eight weeks after medication. Early studies indicated that men were more impaired than were women (Kopala *et al.*, 1989). However, Moberg *et al.*'s meta-analysis found that there was no sex difference, apart from for hedonic response: male schizophrenic patients reported finding odours at higher concentration more pleasant but not more intense than they did lower concentration odours (Moberg *et al.*, 2003).

Patients 'at risk' of developing schizophrenia also demonstrate olfactory deficits, as noted in Doty's summary. Thus, Roalf *et al.* (2006) found that schizophrenic patients and relatives of these patients showed deficits in olfactory identification. At-risk individuals with olfactory impairment are significantly more likely to develop the disorder (Brewer *et al.*, 2003; Corcoran *et al.*, 2005).

There are specific pathological, neural and anatomical features of schizophrenia that correlate with (or might predict) olfactory impairment and these include nasal cavity volume, OB and epithelium volume, dysfunctional neuronal signaling, OEP amplitude and latency, and neural activation as measured by neuroimaging (Smutzer *et al.*, 2003; Turetsky *et al.*, 2009b). In terms of nasal cavity, there are no consistent differences in airflow resistance in the left or right naris in schizophrenia. However, a reduction in the posterior cavity has been found in schizophrenic men (Moberg *et al.*, 2004; Turetsky *et al.*, 2007) and it is this region that contains the olfactory receptor neurons. Moberg *et al.*'s study of forty men also found that this part of the cavity was smaller in 31 per cent on the right and 11 per cent on the left.

OEP studies suggest that decreases are found in N1 and P2 amplitude and increases in P2 latency to H_2S in schizophrenia (Turetsky *et al.*, 2003) and healthy family members (Turetsky *et al.*, 2008). In Turetsky *et al.*'s (2003) study, in which three concentrations of the odour was administered to twenty-one patients, the stronger odour was associated with greater OEP deficit.

Cytoarchitectonically, Arnold *et al.* (1991b) have reported a variety of abnormalities in the entorhinal cortex in schizophrenia, including inadequately formed neuronal clusters in layer 2 of the cortex, a displacement of layer 2 neurons into layer 3, poor lamination and poor differentiation between cortical layers. Others, however, report no such abnormalities (Krimer *et al.*, 1997). The entorhinal cortex is one of two areas in the anterior ventral medial temporal lobe to receive olfactory afferents from the olfactory apparatus (bulbs) directly (the other is the perirhinal cortex, BA35/36, which includes the pirifrom cortex; the area that does not is the temporopolar area, BA38). Turetsky *et al.* (2003) found that volume was smaller in the two regions receiving the afferents in schizophrenia. Olfactory thresholds were related to volume in the perirhinal grey area. Reduced volume in the entorhinal cortex has been reported by others (Prasad *et al.*, 2004; Baiano *et al.*, 2008) as has a reduction in hippocampal volume (Turetsky *et al.*, 2003; Sim *et al.*, 2006) although, as Turetsky *et al.* (2009a) note, these are the only two studies to date to do so.

At the level of the OB and OE, Smutzer *et al.* (2003) note that the number of basal cells has been found to be 37 per cent lower in schizophrenia and the density of postnecrotic immature neurons 316 per cent higher. OB volume is reported to be reduced, usually by 20 per cent or so bilaterally (Turetsky *et al.*, 2000, 2003), a much larger reduction than is found in the hippocampus (Gur *et al.*, 2000). Larger bulbs were associated with better detection thresholds. Turetsky *et al.* (2003) also found a reduction in bulb volume and in the right side in the first-degree relatives of schizophrenic patients, suggesting that there may be a 'genetically mediated vulnerability factor' (Moberg and Turestsky, 2006: p. 307). No olfactory tract reduction investigations have been reported (Turetsky *et al.*, 2009a). The reason for the reduction is unknown – there may be apoptosis of the mitral cells, which leads to faulty innervation of regenerative olfactory neurons (Chazal *et al.*, 2000).

In neuroimaging, there are very few studies of olfactory function in schizophrenia. Clark *et al.* (1991) reported lower frontal metabolism in schizophrenia, as have Bertollo *et al.* (1996), specifically in the right lateral posterior part of the OFC, an area that recives projections from the OB via the perirhinal and entorhinal cortex. Hypoactivation was more symmetrical in the medial OFC.

In a departure from the normal protocol, Turetsky *et al.* (2009a) used a recording technique called the electro-olfactogram (Knecht and Hummel, 2004) which is used to record the summed depolarizarion of olfactory receptor neurons directly via fine wire electrodes inserted into the nasal cativity. In a comparison of twenty-one schizophrenic patients and eighteen controls, they found larger responses in the clinical group to all the odorants used. Why would this group show increased responsitivity if their olfactory performance such as threshold detection is low and bulb volume is also smaller? One suggestion is that the total number of olfactory neurons is greater in schizophrenia but that their specificity is lost or that they are dysfunctional. Turetsky and Moberg (2009) posit an intesting hypothesis. They found that citralva produced large increases in adenylyl cyclose and electro-oculogram (EOG) depolarization but lyral did not. In schizophrenia and unaffected first-degree relatives of schizophrenic patients, there was impaired detection of lyral but not citralva. They suggest that the mechanism of impairment in schizophrenia is not general, but may be odour-specific.

5.7 Epilepsy

Another well-known disorder associated with olfactory anomalies is epilepsy. This neurological disorder is characterized by abnormal electrical cerebral seizures and there are various types and, depending on age, divergently different causes. The primary symptoms of any type of epilepsy, however, are seizures and excessive discharges of neurons. A seizure is a single and sudden event caused by these discharges (Lee, 2004). The discharges can produce changes in sensation, consciousness, cognition and convulsions, and physical symptoms can vary depending on the focus and location of the discharge – a focus in the postcentral gyrus, for example, will produce somatosensory sensations such as tingling or numbness in the tongue or lips (Lee, 2004).

Epilepsy has been commonly associated with distorted olfaction, principally olfactory hallucinations that precede the epileptic aura (the loss of consciousness which, in turn, presages the seizure). Thresholds in epilepsy are normal but there are significant impairments in a variety of other functions and these are specific to the side of the hemisphere that is the focus of the seizure. Thus, right-sided focus has been associated with impaired odour matching (Abraham and Mathai, 1983) and decrease in immediate memory for common odours such as vinegar, coconut and coffee (Carroll *et al.*, 1993); those with left-sided focus did not exhibit this impairment. Kohler *et al.* (2001) found that the identification scores of right temporal lobe epilepsy patients and schizophrenic patients were similar and lower than that of a group of controls and patients with left-sided epilepsy.

More dramatic are the olfactory anomalies reported during an aura. These hallucinations – in fact, any hallucinations – have traditionally indicated some underlying pathology. An olfactory hallucination is a 'subjective experience of olfactory phenomenon in the absence of a stimulus' (Velakoulis, 2006: p. 322) but is normally described as an olfactory aura in epilepsy. One of the earliest studies was that of Hughlings-Jackson (1899) which reported a patient with right temporal lobe epilepsy who experienced a strong camphor – or ether-like smell. Hughlings-Jackson proposed that the fits following olfactory or gustatory auras should be called uncinate fits, a term taken up by Daly (1958) who proposed that the uncus was involved in these fits and that this structure was central to olfaction. In his study, twenty out of fifty-five patients experienced olfactory and gustatory aura prior to seizures, largely involving odours that were unrecognisable but pleasant and unpleasant.

Later studies showed that the percentage of temporal lobe epilepsy patients who experienced olfactory auras varied between 1 and 12 per cent (Gupta *et al.*, 1983; Taylor and Lochery, 1987; Fried *et al.*, 1995; Chen *et al.*, 2003). In Chen *et al.*'s study, all odours were perceived as unpleasant – foetid, rotten, burning, charing or medicinal. In one of the most extensive studies, Acharya *et al.* (1988)

found only fourteen out of 1,423 patients experienced auras and nine of these reported they were a mix of pleasant and unpleasant familiar odours.

Olfactory hallucinations can also accompany cluster headaches or migraines. Silberstein *et al.* (2000) reported a patient with a cluster headache who smelled a 'bad citrus fruit'. Reported unpleasantness of an odour is a leitmotif in severe headache (Fuller and Guiloff, 1987). At the cortical level, Demarquay *et al.* (2008) found that exposure to odour resulted in increased left temporal lobe activation and decreased activation in the left inferior and left and right middle frontal gyri, left and right angular and right posterior superior temporal gyri, posterior cingulate cortex and right locus coeruleus, in a group of eleven migraineurs.

5.8 Other illnesses

In addition to KS, other endocrine disorders that present with problems in olfaction include Addison's Disease, Turner Syndrome and Cushing's Disease. Non-endocrine disorders that appear to show olfactory deficits including Down's Syndrome, Huntington's Disease, progressive supracranial palsy, amyotrophic lateral sclerosis, Post Traumatic Stress Disorder, depression and Obsessive–Compulsive Disorder (OCD). The reliability of the deficits in each is a source of controversy.

Down's Syndrome, which may be a risk factor for AD because plaques, tangles and granulovacuolar degeneration are found in both, is a chromosomal disorder (chromosome 21). In Down's Syndrome deficits in the UPSIT have been reported (Warner *et al.*, 1988; Hemdal *et al.*, 1993; McKeown *et al.*, 1996).

The results of studies with depressed individuals are not consistent. Thus, some studies find normal sensitivity and identification or increased sensitivity to odours such as isoamylacetate after anti-depressant medication (Serby *et al.*, 1990; Gross-Isseroff *et al.*, 1994; Solomon *et al.*, 1998). Clepce *et al.* (2010) found that odour identification, measured with Sniffin' Sticks, declined during the depressed state but there was no difference in valence rating between depressed, remitted or control participants. Pause *et al.* (2001), however, reported that depressed individuals found citral odour as more pleasant during depression. A variant of depression, Seasonal Affective Disorder, has been associated with impaired identification of UPSIT odours when these are presented in the right nostril (Postolache *et al.*, 1999).

Odours related to the trauma prompting Post Traumatic Stress Disorder has been associated with symptoms provocation. Thus, the scent of blood, napalm and so on can activate cognitive schema and provoke emotional reinstatement (Kline and Rausch, 1985) but very few systematic studies exist. Other reports are anecdotal (e.g. Vermetten and Bremner, 2003). Thus, one case study of a 34-year-old paramedic describes how he had to provide mouth to mouth with a patient whose face had been blown off in a factory explosion. He vomited and when eating, this context was immediately reinstated. When asked to think about the incident, he reported being able to smell clearly the odour of vomit.

The findings for OCD are mixed with most studies reporting normal identification (Locatelli *et al.*, 1996; Fenger *et al.*, 2005). However, a recent study of fifty-five patients found greater impairment on the UPSIT (Dittrich *et al.*, 2010). However, when anxiety level was controlled for, the deficit only remained for the odours rated as disgusting (not pleasasant) and they rated these as less disgusting and intense.

5.9 Neurodegenerative disorders

5.9.1 Alzheimer's Disease

Chapter 2 described the significant effects of ageing on olfactory ability. It is essentially a story of relentless decline. Impairment in olfactory function, however, can also accompany major pathological changes in an individual's brain with age but independently of age. PD is one such pathology, AD is another and what has aroused great interest in olfactory pathology in AD is the possibility that it may be a better predictor of the illness than are the more well-known cognitive symptoms.

It has been estimated that approximately 700,000 people in the UK suffer from dementia and half of these have a dementia rising from AD – Dementia of the Alzheimer Type (DAT) (Knapp *et al.*, 2007). It is estimated that 24.3 million people suffer from DAT wordlwide and that the number of people diagnosed with the illness is expected to double every 20 years until 2040 (Ferri *et al.*, 2005). US studies show a 50 per cent incidence in people over 85, and 100 per cent incidence in one Dutch study of nineteen centenarians (Evans *et al.*, 1989; Blansjaar *et al.*, 2000). Diagnosis is only confirmed with certainty post-mortem.

DAT, according to DSM-IV-TVR, is characterized by cognitive decline exemplified by memory impairment (learning new information and recalling previously learned information) and one or more of: aphasia, apraxia, agnosia and executive function problems; symptoms that cause significant decline from previous levels of functioning; gradual onset and continuing cognitive decline; symptoms that are not due to other progressive CNS diseases or conditions causing dementia.

Neuropathological characteristics of AD are neurofibrillary tangles – abnormal proteins that are found in various parts of the person's brain, especially the temporal, parietal and frontal cortices, neuritic senile plaques – abnormal nerve cell processes that surround the protein and are found in the cortex, and granuovacuolar degeneration (Nelson *et al.*, 2010; Hyman *et al.*, 2012). Animal models suggest that the specific protein contributing to the cognitive decline in AD may be an assembly called AB*56 which is found outside cells (Lesne *et al.*, 2006).

There is significant neuron loss in AD. The frontal and temporal gyri are thought to shrink by approximately 20 per cent and atrophy is found in the hippocampus, amygdala and other subcortical areas such as the raphe nuclei and nucleus basalis of Meynert. The most widely accepted model of pathology development in AD suggests that the disease progresses through six stages (Braak and Braak, 1991). In stages 1 and 2 the pathology is in the early stage of development and no clinical symptoms (e.g. cognitive dysfunction) are present. During stage 1, plaque tangles develop in the 'transentorhinal' cortex in the temporal lobe and in stage 2, these extend into the entorhinal cortex (an area that forms part of the olfactory cortex). Stages 3 and 4 see the appearance of the first symptoms of AD. In stage 3, pathology occurs in the hippocampus and this reaches the association cortex in stage 4. At stages 5 and 6, the disease is progressing relentlessly and pathology becomes more widespread, extending to the neocortex. The E4 allele of the apolipoprotein gene appears to be a risk factor in AD and for cognitive impairment (Farrer *et al.*, 1997) and gene loci including CLU, CR1 and PICALM have been identified (Weiner *et al.*, 2012).

5.9.2 Olfactory impairment in Alzheimer's Disease

Of interest to olfactionists and clinicians is the finding that many of the brain regions affected by Alzheimer pathology are involved in olfactory processing – the OE, AON, entorhinal cortex, piriform cortex and hippocampus. There is usually no nasal pathology reported and patients are not anosmic. The brain areas identified are some of the first to be affected and are some of the regions most severely affected by pathology (Murphy, 1999). Tangles in the AON, entorhinal cortex and amygdala have been observed in moderate and mild dementia (Price *et al.*, 1991b). It is this pathology that has suggested to Murphy that the dissociation of temporal lobe structures from the hippocampus and OFC prevents olfactory information from flowing to these areas and produces the deficits. As the OBs project to the temporal regions, a decline in olfactory performance might be evidence of early pathology. Very early studies had suggested that the ability to smell was compromised in AD and PD (Ansari and Johnson, 1975). Later studies administering well-controlled measures found that olfactory discrimination, identification and recognition memory was impaired in AD in as many as 85–90 per cent of patients (Serby *et al.*, 1985; Doty *et al.*, 1987) and birhinally. Identification impairment appears to be the first of the olfactory impairments to present itself in AD and some estimates suggest that the loss may be as large as 90 per cent (Smutzer *et al.*, 2003). In a meta-analysis of forty-three studies examining identification, threshold and recognition memory ability in AD and PD, Mesholam *et al.* (1998) found severe impairments in all three tests in both illnesses. There was little that could discriminate between the two. Threshold performance was better than identification or recognition memory in both groups (Figure 5.4).

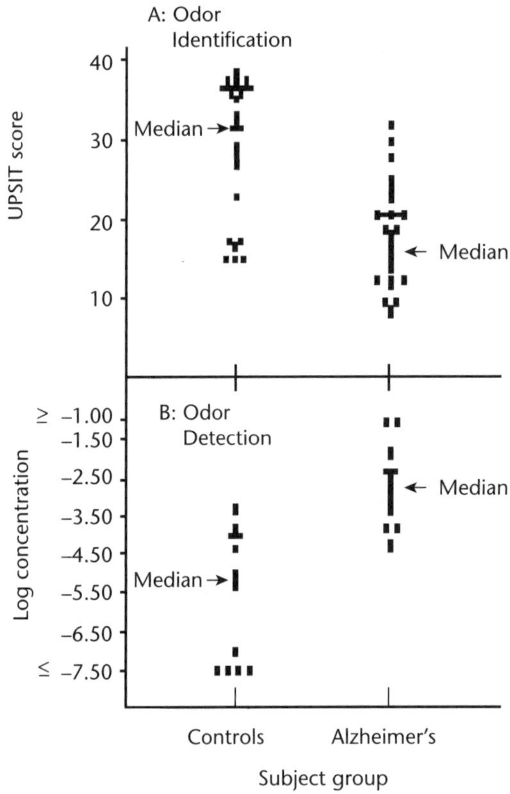

FIGURE 5.4 Odour identification and detection differences in healthy controls and patients with Alzheimer's Disease

Threshold detection appears to be a significant predictor of those at risk, however. In a study of taste recognition, odour detection threshold and visual recognition in sixteen individuals with questionable AD and sixteen age-matched controls (77 years old), Nordin and Murphy (1996) reported declines on all three tests in the AD group, especially olfactory threshold (which did not correlate with recognition memory). Bacon *et al.* (1998) found that individuals who were non-demented but were at risk of developing AD because they had mild cognitive impairment (MCI; and who later did develop the disorder) had higher olfactory thresholds a year before the diagnosis of AD.

Indentification is also impaired in an 'at risk' group but this impairment does not appear to be mediated by verbal ability (or decline). For example, Morgan *et al.* (1995) administered the lexical UPSIT and the less verbal San Diego Odour Identification Test to a group of at-risk/questionable AD individuals and found impairments on both tests, impairments the same research group also found in those carrying the APOE4 (apolipoprotein E) allele. Murphy *et al.* (1990) also found a decrease in n-nutanol threshold but not sucrose threshold in a group of nine e4+ carriers. Others have found that e4 allele carriers, those with familial AD and those who show cognitive decline that is not commensurate with their age, show impaired olfactory performance (Bacon *et al.*, 1998; Schiffman *et al.*, 2002). Morgan *et al.*'s results do not rule out the possibility of any verbal mediation completely. It is possible that the act of sniffing and identifying involves initially applying a label – whether a descriptor of intensity or familiarity or a description of its hedonic tone or an actual name – to that odour and then retaining this word in mind as a means of providing a template next to which other odours are matched. Even when the number of items are reduced, to alleviate the load on memory and to reduce time taken to fend off fatigue in easily fatigued patients, performance on short versions of the UPSIT predict those at risk for AD (Tabert *et al.*, 2005).

The association between olfactory impairment and at risk status for AD is consistent. Several studies have now shown that threshold deficits or identification impairment appear before typical AD symptoms emerge (Morgan *et al.*, 1995; Bacon *et al.*, 1998; Devanand *et al.*, 2000). Graves *et al.* (1999) have gone as far as to suggest that olfactory impairment may be a better predictor of cognitive decline than is cognitive test performance studied concommitantly and there is less to dismiss in this idea than appears. A study of Scandinavian participants found a significant relationship between olfactory identification ability, the e4 allele and cognitive decline. Olofsson *et al.* (2009) administered the Scandinavian Odour Identification Test to 501 65–90-year-old participants. Impairments on this test are associated with cognitive decline in Swedish participants (Nordin *et al.*, 1998). The test uses odours that are very familiar to Scandinavians – pine needles, juniper, violet, anise, clove, vanilla, almond, orange, cinammon, lemon, lilac, tar and apple. Participants are given four labels and asked to choose which describes the odour. The presence of the e4 allele and a decline in odour identification score predicted global cognitive decline as measured by the Minnesota Mini-State Examination whereas vocabulary performance did not. Of interest is the finding that five years later, old age and e3 allele carrying predicted cognitive decline but olfactory performance did not, suggesting that by this later stage the sensitivity of olfactory identification is swamped by other factors.

In a comprehensive study, Djordjevic *et al.* (2008) examined detection threshold, odour discrimination ability, colour discrimination ability and UPSIT and picture identification performance in thirty-three normal, elderly individuals, fifty-one individuals with MCI and twenty-seven patients with probable AD. The inclusion of tests examining visual performance was used as a useful internal, sensory control. The authors used the forced-choice ascending single-staircase procedure – where PEA was presented from weak to strong concentrations and after five correct responses, was presented in reverse strength order (stronger to weaker). Participants were asked which of a pair of stimuli (blank and odour) smelled stronger. In the discrimination task, participants made judgements about the intensity of pairs of odours, after a procedure adopted by Zatorre and Jones-Gotman (1991), deciding whether the stimuli were the same or different (some were similar to each other, some were very different). An analogous visual task employed colours. In the pictorial equivalent of the UPSIT, the participant was presented with a drawing (forty in total) and asked to identify it from four alternative names.

Significant impairment on all three olfactory tasks was found in the MCI group. Impairments were also found in the AD group and these were significantly worse. The discrimination impairment in the MCI group appeared to be accounted for by the threshold deficit. The impaired identification performance, however, was not accounted for by threshold. Tests of identification and discrimination were most highly correlated with cognitive symptoms and threshold test performance produced with lowest correlation. As the impairment (cognitive) in MCI is memorial, with preserved verbal, reasoning and attention ability, Zatorre and Jones-Gotman (1991) suggest that their data provide strong evidence for the hypothesis that olfactory symptoms predate cognitive symptoms in AD. They suggest that identification and threshold performance can predict/classify 91 per cent of individuals with AD and 73 per cent of individuals with MCI. They also found that AD patients who reported detection problems showed the highest thresholds, suggesting that some AD patients are self-consiously hyposmic and that the hyposmia can interfere with everyday life. Research normally indicates that patients are unaware of this deficit until it is formally assessed (Doty *et al.*, 1987). Three out of twenty-four patients indicated that they suffered from a chemosensory complaint; with formal testing with the UPSIT, 90 per cent scored below normal.

In a larger study of 471 cognitively intact individuals (with a mean age of 79 years), Wilson *et al.* (2009) examined the relationship between memory performance and olfactory identification (B-SIT). Olfactory performance was correlated with episodic memory and risk of MCI, and the correlation was positive. After controlling for the e4 allele, this relationship remained. Thirty-four of the participants later died and it was found that these participants scored at the lower end of the B-SIT range.

5.9.3 Neuropsychological mechanisms and olfactory impairment in Alzheimer's Disease

A cognitive feature of AD is semantic memory impairment but it has been unclear in the liteature whether these problems are due to a difficulty with retrieval or with the organization of knowledge (Chertkow *et al.*, 2008). Tasks that rely on semantic networks, such as category tasks, are more seriously affected than are letter fluency tasks suggesting that the problem may be organizational. There has been a debate in olfaction over whether we process odours perceptually or semantically (or both). An early view was that there was no semantic component involved in odour perception, a view challenged by data presented by Lyman and McDaniel (1986, 1990) which showed that odours could be better remembered when associated with their names, a life event or a visual stimulus. This finding suggests a degree of semantic encoding and retrieval. Would such encoding be defective in AD and, if so, how?

Au *et al.* (2003) reported that patients with AD enagage in unusual clustering when asked to place classes of objects into subordinate classes. For example, if asked to cluster animals, they would do so according to the – rather superficial – criterion of size, rather than (the more abstract concept of) domesticity. They intrepret this finding in the following way: in these patients the visual system is intact and they can therefore organize stimuli according to perceptual features; the temporal lobe, however, is not intact and this affects the ability to think abstractly. To examine whether this reasoning was correct and to determine whether a similar phenomenon occurred in olfaction, Razani *et al.* (2010) undertook the following experiment. They recruited twelve patients with probable AD and twelve controls (mean age 73 years) and presented them with three colours or odours. Participants were asked to indicate which two were alike. They were also asked to match a written label to the appropriate colour or odour. Participants were then asked to sort ten odours and ten colours by attribute – edibility, fruitiness, pleasantness and intensity (for odour) and warmth, likeablity, brightness and red-purple (for colour).

When multi-dimensional scaling maps (see Chapter 1) were created, the control group sorted the odours more by edibility but the AD group sorted by fruitiness. In addition, the grouping appeared more logical in the control group – chocolate, coffee and cinnamon were grouped together – whereas it was much looser and and disparate in the AD group. Another example illustrates this. The controls were more likely to group pineapple, banana and orange according to fruitiness whereas the AD group did not – orange was associated with baby powder as was banana. There was no difference between the groups in terms of colour classification/clustering (presumably, the authors suggest, because the striate cortex was intact) (Figure 5.5).

The study is interesting because it suggests a semantic impairment in AD that extends to the olfactory modality. What the study did not do, however, which the authors acknowledge, was to administer an identification test (to ensure that the identification performance of the groups was the same or different) or to match the groups for olfactory sensitivity (threshold).

Patients with semantic dementia show similar symptoms to those with AD but, because the lesions are localized to the fronto-temporal region, the deficits are limited to word comprehension problems and impairments in non-verbal semantic memory (Hodges and Patterson, 2007). Initially, the focus of pathology is in the left anterior and inferior temporal lobe and this extends to the OFC as the illness progresses. Snowden *et al.* (2001) and Rosen *et al.* (2006) have reported an unusual symptom of semantic dementia: the development of preference for unusual flavour combinations and the development of food fads. The role of the OFC in smell and taste, as Chapters 3 and 4 showed, is well reported and frontal lobe injury has been associated with impairment in food-related behaviour, such as the ability to follow recipes. The temporal and frontal cortices also contain all of the regions involved in smell and taste perception, including the primary olfactory and gustatory cortices, the secondary smell and taste areas, and the amygdala. Like patients with AD, patients with semantic/fronto-temporal dementia also

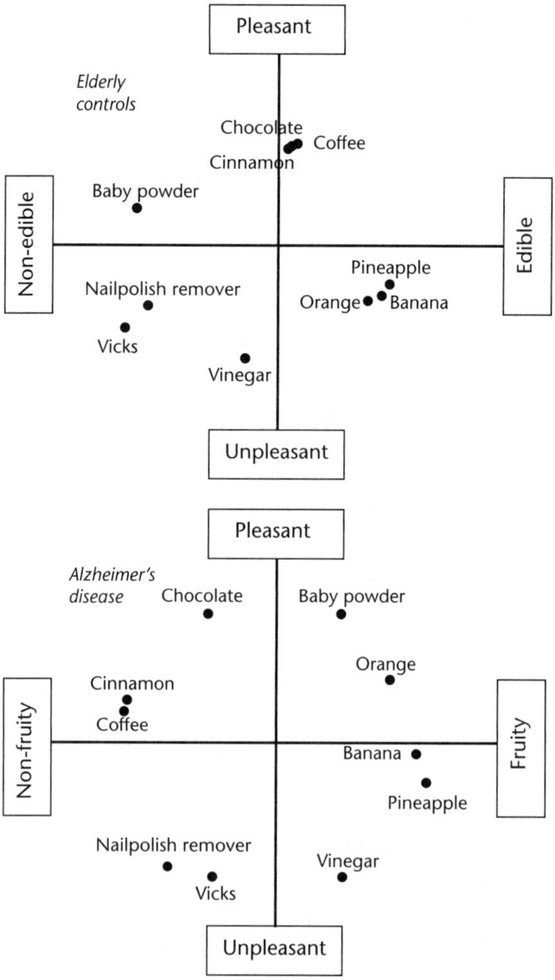

FIGURE 5.5 Spatial odour maps for the demented and healthy participants in Razani *et al.*'s study

show impairments in odour identification and also impairment in identification of food flavour (Gorno-Tempini *et al.*, 2004; Luzzi *et al.*, 2007; Rami *et al.*, 2007).

To examine the effect of semantic dementia on flavour identification and semantic decision-making related to smell and taste, Piwnica-Worms *et al.* (2010) asked three patients at various stages of semantic dementia (including one with aphasia) and six controls to complete the UPSIT and two other tasks. In one of these, participants were asked to assess whether two flavours were congruent (e.g. watermelon and mango or chocolate and coffee), asked whether they were the same or different, and whether the combination was normal or unusual. They were also given a word matching task where, for example, chocolate would be matched with its label in the following list: chocolate, coffee, watermelon. The semantic dementia group was unable to identify flavours or to assess the congruence between two flavours/tastes. The authors suggest that this represents a form of associative agnosia. In the visual form of this disorder, patients have difficulty in integrating several elements of stimulus in order to create and identify a percept. In this study, the authors argue, the patients have an inability to integrate different elements of food flavour – the integration of which gives rise to food flavour – as demonstrated by their inability to indicate whether two flavours/tastes are compatible.

5.9.4 Peripheral neuropathology in olfactory structures in Alzheimer's Disease

One of the most exciting possibilities in experimental molecular work is that biomarkers in the OE may allow the early detection of AD. Talamo *et al.* (1989) reported that the autopsied neuroepithelium of eight patients with AD had reduced olfactory neurons but that similar neuropathology (dystrophic neurites) was found in other parts of the brain suggesting that the neuronal loss was not specific, nor predictive of later pathology. Of note is the finding that plaques and tangles are not found in the OE (Lee *et al.*, 1993). However, the epithelia of patients with AD show reactivity to antibodies – there is increased immunoreactivity to tau (the protein that comprises the helical filaments in tangles) and ubiquitin (a stress protein essential for cell vitality) in the olfactory mucosa (although there are no plaques and tangles there) (Tabaton *et al.*, 1991). Other studies have suggested that this feature is not specific to AD but is also found with normal ageing (Trojanowski *et al.*, 1991). The dystrophic neurites found in the OE of patients with AD have also been found in healthy adults.

Lee *et al.* (1993) reported an interesting finding: dystrophic neurites, while not containing the paired helical filaments seen in tangles in AD, did express themselves as 10–15 micrometre long filamented bundles in AD. Other studies have shown abnormal tau protein processes immunoreactivity in the dendrites and perikarya of olfactory neurons (Yamagishi *et al.*, 1994); a post-mortem study of AD found B-amyloid in ten out of twelve patients, particularly in the ventral third of the OE. Hock *et al.* (1998) also reported tau pathology in the OE but this was not the primary symptom of AD.

Of the proteins thought to be involved in pathology in AD and PD, alpha-synuclein appears to be important (Kruger *et al.*, 1998). It is a soluble protein containing 140 amino acids and can be isolated from the plaques of AD patients (it is also a contributor to the filaments of Lewy bodies). It is mostly found in presynaptic neuron terminals. It is widespread in the OE and there is evidence that the alpha, beta and gamma variants are expressed in the Bowman's glands and basal cells in the OE (Duda *et al.*, 1999) and have been reported in PD and Lewy body disease. Other molecular, neurophysiological processes or abnormalities in the olfactory system of the AD include an increase in superoxide dismutaste, which protects cells against oxidation (Kulkarni-Narla *et al.*, 1996), the down regulation of cells and a reduction of heat shock proteins, which help cells after temperature shock and injury (Getchell *et al.*, 1995), a decrease in calbidin D28K in olfactory cells (Yamagishi *et al.*, 1998), which helps cells to buffer positive calcium ions, and an increase in the amount of APOE in olfactory receptor cells (Yamagishi *et al.*, 1998).

Tangles have been reported in the next stage of olfactory processing – the OB and AON – in the elderly and a loss of glomeruli in the OB has also been reported (Kishikawa *et al.*, 1990). A reduction in the left, right and total OB volume is correlated with Mini Mental State Examination (MMSE) scores in 72-year-old individuals with AD (Thomann *et al.*, 2009). Kishikawa *et al.* found that the number of tangles in the OB increases with increasing age: 17.7 per cent when people are in their 50s, 25.6 per cent in their 60s, 57 per cent in their 70s, 86.7 per cent in their 80s and 100 per cent in their 90s.

This loss is thought to be accelerated in AD. In addition, there are thought to be 62 per cent fewer cells in the AON in AD and 50 per cent fewer myelinated axons (Davies *et al.*, 1993). The bulbs may also be the first structure affected by tangles – or show evidence of tangle formation (Kovacs *et al.*, 2001). Bundles in these areas may be responsible for the olfactory loss (Wilson *et al.*, 2007) and tau pathology here correlates with MCI (Attems and Jellinger, 2006). Work with transgenic mice has shown that the over-expression of mutant amyloid precursor protein was associated with impairments in short-term memory for odour and olfactory discrimination (Wesson *et al.*, 2010). Whether these abnormalties are specific to AD, however, is questionable. As the next section shows, hyposmia is also found in PD and it is also found in Lewy body disease and fronto-temporal dementia (McLaughlin and Westervelt, 2008; Williams *et al.*, 2009). Furthermore, total anosmia is not commonly reported in AD nor PD. Olfaction, therefore, may not be uniquely predictive of AD (Wesson *et al.*, 2010).

5.9.5 Pathology in olfactory cortex in Alzheimer's Disease

The cortical regions affected in AD invariably include the olfactory areas – the medial temporal lobe, priform cortex, prepiriform cortex, olfactory tubercle and entorhinal cortex, all of which have connections to (secondary) olfactory areas – the OFC, insula and dlPFC. Some of the highest densities of plaques and tangles and the greatest pathology, for example, are found in the entorhinal cortex, subiculum, temporal pole, piriform cortex, amygdala, OFC and prepiriform cortex (Kromer Vogt *et al.*, 1990; Arnold *et al.*, 1991a; Chu *et al.*, 1997). The cortical link between the cortex and the subcortex is the entorhinal cortex, which sends projections to the hippocampus (van Hoesen *et al.*, 2000). Kesslak *et al.* (1991) reported an association between loss of volume in the hippocampus and olfactory test performance. However, Koss *et al.* (1988) found no correlation between detection threshold and identification performance, and brain metabolism. Buchsbaum *et al.* (1991) reported less activation in the hippocampus in six patients with AD who completed an odour memory test (they were also poorer at this than were controls) but increased activation in the lateral temporal lobe. Kareken *et al.* (2001) examined UPSIT and odour threshold (for PEA) performance in seven patients with AD (average age of 73 years) and eight age-matched controls while brain blood flow was also investigated. The control group showed significant activation at the fronto-temporal junction bilaterally – an area that is the equivalent of the piriform cortex (or POC) in non-human brains – whereas the AD group showed less odour-evoked activation in the right piriform cortex and right anterior ventral temporal regions. These effects occurred for odour identification (which was poorer in the AD group); there were no significant group differences for threshold (Figure 5.6).

A study comparing areas of damage in patients with corticobasal syndrome and fronto-temporal dementia and patients' performance on the UPSIT noted that UPSIT scores correlated with right midfrontal gyrus in the latter group and right insula, midfrontal gyrus and bilateral inferior frontal gyrus for the former (Pardini *et al.*, 2009). UPSIT scores predicted general memory performance on the Wechsler Memory Scale and Boston Naming Test in the fronto-temporal group. Similar involvement of the frontal gyrus and inferior frontal gyrus was reported in a PET study of patients with early AD (Forster *et al.*, 2010). It found that olfactory identification scores (measured via Sniffin' Sticks) correlated with right superior parietal lobule, frontal gyrus, inferior frontal gyrus and precuneus activation, discrimination with left postcentral cortex and threshold with the right thalamus and cerebellum.

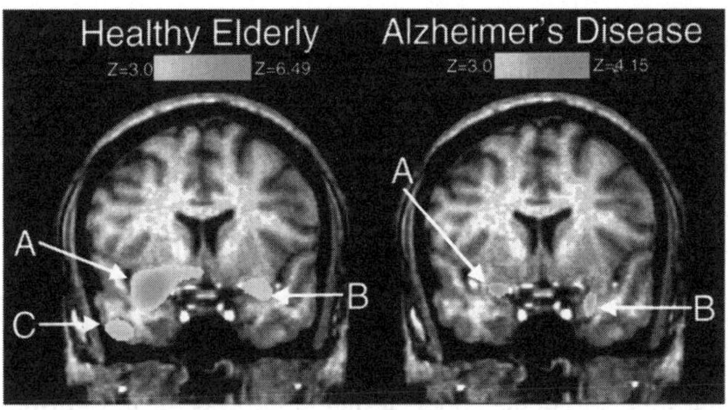

FIGURE 5.6 Areas of activation or hypoactivation in a healthy elderly individual and a person with Alzheimer's Disease seen during olfactory stimulation

5.9.6 Parkinson's Disease

PD is a disorder of motor behaviour whose classical features include akinesia (general loss of movement), rigidity (resisting passive movement) and tremor at rest. Tremor at rest arises from alternating agonist and antagonist contraction in distal muscles of the arm. The most common manifestation of this is 'pill-rolling', where the fingers appear to imitate the rolling of a pill. Akinesia can take the form of slowness of movement (bradykinesia), reduction of movement (hypokinesia) or lack of spontaneous and automatic voluntary movement (akinesia itself). Axial akinesia refers to impaired movement of the trunk or proximal muscles; this makes turning difficult for the Parkinsonian patient. The most prominent neuropathological feature of PD is a loss of the striatal dopamine pathway which runs from the substantia nigra to the neostriatum (caudate nucleus and putamen) and globus pallidus. There is degeneration of the ventrolateral layer of the substantia nigra projecting to the striatum (Fearnley and Lees, 1991). Lewy bodies can also be found in the substantia nigra and locus coeruleus which may be a diagnostic marker for the disease (although other diseases with parkinsonian symptoms also show evidence of these Lewy bodies and Lewy bodies are also found in other parts of the brain such as the cortex and raphe nuclei).

According to Braak *et al.* (2002, 2003), the first regions affected in PD are the dorsal motor nuclei of the glossopharyngeal and vagus CNs and AON and the pathology begins in the brain stem and extends to the subcortex which is affected later. The most common form of Parkinsonism occurs in idiopathic PD where the aetiology of the disorder is unknown. There is also a familial form and variants of the disease (Guam-PD dementia complex, multiple system atrophy, Progressive Supranuclear Palsy and corticospinal degeneration).

5.9.7 Olfactory deficits in Parkinson's Disease

The first report of olfactory dysfunction in PD was published in the mid-1970s when increased threshold to amyl acetate was associated with progressive degeneration in a sample of twenty-two patients (Ansari and Johnson, 1975). Since then a raft of studies have reported, as they have for AD, deficits in olfactory detection, identification and recognition in PD (Ward *et al.*, 1983; Doty *et al.*, 1988, 1992; Zucco *et al.*, 2001; Chou and Bohnen, 2009). Quinn *et al.* (1987) reported reduced thresholds to the same odour in seventy-eight PD patients and Doty *et al.* (1988, 1992) found a decrease that was independent of age and disease duration. Hawkes *et al.* (1997) found that UPSIT scores were lower in 155 34–84-year-old patients with PD; 19 per cent had normal scores and 42 per cent were identified as anosmic. Of all the odours tested, pizza was the most difficult to identify. Another study found that this odour and that of wintergreen were best at discriminating between disease and non-disease (Hawkes and Shepard, 1998). The short version of the UPSIT and Sniffin' Sticks also discriminate between PD and healthy individuals with 90 per cent sensitivity (Double *et al.*, 2003). Problems in identification and discrimination may occur independently of threshold problems (Boesveldt *et al.*, 2008) (Figure 5.7).

There is also a susceptibility to olfactory impairment in relatives of patients with PD (Markopoulou *et al.*, 1997). Sons and daughters of fathers with PD (especially) had lower UPSIT scores than those of unaffected parents (Montgomery *et al.*, 1999). However, it is rare for these relatives to develop the disorder although they are more likely to if they are hyposmic than normosmic (Ponsen *et al.*, 2004). One explanation for this phenomenon is that there is lower striatal dopamine binding in these hyposmic non-Parkinsonian relatives (Berendse *et al.*, 2001). There is also a curious PD mutation, PARK2 (Parkin) which is an autosomal recessive form of juvenile PD and which is associated with lower UPSIT scores but not significantly lower than those of controls (Khan *et al.*, 2004).

Smutzer *et al.* (2003) conclude that the olfactory deficit in PD appears before the motor symptoms and several researchers have noted that the olfactory system is the first affected by impairment and pathology in PD (e.g. Del Tredici *et al.*, 2002). A study of B-SIT performance in 2,207 men in Honolulu who were part of the Asia Aging Study found that identification deficits were present at least four years prior to onset of the movement disorder of PD (Ross *et al.*, 2007). A similar finding has been reported by Haehner *et al.* (2007). One researcher (Hawkes, 2006) has taken this observation a step further (see section 5.9.8).

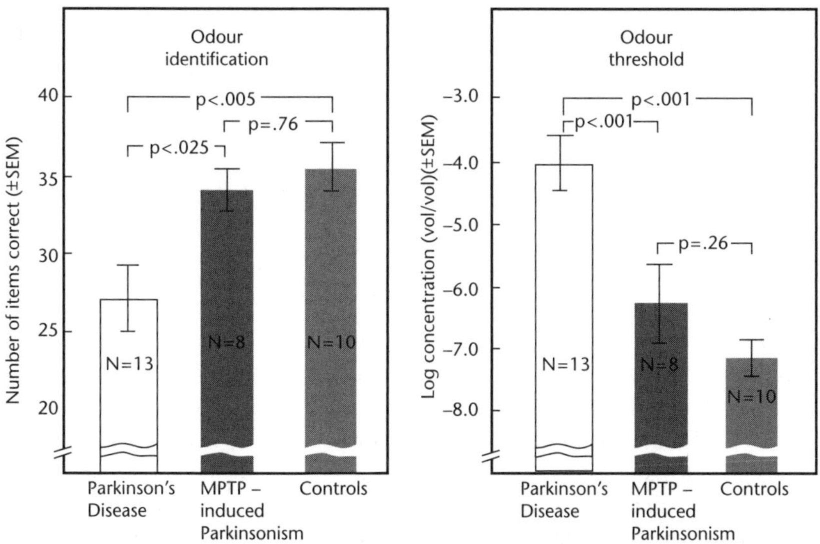

FIGURE 5.7 Odour identification and threshold scores for a group of healthy controls, a group with Parkinson's Disease and a group with induced Parkinsonism

5.9.8 Pathology in peripheral olfactory areas in Parkinson's Disease

In his review of the neurophysiology and anatomy of PD and olfaction, Hawkes concluded: 'It remains possible that the olfactory system is the site of the initial damage in idiopathic PD, along with the dorsal medulla, and that the motor component is a late manifestation of what is a primary olfactory disorder' (2006: p. 287). There is some evidence for this view. The first conclusive evidence of pathology in the OE of PD (B amyloid protein) was found in four individuals post-mortem (Crino *et al.*, 1995). Dystrophic neurites and synnuclein (especially the alpha variant) but not Lewy bodies have been reported in the OE (Crino *et al.*, 1995) – two mutations of the gene have been associated with a type of familial Parkinsonism. It is also the most common abnormality in the neurites of OE in PD (more common than in AD; Duda *et al.*, 1999). Daniel and Hawkes (1992) compared the OBs and tracts of brains taken from the UK Parkinson's Disease Brain Bank and found that Lewy bodies were present in the AON and mitral cells. The loss of the AON appears to correlate with the duration of the illness and is observed in PD and in the Guam variant of PD (Pearce *et al.*, 1995). Braak *et al.* (2003) reported the presence of alpha synnuclein and Lewy bodies in the dorsal motor nuclei of CNIV and CNX and in the AON and noted that these were the earliest neural changes observed. Neural degeneration and alpha-synnuclein presence has been consistently reported in the OB and AON (Hawkes *et al.*, 2009) and also occurs in the limbic system (Silveira-Moriyama *et al.*, 2009), especially the amygdala (Sengoku *et al.*, 2008) and the corticomedial nucleus of the amygdala specifically (Harding *et al.*, 2002). The appearance of Lewy bodies in PD might also suggest that individuals with Lewy Body Disease demonstrate olfactory impairment and there is some evidence for this (McShane *et al.*, 2001); they also show tau and alpha synnuclein pathology in the AON (Tsuboi *et al.*, 2003).

5.9.9 Cortical and subcortical abnormalities in Parkinson's Disease

A number of studies have correlated brain volume and anomalous brain activation with olfactory integrity in PD. An MRI study of fifteen patients with early-onset PD and twelve with moderate PD found a correlation between olfactory performance and cell loss in the right piriform cortex (early PD) and right amygdala (in the moderately advanced) (Wattendorf *et al.*, 2009). Entorhinal and piriform cortex abnormalities have been reported (Braak and Del Tredici, 2008) – specifically, tangles and Lewy bodies in the second layer of the entorhinal cortex. Tangles have been reported in 98 per cent of cases (Mattila *et al.*, 1999) and Lewy bodies in the amygdala, hippocampus and cortical gyrus, the three subcortical areas most usually found to be affected (Mattila *et al.*, 2000). UPSIT scores, for example, have been found to correlate with acetylcholineesterase activity in the hippocampal formation, amygdala and neocortex (Bohnen *et al.*, 2010). As PD is a disorder characterized by cholinergic and dopaminergic dysfunction, Bohnen *et al.* investigated the relationship between these neurotransmitters and olfactory dysfunction. They found that cholingeric denervation in the limbic system was a better predictor of smell loss than was nigrostriatal dopamine denervation and suggest that cholingeric denervation is associated with the most severe smell loss (Figure 5.8).

Other studies have also downplayed the importance of dopamine and dopamine transporter (DT) activity to olfactory impairment (Chou and Bohnen, 2009). Postuma and Gagnon (2010) correlated UPSIT performance with various cognitive measures in seventy-seven individuals with idiopathic PD. These measures included executive function measures, episodic verbal memory, non-verbal memory and visuoconstructive ability (copying a complex geometric shape, the Rey-O). Impaired episodic verbal memory was correlated with impaired UPSIT performance as was non-verbal memory. 'Both olfactory and cognitive dysfunction', the authors suggest, 'are markers of a disease subtype (perhaps a "diffuse" degenerative process) that can impair multiple non-motor domains' (p. 2).

In terms of brain activation, whereas increased activation in the precentral gyrus and middle temporal gyrus is seen in controls, little activation in these areas is found in naturally breathing PD patients smelling the peach-like alpha-undeclactone (Takeda *et al.*, 2010). Hummel *et al.* (2010a) found decreased

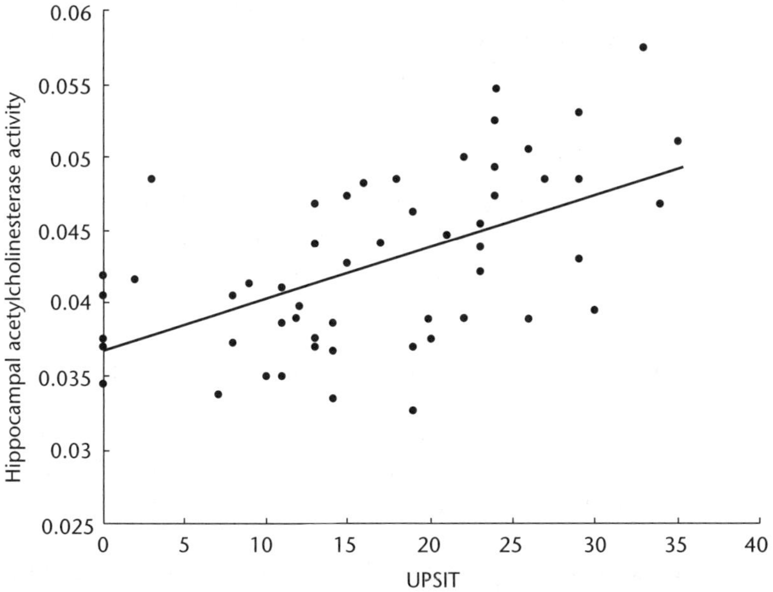

FIGURE 5.8 A scatterplot showing the correlation between hippocampal acetylcholinesterase integrity and UPSIT score in a group of patients with Parkinson's Disease

activation in the amygdalo-hippocampal complex when PD patients smelled H_2S and PEA and an increase to PEA in the striatum and left inferior frontal gyrus. Unpleasant stimuli were associated with a decrease in the striatum in the PD group.

The ability to inhale naturally is important because sniffing may be problematic in PD patients as it requires motor effort and control. One possible explanation for the depressed detection thresholds in olfaction has been this inability to sniff, an act that clearly enhances olfactory stimulation and also increases activation in the piriform cortex and OFC (Sobel *et al.*, 1998). Sobel *et al.* (2001) found that increased sniffing improved UPSIT scores in PD and this is not a variable that is considered in PD studies. In the midly affected, Sobel *et al.* suggest, this may not be such a great problem; in more advanced cases, the effect of the inability to sniff may have more pronounced effects.

Neurophysiologically, 46 per cent of PD patients in one OEP study showed no or abnormal OEPs (Hawkes *et al.*, 1997). Of the thirty-seven patients with some trace of OEP to H_2S, latency was longer, a finding that has also been reported with vanillin (Barz *et al.*, 1997). In a study of OEPs, DT, UPSIT and PD duration, Deeb *et al.* (2010) followed up forty-nine patients with early PD. At follow-up, they were classified as PD or non-PD. Deeb *et al.* found that a DT scan was 92 per cent sensitive in predicting PD. However, the UPSIT also scored particularly highly, predicting 86 per cent of PD patients. Increased OEP latency was found in the PD individuals but there was no difference in amplitude – OEP changes did not correlate with the DT scan. An example of the OEP response of a patient with PD can be seen in Figures 5.9a and 5.9b.

The evidence, therefore, makes an increasingly persuasive case for olfactory impairment as a significant predictor of PD.

5.10 Disorders of taste

Unlike olfaction, taste appears to present relatively few problems of impairment with age or with disease, neurodegenerative disease in particular. However, there are specific disorders of taste perception that are reported as a consequence of brain injury, illness and genetics. Olfactory disorders have gustatory equivalents. Thus, patients may report taste loss (ageusia), decreased sensitivity (hypogeusia), distorted taste, usually in the form of unpleasant taste sensations (dysgeusia) and a form of taste agnosia in which an individual is unable to recognize or identify a taste despite intact sensory or linguistic ability. There is a rare condition, Burning Mouth Syndrome (glosodynia/alicia), which is characterized by abnormal burning, oral sensation in the mouth and altered taste perception.

According to the 1993 AMA Guide to the Evaluation of Permament Impairment, cited earlier, a bilateral loss of taste is considered to be a 3 per cent impairment of the whole individual. In general, individuals with taste loss present with a perceived reduction in taste sensation or abnormal taste sensations and, as Chapter 2

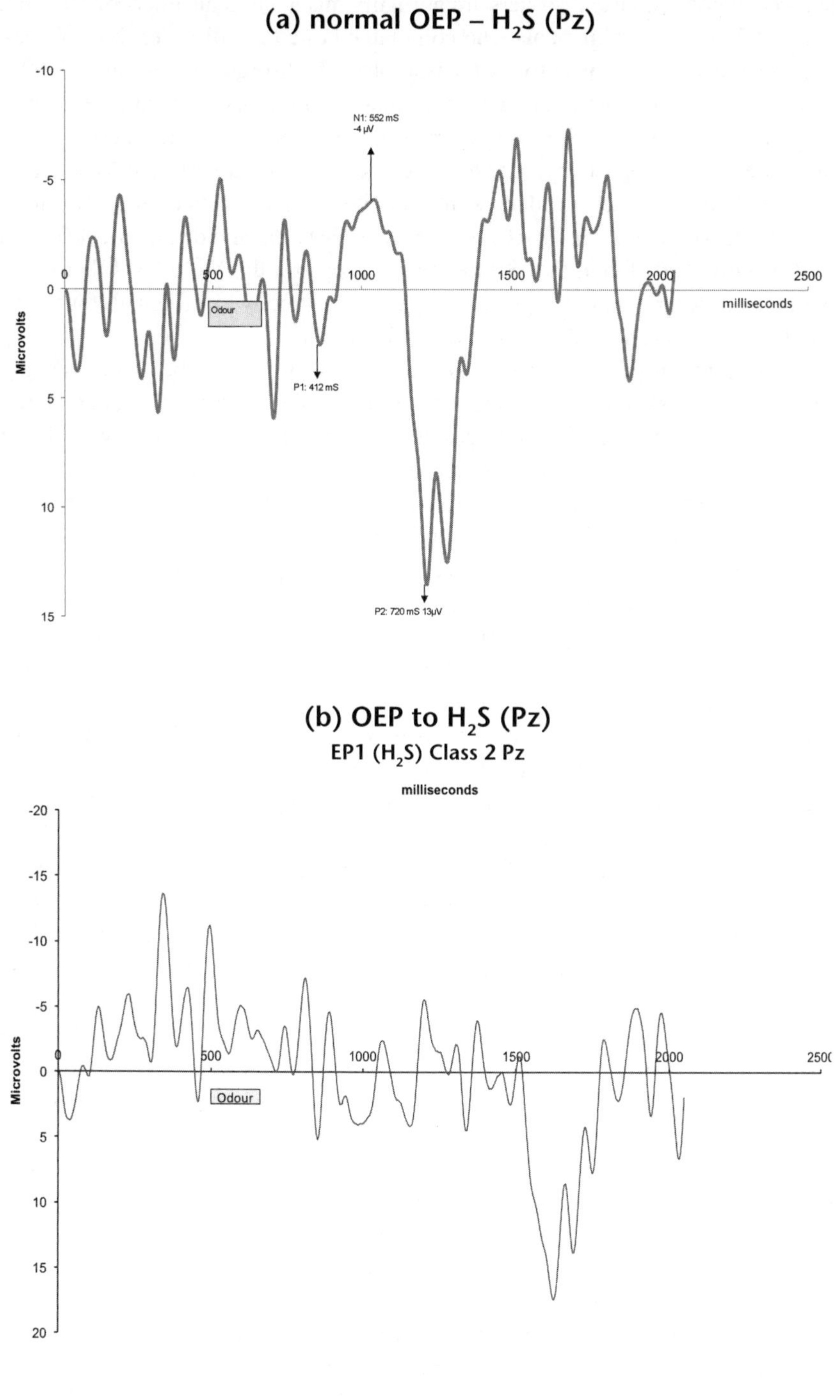

FIGURE 5.9 (a) normal OEP to H_2S; (b) OEP to H_2S in a patient with Parkinson's Disease

showed, reports of taste disorders are normally mistaken as disorders of olfaction. Thus of 74 per cent of patients who complained of a taste disorder (N=438), less than 4 per cent had an objectively measurable taste disorder (Deems *et al.*, 1991). Deems *et al.* found that total ageusia in major head injury was extremely rare – less than 1 per cent. Patients with injury to nerves innervating the anterior lingual area often do not report any taste loss, i.e. they are unaware of their hypogeusia. The incidence of taste disorders is thus smaller (and rarer) than olfactory disorders. Welge-Lussen *et al.* (2010) in a study of taste thresholds and identification in 761 individuals found that 5.3 per cent scored at a threshold level that would warrant a description of hypogeusia, 83 per cent correctly identified all four tastes and no individual showed evidence of total ageusia.

Measurements of taste thresholds and recognition need to be objectively and formally undertaken before a diagnosis of a disorder is warranted. As different areas of the tongue receive projections from, and are innervated by, different CNs, regional testing of the tongue can help determine whether there is damage or dysfunction to one or more of these nerves. (This testing, as described in Chapter 1, might include regional liquid stimulation or electrogustometric stimulation of parts of the tongue.) For example, various concentrations of a tastant will be administered to the back and the front of the tongue. A taste disorder might indicate an underlying brain pathology, such as infarct or ischemia. As with olfaction, the loss of a sense of taste is important because this may affect the prevention of ingestion of material that is harmful – toxins, as Chapter 3 showed, tend to be bitter-tasting and it is one of the sense of taste's functions to act as a gastrointestinal guardian in this respect. Taste is also required for the process of salivation and, without salivation, ingestion is difficult (Giduck *et al.*, 1987).

There are various causes of taste disorders and a comprehensive list would include: oral cavity problems or disorders, problems with salivary transport, destruction, inflammation (Wang *et al.*, 2009) or damage to the taste buds, damage to the CNs innervating the taste bud fields, lesions or dysfunction in CNS areas such as the insula, disturbances in metabolism (Bromley and Doty, 2003) and also other illnesses such as cancer, especially when patients are undergoing chemotherapy (Epstein and Barasch, 2010; Sanchez-Lara *et al.*, 2010), and diabetes (Naka *et al.*, 2010). In the latter, for example, there is progressive hypogeusia for glucose and sweeteners, then salty tastants and then all tastes (Hardy *et al.*, 1981). Medication use can also affect taste and a variety of drugs have been found to alter taste (Naik *et al.*, 2010), including amphetamine, baclofen, levodopa and amitriptyline. Schiffman (1983) found that the drug group sulfhydryl impaired taste perception and suggested that this contributed to taste loss in older people (Schiffman *et al.*, 1999).

However, it is also noteworthy that altering the physiology of CNs can enhance taste perception, although not in a particularly positive way. Thus, if the chorda tympani and CNVII is anaesthetized unilaterally, perception of taste at the back of the tongue is enhanced and individuals report finding unpleasant tastes as more unpleasant (Kveton and Bartoshuk, 1994). Damage to the tongue can also cause phantageusia or a distorted sense of taste. For example, increased bitter sensation at the rear of the tongue has been reported (and the opposite effect for NaCL; perception decreases; Yanagisawa *et al.*, 1998). Anaesthetic applied to both chorda tympani resulted in taste phantoms being reported in 40 per cent of patients when both sides of the tongue are stimulated or to the back of the tongue contralateral to unilateral anaesthetic.

Damage to any of the CNs innervating the tongue and nasopharynx can produce taste impairment. The facial nerve, because it is long and sinuous, is susceptible to injury because of this physical characteristic. There are also specific disorders associated with CNVII damage, such as Bell's Palsy which can result in impaired functioning of the anterior tongue. Because of the course of the CNVII through the middle ear, damage to the middle ear, commonly caused by head injury, results in ipsilateral impairment in taste and also impairment in the ability to produce saliva. Another nerve close to CNVII is the lingual branch of the mandibular division of the trigeminus and this, when injured, can also cause taste impairment, including numbness and whole-mouth taste loss (Scrivani *et al.*, 2000). As the third molar area is also innervated by this nerve, deeply impacted wisdom teeth and their removal has also been associated with short-term taste loss (Shafer *et al.*, 1999).

There is no systematic treatment for taste loss following injury – the condition is considered to be so rare and the recovery of the function so spontaneous that none has been seriously proposed. It is noteworthy that the ability to perceive sweetness recovers before the ability to perceive bitterness.

Taste neglect and taste ageusia are rare disorders. The former is the gustatory equivalent of the more well-known unilateral visual neglect. In buccal hemineglect, individuals might neglect one half of the mouth or food in the left half of the mouth (Andre *et al.*, 2000). Small *et al.* (2005a) report the case of a 38-year-old epileptic woman who underwent left medial temporal lobe and amygdala surgery and was able to detect and name tastes, as well as determine the intensity of taste, but was unable to recognize tastes. Two-to-five years later, there was no change in her behaviour.

In terms of cortical activation in patients with smell loss, studies are few. Hummel *et al.* (2007) used fMRI to study gustatory stimulation in twelve normogeusic and eight hypogeusic individuals. The OFC and insula was activated in both groups. However, there was higher activation in these areas in the patient group, which the authors interpret as reflecting enhanced neuronal recruitment to compensate for the functional impairment.

Although apparently trivial in nature, smell and taste disorders can have serious consequences for well-being and psychological functioning on the one hand, and may be predictive of more serious disorders on the other. The work in PD which has revealed and may reveal further neuroanatomical markers for the disorder in the olfactory system is exciting. If replicated and found consistently, the findings regarding the sense of smell may provide the earliest indicator of a potentially devastating disorder and help prepare for this eventuality. To a lesser extent, such a possibility may also exist for AD but the data are not as consistent. So much for Cinderella.

She makes her final entrance in the next chapter. One of psychology's most complicated conundra is: how do we perceive flavour and how does the brain allow us to do this? An attempt at answering these two questions is presented next.

6

THE NEUROPSYCHOLOGY OF FLAVOUR

Multisensory interaction at the behavioural and neural level

6.1 Flavour: a starter

There is an observation made by the functional psychologist, Edward Titchener (1909), about the nature of the peach which encapsulates many of the puzzles, intrigues and uncertainties concerning the role of smell and taste and other senses in food flavour. 'Think of the flavour of the ripe peach,' he writes,

> The ethereal odour may be ruled out by holding the nose. The taste components – sweet, bitter, sour – may be identified by special direction of the attention upon them. The touch component – the softness and stringiness of the pulp. The puckery feel of the sour – may be singled out in the same way. Nevertheless, all these factors blend together so intimately that it is hard to give up one's belief in a peculiar and unanalyzable peach flavour.
>
> (p. 135)

And this is the challenge – almost seemingly insurmountable problem – of studying flavour: understanding how the individual sensory, kinetic and hedonic components combine and interact to produce something identifiable and identifiably unitary, in McBurney's (1986) word 'fusion'. If the psychology leaves us defeated, as it does when defining emotion, then perhaps neuroscience can provide us with a soupçon of hope. For example, as the previous chapters have described, the three major senses involved in the hard graft of flavour perception – smell, taste and touch – have separate systems at the anatomical, physiological, neural and neurochemical level. The pathway of the signal from olfactory, gustatory and somatosensory receptors is well delineated and well studied at the periphery and the evidence is that they are distinct. This much we know. What is less clear is whether these pathways interact at some point or at various points or whether, as some authors have suggested, citing neurophysiological evidence, there exists a convergence zone in the brain that receives all of these sensory inputs and collates the information, assimilates it and produces the flavour percept (Rolls and Baylis, 1994). As Titchener went on to expound: 'It is a univeral rule in psychology, when sense qualities combine to form what is called a perception, the result of their combination is not a sum but a system; not a patchwork but a pattern' (p. 135). According to Rolls, this system occurs at the level of the cortex, where the individuality of the sensory pathways becomes more blended. The specific region thought to be responsible for this blending is the posterior OFC although, as a later section will show, this may not be the only site.

Rolls argues strongly that the individuality and separateness of the sensory systems pre-cortically results in more sophisticated processing later in the system. For example, when taste fibres enter the brain stem and synapse with the NST, these make subseqent projections to specific areas of the thalamus (note that Rolls's assumptions and descriptions are based on non-human experimentation) such as the parvocellular and ventroposteromedial nucleus (VMPN). The subnucleus, Rolls argues, responds exclusively to taste. The VMPN does not respond to auditory or visual stimulation (Pritchard *et al.*, 1986). The insula and opercula receive projections from the VMPN that are gustatory in nature and the structures themselves appear to contain cells that respond to tastes (Scott *et al.*, 1986; Yaxley *et al.*, 1990). In macaques, Scott *et al.* (1995) note that cells in the OFC are multisensory in nature. That is, the region responds to olfactory, gustatory and visual stimulation. However, a part of this region – the caudolateral OFC – appears to be selectively responsive to gustatory stimulation. In contrast to the PTC, the posterior OFC is thought to contain cells that are bimodal – they respond to more than one type of sensory input. In the PTC, 85 per cent of neurons have been found to be selectively activated by taste, with 66 per cent showing unimodal properties. The anterior portion shows a slightly different proportion of responsiveness: 60 per cent gustatory and 89 per cent unimodal. The posterior region contained 4 per cent of neurons that responded to smell and taste and were, therefore, bimodal. Rolls and Baylis (1994) proposed that this general area represented a convergence zone because almost 30 per cent of neurons in the taste cortex were also bimodally responsive to visual stimuli and taste and the anterior region contained 39 per cent of cells that were similarly bimodal. They note that this region receives inputs from the inferior temporal visual cortex. Overall, they note that 13 per cent of all cells in this are bimodal and argue that olfactory/gustatory (and visual) convergence does not occur before the OFC. Furthermore, they argue that unimodal neurons mingle with bimodal neurons en route to the OFC. This proposition is reviewed in the last sections of the chapter.

6.2 Odour–taste interactions

The dominant senses in flavour are smell and taste, but both are confused. We think that the dominant flavour sense is taste when, in fact, it is smell. There are five basic tastes but hundreds and thousands of potential odours and it is these odours that help us to identify the food that we ingest. Rozin (1982) has argued that the two types of breathing we undertake allow two types of odour identification – one distal (orthonasal, though the nares), one proximal (retronasal). As noted in Chapter 1, when the olfactory receptors are impaired – as they are when individuals have a cold or influenza – people report (incorrectly) being unable to 'taste' food although taste is relatively well preserved. People are able to determine whether what they ingest is sweet, salty and so on, but are unable to detect the aroma of food, that which gives the food its identity. Prescott (1999) describes this as the 'orthonasal effect' because we attribute taste to odour. It is a misunderstanding that is entrenched and difficult to dislodge presumably because we do not consider what we ingest to have an odour, but a taste: food is placed on the tongue and not presented to the nose. This geographical property of food results in a very strange, distorted map of flavour being drawn. Gastronomists such as Brillat-Savarin were aware of the predominant importance of smell almost two centuries ago, as noted in Chapter 1.

Chapters 3 and 4 described the current state of our understanding of the neuroanatomy and neuropsychology of the two major flavour senses – smell and taste – and suggested some ways in which these senses interact. The emphasis was on describing their unitary functions rather than how they influenced each other. This section examines in a little more detail some of the unusual interactions between smell and taste – how each can influence each other – at the behavioural level and how this interaction might be expressed at the neural/cortical level.

The senses interact with each other constantly and almost unstoppably in life. In the flavour domain, some of these interactions are learned and some appear to be more innately determined, although the former is easier to determine than the latter (Delwiche, 2004; Verhagen and Engelen, 2006). The interactions involve those between taste and smell, smell and taste, sight and taste, sight and odour, somatosensation (e.g. temperature) and taste, somatosensation and odour, sound and taste, taste and sound, odour and sound, and sound and odour. Lest this be misconstrued as word salad, what is meant is that sound might affect the way in which taste is perceived and taste may influence the way in which sound is perceived so that the interactions may not be unidirectional but bidirectional.

The most well-studied interactions involve those between smell and taste (and within these domains, sweetness is the most well-studied quality) and there is considerable evidence that one can influence, enhance or impair the preception of the other. Detection of flavour is invariably better when information is available from both senses. Early studies had shown that the effects of sweet taste and smell were additive: each made the other sweeter. Murphy *et al.* (1977) found such an additive effect when fruit-smelling ethyl butyrate was presented with saccharin. The presence of the odours of strawberry, vanilla, lemon, almond, caramel and lychee has been found to enhance the perception of the sweetness of aspartame in a variety of studies (Frank and Byram, 1988; Lawless and Clark, 1992; Frank *et al.*, 1993; Schifferstein and Verlegh, 1996; Stevenson *et al.*, 1999; Sakai *et al.*, 2001). Aroma associated with salt can enhance the perceived saltiness of food (Nasri *et al.*, 2011). Other odours – peanut butter, chocolate, ham and wintergreen – leave the perception of saltiness unchanged (Frank and Byram, 1988; Frank *et al.*, 1993; Schifferstein and Verlegh, 1996). However, some authors have noted that sweet odours may not necessarily enhance the taste of sweetness but affect sweetness ratings, two very different things (Clark and Lawless, 1994).

Some odours suppress sweetness – maltol and angelica oil, for example – whereas others (such as chocolate and caramel) suppress perception of sourness (Frank *et al.*, 1993; Stevenson *et al.*, 1999). Other odours – soy sauce or vegetable – can enhance the perception of salty taste (Djordjevic *et al.*, 2004a; McCabe and Rolls, 2007) as can those of bacon, sardine, anchovy, ham, chicken and tomato (Lawrence *et al.*, 2009; Nasri *et al.*, 2011). These 'savoury/salty' odours can also enhance the perception of cheese (Lawrence *et al.*, 2011). The odour of lemon and strawberry can enhance the sourness of citric acid (Frank, 2002). The addition of almost undetectable levels of saccharin, but not umami, can lead to enhanced sensitivity to cherry almond taste (Dalton *et al.*, 2000; Lawrence *et al.*, 2009) and the presence of salt can increase the odour and flavour intensities of beef broth (Ventanas *et al.*, 2010). Trigeminal CO_2 presented with seven drinks was associated with an increase in perceived sourness and enhanced aroma perception. The presence of mint enhanced the 'tingling/fresh' sensation of the drink (Saint-Eve *et al.*, 2010). A similar finding was reported with lemon and orange scents: these odours were associated with more intensely refreshing feelings (Zellner and Durlach, 2003). Olfaction can also seriously impair correct somatosensory evaluations. De Wijk *et al.* (2004) presented participants with cups of two halves. They could see the top half but not the bottom. Both halves were filled with custard but the straw through which they could consume the custard was linked to the bottom rather than the top of the cup. Participants were asked to judge the various qualities of the custard they smelled (and thought they tasted) and which they could see. The odour of the top of the cup affected the ratings of the custard's creaminess, melting quality and thickness.

The ability of one sense to evoke a sensation in another is common in chemoreception. Stevenson and Boakes (2004) suggested that as the odours of amyl acetate and vanilla were perceived as 'sweet' and hexanoic acid as 'sour', that this reflected a genuinely true synaesthesia. As odour influences taste so taste influences odour: the presence of sugar can enhance the 'fruitiness' of a perceived odour, for example (Bonnans and Noble, 1993).

The interaction between these senses extends to imagination – people who imagine an odour rate the imaginary sensation of a congruent taste as stronger (Algom and Cain, 1991). Djordjevic *et al.* (2004a) asked participants to detect sucrose while imagining the odour or taste of ham or strawberry. Both imagination conditions led to better detection when strawberry was the subject.

Smell and taste interact with somatosensation, specifically temperature and viscosity (an aspect of texture). The quality of food texture that is most commonly associated with flavour (Vickers and Wasserman, 1980), in that participants are more aware of it, is crispness although this is not an aspect of texture per se but a combination of somatosensory and auditory sensations, the psychological basis of which is poorly studied (Vickers and Wasserman, 1980). According to Stevenson and Mahmut (2011) there are three distinct texture qualities – mechanical (hardness, viscosity), geometrical (grittinesss) and chemical (fatiness). Creaminess may be a combination of the first and last of these. Christensen (1977, 1980) found that an increase in a tastant's sweetness enhanced perceived viscosity whereas the presence of citric acid reduced it and caffeine had no effect. Usually, viscosity tends to reduce perceived taste sensitivity and intensity (Mackey and Valassi, 1956), an effect that has been found for tomato juice, orange juice and coffee. It also appears to reduce olfactory intensity (Bult *et al.*, 2007). Odour can change perceived viscosity, however. Stevenson and Mahmut (2011) found that adding an odour to a viscous solution (carboxymethylcellulose dissolved in water) increased the perceived viscosity and sweetness of that solution, thereby suggesting that odours can aquire 'tactile-like somatosensory qualities'.

But perhaps more than viscosity, temperature is the non-chemosensory quality of flavour that most closely affects taste. Temperature refers to the mechanical change in heat/cold in the mouth or to the chemical changes producing heat or coolness (such as chilli and menthol). Mechanically, our ability to detect tastants is at its best when the tastant is between 22 and 37 degrees (Pangborn *et al.*, 1970). Temperature is important to flavour as foods are preferred at certain temperatures and are only liked at those temperatures – ice cream needs to be ice-cold, carbonated drinks need to be chilled and custard and gravy need to be hot in order to maximize their hedonic reward. Cold foods, unlike hot, also encounter the modifying context of the tongue; the heat from the mouth and tongue increases the temperature of the cold food. The tongue has a temperature of approximately 35 degrees and the intensity of a food tasted, particularly sucrose (Bartoshuk *et al.*, 1982), is enhanced as the food reaches this temperature. Reducing the temperature of the tongue tends to result in a decrease in intensity ratings of some tastants such as caffeine and sucrose but not citric acid and salt (Frankmann and Green, 1987). Altering the temperature on the tongue without the presence of tastants can also generate taste sensations (essentially, taste phantoms) although this phenomenon is not consistent across participants nor across all parts of the tongue (Cruz and Green, 2000). One mechanism for this phenomenon might be that warming initiates G-protein coupled receptors wheras cooling causes gating of NaCl and H+ ions (Delwiche, 2004).

Other tastants can interfere with taste (and odour) perception. Capsaicin, for example, is a well-known irritant and can affect the perception of any food eaten after it. Cream has a similar effect – as anyone who has drunk a glass of wine after eating cream/ice cream can attest (the wine tastes sour). Chilli depresses the intensity of almost all taste qualities – sweet, sour, salty and bitter (Lawless and Stevens, 1984; Gilmore and Green, 1993). Piperine, which gives pepper its distinctive quality, similarly depresses the intensity of these four tastes. Some researchers have found that the effects of chilli are localized on the tongue. At the side, the intensity of sweet, umami and bitter tastes becomes reduced but salty and sour are not affected (Simons *et al.*, 2002). Capsaicin and menthol can also elicit bitter sensations in some people when stimulation is to the circumvallate papillae (Green and Schullery, 2004). Capsaicin's effect is also seen in olfaction where it impairs the intensity of orange and vanilla scent but not strawberry (Prescott and Stevenson, 1995). An increase in temperature tends to increase odour intensity. In one study which compared flavour ratings of custard and mayonnaise heated to 27, 35 and 43 degrees by the tongue, the odour of vanilla and the taste of sourness, respectively, increased with increasing temperature (Engelen *et al.*, 2002).

6.3 Odour, taste and sight interactions

One of the most intriguing interactions that exists in flavour is that between the chemosenses and vision. The principal effect occurs at the level of colour. Increases in the colour of cabonated drinks leads to higher ratings of the flavour's intensity, especially sweetness (duBose *et al.*, 1980), an effect that is also found with yoghurt and cakes. The greener and yellower a lime and lemon drink, the greater the sweetness rating (Roth *et al.*, 1988) and when white wine is coloured red, it is described using terms used to describe red wine (Morrot *et al.*, 2001). Similarly, red wine made white is described using terms normally used for white wine (Ballester *et al.*, 2009). The better the congruence between the colour and the source of the food (yellow and lemon, for example), the greater the intensity of the rated taste. When colours are inappropriately added to drinks, the ability to identify them is impaired. Cakes coloured yellow that are flavourless but described as lemon cakes are considered more acceptable/palatable as the intensity of the colour increases (duBose *et al.*, 1980). Brown M&M's are judged to be more chocolatey – but no less likeable – than are green ones (Shankar *et al.*, 2009) and red added to drinks increases their ratings of sweetness (Johnson *et al.*, 1982, 1983). When drinks are colourless, the ratings of the acceptability decline (duBose *et al.*, 1980). Colour, therefore, assists us in making judgements about food and its acceptance, palatability and flavour. Other examples can be found in Martin (2013).

6.4 Food-related visual stimulation and brain activation

If we eat with our eyes, neuroimaging has not been slow in examining which regions of the brain are recruited when we look at food. 'The first taste is always with the eyes', as Apicius had it. Studies have examined responses to various types of food cues, the type of participant (obese vs. lean) and the physiological state (hunger, satiety, exercise) and age of the participant. All of these variables have been shown to affect brain activation.

One of the earliest neuroimaging experiments presented thirteen adult women with colour photographs of high- and low-calorie food and non-edible utensils relating to eating as fMRI measured brain activation (Killgore *et al.*, 2003). Foods activated the bilateral amygdala and ventromedial PFC but there were differences between the type of food: high-calorie photographs were associated with increases in the medial/dorsolateral PFC, thalamus, hypothalamus, corpus callosum and cerebellum, while low-calorie stimuli were associated with (smaller) activation in the medial OFC, PTA and the temporal lobe. This finding was replicated in a group of 9- and 15-year-old girls whose responses to images of high-calorie food (such as ice cream) were compared with a group of 22–28-year-old women. With increasing age, there was increasing activation in the OFC and cingulate cortex. In the adolescent and child sample, there was more activation in the visual areas and the fusiform gyrus. For low-calorie food images (such as vegetables), there was increased activation in the amygdala with age (Killgore and Yurgelun-Todd, 2005). Adults showed greater activation in the cingulate gyrus and in the younger sample, the fusiform gyrus (suggesting to the authors a degree of lower-level processing). Common to all foods was activation in the left inferior orbitofrontal gyrus and bilateral parahippocampal and hippocampal areas. Utensils also showed increased activation with age in the left inferior frontal gyrus, left anterior and middle cingulate gyri, bilateral medial frontal gyri, insula, righ thalamus and cerebellum. On the basis of this Killgore *et al.* (2003) suggested that there was a shift in processing from vision to cognition with age.

Further studies examining activational effects to appetizing foods have found activation in the right insula and left OFC and the visual areas (Simmons *et al.*, 2005); in the left OFC and bilateral insula/operculum in lean participants who had fasted (Porubska *et al.*, 2006) with these responses modulated in the insula, operculum and putamen when participants rated their appetite; increased activation in the ventral striatum, amygdala, anterior cingulate and premotor area (Passamonti *et al.*, 2009), especially in those for whom the sight of food increased the desire to eat; increased activation in the medial OFC and insula in women who saw high-calorie food and had been deprived of food (Siep *et al.*, 2009); increases in the dorsolateral and ventromedial PFC, middle posterior cingulate and insula in women (but not men) who viewed high-calorie food (Killgore and Yurgelun-Todd, 2010); and increased activation in the ventroanterior insula to disgusting foods in those who were disgust-sensitive (Calder *et al.*, 2007).

Hunger is a strong modulating variable in these studies and has an effect that is also observed when participants taste food (see, also, the section on sensory-specific satiety (SSS) below). The tasting of sucrose, caffeine, saccharin and citric acid is associated with stronger responses in the insula, thalamus and substantia nigra in hungry participants (especially to sucrose) and greater in the inferior and superior insula and OFC areas 11 and 4 in the satiated (Haase *et al.*, 2009). Satiety has been associated with decreases in the parahippocampal gyrus, amygdala and ACC. Haase *et al.*'s (2011) fMRI study of men and women who perceived four tastants under conditions of hunger or satiety found that greater changes were seen in men than women to all tastants in the middle frontal gyrus, insula and cerebellum, regardless of hunger state. During hunger, the greatest sex difference was found to citric acid where activation was greater in men in the insula, thalamus, posterior cingulate, dorsal striatum and a range of other regions. Activation was stronger to salt in the dorsolateral PFC.

A meta-analysis of neuroimaging studies of response to visual images of food has noted that such cues are not simply behaviourally important – in that they provide visually driven information about a stimulus that can invoke desire or disgust or indifference – but that they are physiologically important in that they ready the body for possible ingestion and the consequent release of insulin and alteration in heart rate (van der Laan *et al.*, 2011). Images provoke the desire to eat. Van der Laan *et al.* note that, of the seventeen studies they assessed, the most common areas of activation during viewing of food images were the bilateral posterior fusiform gyrus, left lateral OFC and left middle insula. When hunger was manipulated (five studies), activation was modulated in the right amygdala and left lateral OFC. When the energy content of the food was assessed (seven studies), activation was modulated in the hypothalamus and ventral striatum. The authors argue that most agreement concerns the role of the occipital cortex (because of visual processing) and the insula, but agreement on the consistent involvement of other areas is less clear. There is also a high degree of methodological variability between studies which means that drawing consistent conclusions is not possible. They also note that, as yet, there are very few studies that have examined the effect of age, mood, dietary restraint and reward sensitivity.

6.5 Hunger, satiety and sensory-specific satiety

Studies of the effect of hunger on the response to food-related stimui using fMRI tend to be food-general. That is, hunger is seen as a unitary variable thought to affect all food types. There is evidence, however, of greater specificity in terms of the effect of physiological state on food acceptance. SSS, for example, is a well-documented phenomenon whereby the pleasantness (and the duration of ingestion) of ingested foods declines over time but the pleasantness and ingestion of foods that have not ingested are not affected (Rolls, 1986). For example, the pleasantness of chicken and banana odours (but not those of other foods) declines significantly after these foods are eaten to satiety (Rolls and Rolls, 1997). The specificity of the hedonic response may not be limited to the food's taste or smell, but can extend to its shape, its variety and its colour. For example, people who eat the same colour of smartie, or the same shape of pasta, or one type of sandwich filling eat less of this one type of food than when more than one colour, shape and filling is available. They also find the one shape, colour and filling of food they eat to be less pleasant than the food they have not eaten.

At the cortical level, it has been proposed that the taste areas become more or less involved during satiety to specific foods. The OFC appears to be particularly susceptible. Small *et al.* (2001a) found that activation here decreased when people ate chocolate or tomato juice to satiety and this activation correlated with reduced pleasantness ratings (Kringelbach *et al.*, 2003). Decreased activation in the OFC has also been observed when people were exposed to the odour of banana after they had eaten bananas to satiety (O'Doherty *et al.*, 2000). No such decrease was observed when people sniffed the odour of vanilla.

6.6 The neuropsychology of flavour

Which brings us to the neuropsychology of flavour and whether such a thing is possible. There are many variables that would militate against this proposal. For example, although certain assumptions are made about flavour and what it means, there is no satisfactory definition that would meet strict scientific criteria. In this sense it is a little like love.

We assume that flavour involves the interaction of at least two senses – smell and taste – and flavour cannot occur without both of these senses becoming stimulated. However, flavour also involves other variables such as texture, temperature, sound and vision. When we attempt to disentangle what happens at the level of the brain – what we assume translates the sensations into psychology – we are faced with an interesting conundrum. If it is possible to identify a flavour system in the brain, is this system activated similarly to all flavours or are different parts of the system activated by different flavours or are different systems responsible for mediating different flavours? Is the neurophysiology, neurotransmission and neuroanatomy that allows us to distinguish an eggs Benedict from a hamburger, ice cream from Durian fruit, a tandoori chicken from a bouillabaisse or a cheese and onion crisp from a fondue the same? Are the central pathways identical in the perception of all of these foods but are influenced by sensory-specific pathways (mediating heat or coldness, or softness and hardness, or pleasantness or unpleasasntness) or are all of these assumptions correct with the addition of a degree of stimulation of these pathways? Are there different classes of receptors and number of receptors that contribute to the process, as you have in olfaction, that allows us to perceive thousands of very different odours? In short, are there neuropsycholog*ies* of flavour? Let us look at what we think may occur in the brain when we perceive what is commonly known as a 'flavour'.

6.7 An anatomy of dinner

You are at a restaurant, perhaps *Restaurant Gordon Ramsay* in Chelsea (there is nothing quite like treating yourself and frankly, if you are going to treat yourself, treat yourself well). You have ordered, and are now looking at an appetizing dinner: a lobster ravioli with pea puree, dotted with pools of Lesvos's finest extra virgin olive oil and topped with a parmesan tuile (I am not sure if Gordon does serve Lesvos olive oil but perhaps he should; his dish of ravioli and pea is a signature one). The sight of this highly calorific and immensely desirable dish will activate the primary visual cortex, bilateral posterior fusiform gyrus, left lateral OFC and left middle insula. The sight of the food will also activate those regions of the brain responsible for object recognition, storing visual memories of the elements of the lunch and possibly the 'taste-memory' of this particular dish or elements of it (e.g. pea), thereby reinstating the pleasurable, delicious and rewarding feeling generated the first time it was enjoyed. The insula will have been necessary for novel taste memory and retrieval, and the PBN for the formation, but not retrieval, of this taste memory (Nunez-Jaramillo *et al.*, 2010). The knowledge of the items of the dish will also activate the mental lexicon containing the referents for lobster, pasta, pea, puree and olive oil and the regions responsible for the semantic processing of the food – that the lobster has been boiled, seasoned, flavoured, placed inside two small square sheets of kneaded flour and eggs, and the item boiled again. The words associated with these tastes and smells will probably activate the piriform cortex bilaterally, and the amygdala, but not the OFC (Gonzalez *et al.*, 2006). If the food is familiar, it will activate the superior and inferior frontal gyri, precuneus, angular gyrus, parahippocampal gyrus and hippocampus (Plailly *et al.*, 2007a).

The scent of the food may not be identifiable orthonasally as the lobster ravioli will not be immediately or particularly crustacea-scented – that will come later when the fork cuts through the gelatinous shell of the pasta – but the pea may emit the characteristic sweet scent of verdant pulse and the olive oil may emit that green, grassy aroma of good-quality, first-pressing of olive oil and the parmesan crisp may have the faint aroma of cheese. Thus, orthonasally, families of olfactory receptors will send signals to the OB, in which the spatial array of the signal will be determined by the strength of the odour sensation, and from there will project to the AON which is the bridge between the bulbs and the POC: the prepiriform cortex, lateral entorhinal cortex, ventral tenia tecta, nucleus of the LOT, olfactory tubercle and cortical nucleus of the amygdala.

The motor precision needed for the coordination of knife and fork recruits the PMC/precentral gyrus. This area receives afferent axons from a variety of areas including areas 1, 2, 3, 5 and 6 and the ventrolateral nucleus of the thalamus. A large part of the PMC is devoted to hand movement (especially the finger and thumb), probably owing to the cortical requirements necessary to make precision movements. Stimulation of the PMC elicits movement in other muscles as well, such as the abdomen and back muscle. Unilateral or bilateral stimulation results in movement of these muscles. However, unilateral stimulation of the 'hand area' of the cortex produces movement in the contralateral hand because pyramidal tract fibres are crossed.

The SMA appears to subserve a different type of motor behaviour. Whereas movement of the hand will activate the PMC, sequences of movement will activate the SMA. The ability to imagine also activates this part of the motor cortex and it appears to be involved in motor planning and organization. The premotor area, conversely, is important for the control of visually guided movement such as coordinating hand movement when reaching for an object and guiding it into the mouth. The posterior parietal cortex will be recruited, as destruction of this region produces an inability to execute complex voluntary movements (apraxia). Superior parietal cortex involvement should occur with visually guided movement.

The movement and visuomotor skill required to reach toward and cut the food will in all likelihood recruit two different streams in the parieto-temporal area. In early models, the dorsal stream/pathway was thought to mediate the analysis of spatial relations; the ventral stream, object recognition. Goodale and Milner's (1992) patient, who had damage to area 18 and 19 (inferotemporal cortex), was good at grasping and guiding her hand but poor at object recognition. Goodale and Milner suggest that because other cases of parietal lobe damage show the opposite pattern – patients may recognize objects but are poor at guidance – and because temporal lobe damage is associated with aspects of visuospatial processing, the visual projection system to the parietal cortex represents information about object characteristics and orientation that are related to movement. Jeannerod (1997) proposed that ventral damage impaired grasping but ventral damage impaired judgements regarding object size. The dorsal stream is, therefore, important for visually informed, goal-directed action (such as cutting pieces of food on a plate with a knife and fork, putting them together on a fork, guiding the food towards the mouth and inserting it).

The ravioli, pea and some oil reach the mouth and are deposited there. Once in the mouth and on the tongue, the process of mastication will occur to enable the flavour to be released as molecules are broken down and the gustatory and retronasal receptors become activated and begin sending signals to peripheral parts of their systems. Before mastication, however, there will be immediate stimulation of the tongue in terms of temperature and taste. The combination of increased airflow and mastication releases the odour molecules which will give the food its identity or its 'flavour percept'. It is at this moment that the brain enables the construction of flavour.

The taste of the food and the biomechanics of eating will stimulate the facial nerve (VII), the glossopharyngeal (IX) and the vagus nerves (X), each of which, as described in section 3.17, innervate different parts of the tongue. CNVII, the chorda tympani and superior petrosal nerves innervate the front (anterior) two thirds of the tongue and the soft palate and will be stimulated by taste. Fungiform papillae on these parts of the tongue will produce the action potentials that signal information throughout the rest of the gustatory system. CNIX, via its lingual tonsillar branch, will innervate the back of the tongue, the circumvillate papillae and foliate taste buds, and the parotid gland which allows salivary secretion and the stimulation of saliva produced by the food. The pharyngeal branch of this nerve will innervate the taste buds in the nasopharynx. Finally, CNX, via its superior laryngeal branch, will innervate the pharynx and larynx but will play little role in sensation. It will, however, allow chewing and swallowing as well as mediating the auditory sensation of the crunch of food.

The sweetness of the pea and the lobster will activate sweet receptors on the tongue, probably TAS1R1 and TAS1R2, but we will respond to this sensation more slowly than we would the saltiness of the food (by around 200msecs). The sweeter the pea and the lobster, the stronger the neural response will be in the chorda tympani, NST and parabrachial nerve. Sugar is a reinforcer and so will facilitate the ingestion of more sweet food – it may stimulate cholinergic receptors in the ventral tegmental area of the NA, a region important for reward and reinforcement. If the pea smells sweet, activation will be seen in the insula, piriform cortex and frontal operculum (and the sweeter the odour is judged, the greater the activation) but the sweet taste will only activate the insula (Veldhuizen *et al.*, 2010).

The saltiness of the food will activate a different set of tongue receptors – probably the amiloride-sensitive epithelial sodium channel ENaC. The brothiness of the lobster and ravioli and the parmesan will also, in all likelihood, activate the receptors on the tongue that process umami taste and some of these receptors are shared with sweetness (e.g. the T1R3 receptor subunit). The pepperiness of the olive oil may stimulate another CN, the trigeminus, via a mechanism that is not completely understood. The odour of the oil will also activate BA10 and BA11 bilaterally, and the inferior frontal gyrus.

Signals will be sent to the nucleus terminalis of the medulla where the CNs synapse. From here, signals enter the NST which diverges and then unites to form the insula. The rostral part of this will respond to the sensation of taste; an intermediate part will respond to the stimulation which results in enjoyment of the food and the left side may be more active than the right (Frank *et al.*, 2008). This region and the PTA will also receive input from somatosensory pathways. The middle insula will be activated as the individual is attempting to detect the taste (Veldhuizen *et al.*, 2007). Projections from the NST will synapse with the parvicellular division of the ventroposteriormedial thalamic nucleus or taste nucleus. The taste of the food will, therefore, activate the pathway from transduction at the receptor level on the tongue through peripheral gustatory structures to central structures, especially the anterior insula. If the food is judged on its hedonic merits, there may be activation in the left lateral OFC (Bender *et al.*, 2009). All tastes will activate the anterior insula, frontal operculum and the caudolateral OFC.

The somatosensory elements – mouthfeel, texture – of the food will activate the left and right Rolandic operculum, and possibly the left inferior insula and angular gyrus, activation which probably reflects semantic processing (Cerf-Ducastel *et al.*, 2001). The viscosity of the food will activate the anterior insula, OFC and amygdala – neurons in these areas respond to viscosity in non-human primates (Rolls *et al.*, 2003b; Verhagen *et al.*, 2004; Kadohisa *et al.*, 2005), neurons that also respond to other sensory modalities. Verhagen *et al.* (2004) identified thirty-three 'viscosity neurons' and found that twelve responded to taste and fourteen to taste and other stimuli. These neurons may be selectively multimodal. Rolls *et al.* (2003b) found that almost 70 per cent of multimodal taste neurons responded to glucose but only 13 per cent of those neurons responding to taste and viscosity did: these were more likely to respond to salty and bitter tastes.

The crunch of the parmesan crisp recruits an additional sensory system, audition, and one of the first bits of information this system delivers is the location of the sound (inside the mouth, rather than outside the ear). Auditory information (sound frequences) from the buccal cavity reach the cochlea in the ear and from here, the sound impulse is sent via part of the eighth CN – the cochlear nerve. These efferents reach a collection of structures (the superior olivary complex) in the medulla. The cochlear nuclei form a pathway that reaches the inferior colliculi. Other nuclei reach the olivary complex which then project to the inferior colliculi. From the inferior colliculi, fibres are sent to the medial geniculate body of the thalamus. The efferents from this part of the thalamus finally terminate in specific areas of the primary auditory cortex in the temporal lobe.

Rhythmic mastication to create a bolus will activate the sensorimotor cortex, supplementary motor cortex (MII), insula, thalamus and cerebellum and will probably activate the first three areas more when chewing a pea than the pasta, and the pasta more than the parmesan crisp (Onozuka *et al.*, 2002). If the person doing the chewing is between 63 and 75 years old, activation will be less in MII, the cerebellum and the thalamus but higher in the right PFC (Onozuka *et al.*, 2003). The amydala might also be involved in this process (Sasaguri *et al.*, 2004). If the person has a chewing side preference (say, the left), then increased activation will be observed in the left sensorimotor cortex, even when chewing is bilateral (Ono, 2004). The biomechanics of swallowing then push the bolus towards the pharynx, a process that will activate the lateral and premotor area, frontal operculum and insula.

What then follows is digestion, and an entirely different book.

The unitary (or, sometimes, bi and multimodal) pathways described above allow the perception of the elements of flavour at the macro level. They demonstrate, especially at the level of the cortex, large changes in activation. It has been argued that flavour can be more accurately and sensitively studied at the cellular level but the mechanism by which cells produce flavour is unclear. Rolls's studies were described earlier. Others, in experiments with macaques or other primates, have found that there are cells responsive to smell and taste in the insula and the frontal operculum (Scott and Plata-Salaman, 1999). At the macro level, Small *et al.* (2004) asked eleven adults to perceive either a scent with a congruent or incongruent taste (vanilla with either a sweet tastant or salty tastant) or had them perceiving the stimuli, and tasteless water, individually (unimodal). Responses were greater in the congruent condition in the ACC, dorsal insula, anterior ventral insula, part of the caudal OFC, frontal operculum, ventrolateral PFC and posterior parietal cortex compared to resposes to unimodal stimuli. These effects were not observed when the stimuli were incongruent and compared to the incongruent unimodal parts, and to congruent bimodal parts. In another experiment in which green tea consumers held the liquid in the mouth or sampled green teas and had to retain the memory of the drink in mind for later evaluation with other samples,

Okamoto *et al.* (2006) found activation in the left lateral PFC and right inferior frontal gyrus.

One way in which the integration of smell and taste at the neurophysiological level might be studied could be to investigate activation during conditioned odour or taste preference. It is known that the amygdala is involved in the conditioning of appetitive behaviour and the conditioning of fear (as lesions to the amygdala prevent classical conditioning of fear to tones and a flash of light in rodents). Lesions here also impair conditioned odour preferences but lesions to the insula appear not to have the same effect. Desgranges *et al.* (2010) examined the effect of pairing an odour with a taste and found a subsequent fourfold increase in activation in the number of cells responding to smell and taste in the basolateral amygdala but not the insula. Conditioned aversions, however, may need a part of the insula. Conditioned taste and odour aversions are unaffected by lesioning of the posterior insula and OFC but lesions to the anterior insula do impair conditioned aversion (Lasiter *et al.*, 1985). These studies suggest that there may be at least two sites for the conditioning of odour/taste aversion or preference.

The evidence appears to be strong that the insula is multimodal. In addition to its role in cognitive processing, it undertakes a vital role in chemoreception. In the anterior insula and frontal operculum, neural responding appears to be stronger for taste, temperature and viscosity perception (Verhagen *et al.*, 2004). Of the neurons studied in Verhagen *et al.*'s experiment, none responded to odour. Imaging, however, has found activation in the right insula (amygdala, BA32, BA6 and the ventral lateral thalamic nucleus) and the left insula (and BA44, BA47, amygdala and cerebellum). Activation to smell and taste are found bilaterally in the insula, amygdala and OFC and in the anterior cingulate (BA32). Selective activation to odour has been found in BA21, the cerebellum, BA3 and BA6 but not always. In Verhagen and Engelen's (2006) review of chemoreceptive neuroimaging studies, they concluded that only BA11, BA13 and BA32 were activated in more than one of the twelve gustatory and twelve olfactory studies they examined. De Araujo *et al.* (2003a, 2003b) found that while activation to the taste of sucrose and the scent of strawberry activated the amygdala, insula, frontal operculum, caudolateral OFC and ACC bilaterally, together these stimuli activated the amygdala, right frontal operculum, right anterior insula and right caudolateral OFC. The frontal operculum did not respond to strawberry, but did respond to sucrose suggesting that this might be, the contraints of the odour studied notwithstanding, a unimodal area that responds to taste.

In Small *et al.*'s (2004) study described above, sucrose, vanillin, NaCl and a tasteless stimulus were combined in pairs in congruent and incongruent mixtures. When the results from the imaging during the single sucrose and vanillin presentations were added and subtracted from the sucrose and vanillin joint-presentation condition, activation (superaddition) was found in the anterodorsal insula, anteroventral insula, caudal OFC, frontal operculum, ACC, ventrolateral PFC and posterior parietal cortex. A combination of NaCl and sucrose activated the frontal operculum only, not – surprisingly – the OFC. Thus, Verhagen and

Engelen conclude that multisensory integration could be observed in de Araujo *et al.*'s study, but not Small *et al.*'s when pleasantness and intensity ratings of the various stimuli were compared. Whereas Small *et al.* found superadditive effects in at least six regions (including the OFC, frontal operculum and ACC), de Araujo *et al.* found one, the anterior OFC (a similar, not identical, area was activated in Small *et al.*'s study but on the opposite side). Various reasons have been proposed for these and other differences, including differences in the intensity of stimuli across studies and variations in what participants were asked to do (whether sniff, taste, evaluate for intensity, evaluate for pleasure, etc.). When visual stimuli were congruent with olfactory stimuli, superadditive effects were observed in the left anteriomedial OFC and left posterior inferior parietal sulcus (Gottfried and Dolan, 2003). Verhagen and Engelen note that OFC activation is some distance from the OFC area activated in de Aruajo *et al.*'s study, as well as being on a different side.

6.8 Petits fours?

Which brings us to a number of extant questions, some of which may be unanswerable. A judicious stocktaking of the neuropsychology of smell and taste suggests that there are signs of positive development and promising new avenues for investigation in the field. Given the dismissive conclusions of psychology's great and good a hundred or so years ago, the field of chemoreception is relatively buoyant, boasting at least one Nobel prize, two professional societies and two dedicated journals. There is some sense of coherence, even if much research in the field is driven by a small number of well-known groups and their productive staff, or individual researchers with an interest, ranging from dilettante to sophisticate, in these two senses. Everyone appears to judge themselves competent to research the chemosenses, but it is a lot more difficult than it appears. Control – whether over delivery systems or participant variables – is crucial and can sometimes be overlooked. How many studies indicate that their participants (or their experimenters) were not wearing perfume, cologne or scented deodorant, for example? Control over delivery is less problematic with EP and, to a lesser extent, neuroimaging work but, focusing on olfaction for the time being as this is the sense with stimuli that are most difficult to rein and chaperon, but scent is more problematic to study and control in behavioural studies. Habituation, adaptation and intensity can all affect the independent variable under investigation. There then occurs a trade off between extreme control – where you set up an experiment with the rigorous delivery control of olfactometry, and this would include attaching tubes to your participant – and ecological validity. Ecological validity is not enhanced by the insertion of tubes into nostrils but is demonstrated by the administration of less controllable ambient odour (which, by its nature, may not be directly perceived or may be perceived variably). Taste, as a proximal sense, presents fewer methodological problems of this kind but introduces its own spanner (when studying flavour, for example).

Perhaps the one problem that has dogged and will dog research in this field is the nature of the stimulus. This problem is potentially enormous and will contaminate any conclusion of any study. In short, the problem can be expressed in the following question: how do we determine whether the brain or the person is responding to a specific scent, rather than a class of scent or a category of scent, and how can we ensure that the response is not scent-specific? And this problem is currently insoluble because the problem of the way in which we classify odours is insoluble. Of course, we have our crude attempts at categorization – determining whether a scent is pleasant or unpleasant, for example – but one person's pleasant orange blossom is another person's bête malodorant; one person's delightful cabbage is another's olfactive horror. One way of obviating this problem, and one that is rarely implemented, is to examine hedonic responses to all odours and classify odour conditions based on those responses, regardless of original odour type. For example, if an experiment does use orange and cabbage, then a first analysis would look at the effects of these two scents as discrete, categorical variables. A second, more sophisticated analysis would take the participants' hedonic ratings of these odours and regroup the variables such that you would now have odours that are liked (which could include cabbage) and odours that are not liked (which could include orange). Thus the a priori division of categories of odour is replaced by the genuine hedonic response – whether the participant did or did not like an odour, regardless of the a priori assumptions regarding an odour's quality. This is only the beginning: it does not scratch the surface of the decision-making that determined the categorization of these odours into this type by participants in the first place.

But even this does not circumvent the malodorous elephant in the room – which is that our responses may be scent- rather than type-specific. Research in transduction, chemistry or genetics may help this endeavour along a little, especially in determining scents that may be malodorous. Perhaps it will help identify classes of odour that may produce similar responses. But the often individual, customizable nature of scent and our response to it can overwhelm any supposed generic effect. The scent of custard may send one person into a vanilla-induced orgiastic frenzy and to another, send her to the school dinner table, inside from the rain, a prelude to detested instruction and a shower of taunting. This is a flawed and fanciful example, invoked to make a point – that our response to scent may be inextricably linked to the episodic memories it evokes, in addition to the pleasure or displeasure it causes. There is also the issue of intensity, which is so closely allied to pleasure as to suggest a re-evaluation of the use of the word 'pleasant' in olfaction. Does strength genuinely equate to pleasant or are there different 'types' of pleasant? Floral, spicy, relaxing, alerting? Perhaps the term pleasantness is too all-encompassing a descriptor to be of genuine use and what we need are more specific psychometric categories, such as the above. Thus far, the brain has been able to shed some but not much light on this process, and the processes above. We have, currently, a well-conducted factor analysis – the factors have now become clear, as has the rubble. But what next?

The same problem does not beset taste, with its modest gamut of sensations. But when taste joins smell, this is when neuropsychology meets one of its hardest problems. It may be relatively easy to map the single projections of a single scent in the brain. But mapping the multisensory confusion of flavour is genuinely a challenge. Smells, tastes, textures, temperature/s, biting, mastication… it can theoretically be possible to design a neuroimaging experiment (and it would have to be a neuroimaging experiment) where all of these variables could be controlled for so that the relative contribution of each could be parcelled or subtracted out. But the challenge we meet is very similar to the one encountered in the discussion on smell earlier – there is unlikely to be one locus for flavour, as there are, obviously, different flavours. What animal work can, of course, do in this regard is to set up conditioning experiments in which the pathways activated to a food that is eaten and liked can be mapped and compared against those that have not been eaten or not liked. This would take into account the memorial process involved in flavour, which flavour often depends on both for its appreciation (we enjoy and recognize beans on toast when we eat it) and its anticipation/expectation (we know what the flavour of beans on toast is supposed to be like and anticipate this sensation, hence the desire or craving for the food). If the insula is important (which it is), which sectors, and which cells in which sectors, and which projections to and from the olfactory cortex and limbic system are more important and how strong is the selectivity?

But it is easy to give up and give up easily. One positive outcome of the above problems, challenges and conundra is that there is more to study and understand, specifically in terms of understanding the neuropsychology of flavour and in terms of understanding how the brain processes – senses, recognizes, likes, makes decisions about, makes associations to – smells and tastes and under what circumstances.

In *Regimen*, Hippocrates wrote that: 'Through seven figures come sensations for a man; there is hearing for sounds, sight for the visible, nostril for smell, tongue for pleasant or unpleasant tastes, mouth for speech, body for touch, passages outwards and inwards for hot or cold breath.' Through these come knowledge or lack of it; Hippocrates, unlike his less enlightened contemporaries, was prescient in many ways and on many different subjects. The recognition of the chemosenses as a means of acquiring knowledge was, given the noises of the nineteenth century, very astute. We are now aware of how some parts of the brain react when we acquire this knowledge; we know some of the machinery that allows this to occur; we undertand some of the genetic mechanisms that underpin it. The first neuroimaging study of scent was published twenty years ago; subsequent studies have sometimes provoked more questions than answers. But this is a good thing. It may be that the next development in this field will be the simplest. And as Harold McGee wrote in his *Curious Cook*, 'The simplest morsel leaves much to savour'. The next twenty years should be illuminating, worts and all.

REFERENCES

Abele, M., Riet, A., Hummel, T., Klockgether, T. and Wullner, U. (2003). Olfactory dysfunction in cerebellar ataxia and multiple system atrophy. *Journal of Neurology*, 250, 1453–1455.

Abolmaali, N., Gudziol, V. and Hummel, T. (2008). Pathology of the olfactory nerve. *Neuroimaging Clin. N. Am*, 18, 233–242.

Abraham, A. and Mathai, K.V. (1983). The effect of right temporal lobe lesions on matching of smells. *Neuropsychologia*, 21, 277–281.

Acharya, V., Acharya, J. and Luders, H. (1988). Olfactory epileptic auras. *Neurology*, 51, 56–61.

Adler, E., Hoon, M.A., Mueller, K.L., Chandrashekar, J., Ryba, N.J.P. and Zuker, C.S. (2000). A novel family of mammalian taste receptors. *Cell*, 100, 693–702.

Adolphs, R., Tranel, D., Koenigs, M. and Damasio, A.R. (2005). Preferring one taste over another without recognizing either. *Nature Neuroscience*, 8, 860–861.

Adrian, E.D. (1942). Olfactory reactions in the brain of the hedgehog. *Journal of Physiology*, 100, 459–473.

Adrian, E.D. (1953). Sensory messages and sensation: the response of the olfactory organ to different smells. *Acta Physiologica Scandinavica*, 29, 5–14.

Aggleton, J.P. and Waskett, L. (1999). The ability of odours to serve as state-dependent cues for real world memories: can Viking smells aid the recall of Viking experience? *British Journal of Psychology*, 90, 1–8.

Aglioti, S., Tassinari, G., Corballis, M.C. and Berlucchi, G. (2000). Incomplete gustatory lateralization as shown by analysis of taste discrimination after callosotomy. *Journal of Cognitive Neuroscience*, 12, 238–245.

Aglioti, S., Tassinari, G., Fabri, M., Del Pesce, M., Quattrini, A., Manzoni, T. and Berlucchi, G. (2001). Taste laterality in the split brain. *European Journal of Neuroscience*, 13, 195–200.

Albrecht, J., Kopietz, R., Frasnelli, J., Weismann, M., Hummel, T. and Lundtrom, J.N. (2010). The neuronal correlates of intranasal trigeminal function: an ALE meta-analysis of human functional brain imaging data. *Brain Research Reviews*, 62, 183–196.

Algom, D. and Cain, W.S. (1991). Remembered odours and mental mixtures. *Journal of Experimental Psychology: Human Perception and Performance*, 17, 1104–1119.

Allam, M.D.E., Marlier, L. and Schaal, B. (2006). Learning at the breast: preference formation for an artificial scent and its attraction against the odour of maternal milk. *Infant Behavior and Development*, 29, 308–321.

Allen, J.S., Damasio, H., Grabowski, T.J., Bruss, J. and Zhang, W. (2003). Sexual dimorphism and asymmetries in the grey-white composition of the human cerebrum. *Neuroimage*, 18, 880–894.

Allen, W.F. (1940). Effect of ablating the frontal lobes, hippocampi and occipito-parietal-temporal (excepting piriform areas) lobes on positive and negative olfactory conditioned reflexes. *American Journal of Physiology*, 128, 754–771.

Allen, W.F. (1943). Results of prefrontal lobectomy on acquired and on acquiring correct conditioned differential responses with auditory, general cutaneous and optic stimuli. *American Journal of Physiology*, 139, 525–531.

Allison, T. and Goff, W.R. (1967). Human cerebral evoked responses to odorous stimuli. *Electroencephalography and Clinical Neurophysiology*, 23, 559–560.

Altman, J. (1969). Autoradiographic and histological studies of postnatal neurogenesis. *Journal of Comparative Neurology*, 137, 433–458.

Amaral, D.G., Inausti, R. and Cowan, W.M. (1987). The entorhinal cortex of the monkey I: cytoarchitectonic organization. *Journal of Comparative Neurology*, 264, 326–355.

Amaral, D.G. and Price, J.L. (1984). Amygdalo-cortical projections in the monkey. *Journal of Comparative Neurology*, 230, 465–496.

Amoore, J.E. (1962). The stereochemical theory of olfaction. I. Identification of the seven primary odours. *Proc. Sci. Sect. Toilet Goods Assoc. Suppl.*, 37(1).

Amoore, J.E. (1970). *Molecular basis of odor.* Springfield: Charles Thomas.

Amoore, J.E., Forrester, L.J. and Buttery, R.G. (1975). Specific anosmia to 1-pyrroline: the spermous primary odour. *Journal of Chemical Ecology*, 1, 299–310.

Anderson, A.K., Christoff, K., Stappen, I., Panitz, D., Ghahremani, D.G., Glover, G., Gabrieli, J.D.E. and Sobel, N. (2003). Dissociated neural representations of intensity and valence in human olfaction. *Nature Neuroscience*, 6, 196–202.

Andre, J.M., Beis, J.M., Morin, N. and Paysant, J. (2000). Buccal hemineglect. *Archives of Neurology*, 57, 1734–1741.

Ansari, K.A. and Johnson, A.J. (1975). Olfactory function in patients with Parkinson's Disease. *Chronic Disorders*, 28, 493–497.

Anton, F. and Peppel, P. (1991). Central projections of trigeminal primary afferents innervating the nasal mucosa. *Neuroscience*, 41, 617–628.

Arey, L.B., Tremaine, M.J. and Monzingo, F.L. (1935). The numerical and topographical relations of taste buds in human circumvillate papillae throughout the life span. *Anat. Rec.*, 64, 9–25.

Arnold, S.E., Hyman, B.T., Flory, J., Damasio, A.R. and van Hoesen, G.W. (1991a). The topographical and neuroanatomical distribution of neurofibrillary tangles and neuritic plaques in the cerebral cortex of patients with Alzheimer's Disease. *Cerebral Cortex*, 1, 103–116.

Arnold, S.E., Hyman, B.T., van Hoesen, G.W. and Damasio, A. (1991b). Some cytoarchitectural abnormalities of the entorhinal cortex in schizophrenia. *Archives of General Psychiatry*, 48, 625–632.

Aronsohn, E. (1886). Experimentelle Untersuchungen zur Physiologie des Geruchs. *Archiv. Physiol. Leipzig*, 321–357.

Asai, H., Udaka, F., Hirano, M. and Ueno, S. (2008). Odour abnormalities caused by bilateral thalamic infarction. *Clinical Neurology and Neurosurgery*, 110, 500–501.

Attems, J. and Jellinger, K.A. (2006). Olfactory tau pathology in Alzheimer's Disease and mild cognitive impairment. *Clinical Neuropathology*, 25, 265–271.

Au, A., Chan, A.S. and Chiu, H. (2003). Conceptual organization in Alzheimer's dementia. *Journal of Clinical and Experimental Neuropsychology*, 25, 737–750.

Ayabe-Kanamura, S., Schicker, I., Laska, M., Hudson, R., Distel, H., Kobayakawa, T. and Saito, S. (1998). Differences in perception of everyday odours: a Japanese–German cross-cultural study. *Chemical Senses*, 23, 31–38.

Bachmanov, A.A., Li, X., Reed, D.R., Ohmen, J.D., Li, S., Chen, Z., Tordoff, M.G., de Jong, P.J., Wu, C., West, D.B., Chatterjee, A., Ross, D.A. and Beauchamp, G.K. (2001). Positional cloning of the mouse saccharin preference locus. *Chemical Senses*, 26, 925–933.

Bacon, A.W., Bondi, M.W., Salmon, D.P. and Murphy, C. (1998). Very early changes in olfactory function due to Alzeimer's Disease and the role of apolipoprotein E in olfaction. *Annals of the New York Academy of Sciences*, 855, 723–731.

Badia, P., Wesensten, N., Lammers, W., Culpepper, J. and Harsh, J. (1990). Responsiveness to olfactory stimuli presented in sleep. *Physiology & Behavior*, 48, 87–90.

Baiano, M., Perlini, C. and Rambaldelli, G. (2008). Decreased entorhinal cortex volumes in schizophrenia. *Schizophrenia Research*, 102, 171–180.

Bailey, E.H.S. and Nicholls, E.L. (1884). Preliminary notes on the delicacy of the special senses. *NY Medical Journal*, 40, 325.

Bailey, E.H.S. and Powell, L.M. (1885). Some special tests in regard to the delicacy of the sense of smell. *Transactions of the Kansas Academy of Sciences*, 9, 100–101.

Ballester, J., Abdi, H., Langlois, J., Peyron, D. and Valentin, D. (2009). The odour of colours: can wine experts and novices distinguish the odours of white, red and rose wines? *Chemosensory Perception*, 2, 203–213.

Balogh, R. and Porter, R. (1986). Olfactory preference resulting from mere exposure in human neonates. *Infant Behavior and Development*, 9, 395–401.

Barber, C. (1997). Olfactory acuity as a function of age and gender: a comparison of African and American samples. *International Journal of Aging and Human Development*, 44, 317–334.

Baron, R.A. (1983). 'Sweet smell of success': The impact of pleasant scents on evaluations of job applicants. *Journal of Applied Psychology*, 16, 16–28.

Baron, R.A. (1997). The sweet smell of...helping: Effects of pleasant ambient odours on prosocial behaviour in shopping malls. *Personality and Social Psychology Bulletin*, 23, 498–504.

Baron, R.A. and Thomley, J. (1994). A whiff of reality: positive affect as a potential mediator of the effects of pleasant fragrances on task performance and helping. *Environment and Behavior*, 26, 766–784.

Barr, R.G., Quek, V., Cousineau, D., Oberlander, T.F., Brian, J.A. and Young, S.N. (1994). Effects of intraoral sucrose on crying, mouthing and hand-mouth contact in newborn and six-week old infants. *Developmental Medicine and Child Neurology*, 36, 608–618.

Barry, M.A., Gatenby, J.C., Zeiger, J.D. and Gore, J.C. (2001). Hemispheric dominance of cortical activity evoked by focal electrogustatory stimuli. *Chemical Senses*, 26, 471–482.

Bartocci, M., Winberg, J., Papendieck, G., Mustica, T., Serra, G. and Lagercrantz, H. (2001). Cerebral haemodynamic response to unpleasant odours in the preterm newborn measured by near-infrared spectroscopy. *Pediatric Research*, 50, 324–330.

Bartocci, M., Winberg, J., Ruggiero, C., Bergqvist, L.L., Serra, G. and Lagercrantz, H. (2000). Activation of olfactory cortex in newborn infants after odour stimulation: a functional near-infrared spectroscopy study. *Pediatric Research*, 48, 18–23.

Bartoshuk, L.M. (1979). Bitter taste of saccharin related to the genetic ability to taste bitter substance 6-n-propylthiouracil. *Science*, 205, 934–935.

Bartoshuk, L.M. (1989). Clinical evaluation of the sense of taste. *Ear Nose and Throat Journal*, 68, 331–337.

Bartoshuk, L.M. (1991). Sweetness: history, preference and genetic variability. *Food Technology*, 45, 108–113.

Bartoshuk, L.M. (1993). The biological basis of food perception and acceptance. *Food Quality and Preference*, 4, 21–32.

Bartoshuk, L.M., Duffy, V.B. and Miller, I.J. (1994). PTC/PROP tasting: anatomy, psychophysics, and sex effects. *Physiology & Behavior*, 56(6), 1165–1171.

Bartoshuk, L.M., Duffy, V.B., Reed, D. and Williams, A. (1996). Supertasting, earaches and head injury: genetics and pathology alter our taste worlds. *Neuroscience and Biobehavioral Reviews*, 20, 79–87.

Bartoshuk, L.M., Rennert, K. and Rodin, J. (1982). Effects of temperature on the perceived sweetness of sucrose. *Physiology & Behavior*, 28, 905–910.

Bartoshuk, L.M., Rifkin, B., Marks, L.E. and Bars, P. (1986). Taste and ageing. *Journal of Gerontology*, 41, 51–57.

Barz, S., Hummel, T., Pauli, E., Majer, M., Lang, C.J. and Kobal, G. (1997). Chemosensory event-related potentials in response to trigeminal and olfactory stimulation in idiopathic Parkinson's Disease. *Neurology*, 49, 1424–1431.

Bassareo, V., De Luca, M.A. and Chiara, G. (2002). Differential expression of motivational stimulus properties by dopamine nucleus accumbens shell versus core and prefrontal cortex. *Journal of Neuroscience*, 22, 4709–4711.

Beauchamp, G.K. and Cowart, B.J. (1987). Preference for extremely high levels of salt among young children. *Annals of the New York Academy of Sciences*, 510, 171–172.

Beauchamp, G.K., Cowart, B.J. and Moran, M. (1986). Developmental changes in salt acceptability in human infants. *Developmental Psychobiology*, 19, 17–25.

Beauchamp, G.K. and Moran, M. (1982). Dietary experience and sweet taste preference in human infants. *Appetite*, 3, 319–352.

Beauchamp, G.K. and Moran, M. (1984). Acceptance of sweet and salty tastes in 2-year-old children. *Appetite*, 5, 291–305.

Beauchamp, G.K. and Pearson, P. (1991). Human development and umami taste. *Physiology & Behavior*, 49, 1009–1012.

Beckstead, R.M., Morse, J.R. and Norgren, R. (1980). The nucleus of the solitary tract in the monkey. *Journal of Comparative Neurology*, 190, 259–282.

Beidler, L.M. and Tucker, D. (1956). Olfactory and trigeminal nerve responses to odours. *Federation Proceedings*, 15, 14.

Bell, E.A., Castellanos, V.H., Pelkman, C.L., Thorwart, M.L. and Rolls, B.J. (1998). Energy density of foods affects energy intake in normal-weight women. *American Journal of Clinical Nutrition*, 67, 412–420.

Bembich, S., Lanzara, C., Clarici, A., Demarini, S., Tepper, B.J., Gasparini, P. and Grasso, D.L. (2010). Individual differences in prefrontal cortex activity during perception of bitter taste using fNIRS. *Chemical Senses*, 35, 801–812.

Bende, M. and Nordin, S. (1997). Perceptual learning in olfaction: professional wine tasters versus controls. *Physiology & Behavior*, 62, 1065–1070.

Bender, G., Veldhuizen, M.G., Meltzer, J.A., Gitelman, D. and Small, D.M. (2009). Neural correlates of evaluative compared to passive tasting. *European Journal of Neuroscience*, 30, 327–338.

Bengtsson, S., Berglund, H., Cohen, E., Savic, I. and Guylas, B. (2001). Brain activation during odour perception in males and females. *Chemical Senses*, 12, 2027–2033.

Bensafi, M., Pierson, A., Rouby, C., Farget, V. and Bertrand, B. (2002). Modulation of visual event-related potentials by emotional olfactory stimuli. *Clinical Neurophysiology*, 32, 335–342.

Bensafi, M., Sobel, N. and Khan, R.M. (2007). Hedonic-specific activity in piriform cortex during odour imagery mimics that during odour perception. *Journal of Neurophysiology*, 98, 3254–3262.

Berendse, H.W., Booij, J., Francot, C.M., Bermans, P.L.M., Hijman, R., Stoof, J.C. and Wolters, E. (2001). Subclinical dopaminergic in asymptomatic Parkinson's Disease patients' relatives with a decreased sense of smell. *Annals of Neurology*, 50, 34–41.

Berglund, B. (1974). Quantitative and qualitative analysis of industrial odours with human observers. *Annals of the New York Academy of Sciences*, 237, 35–51.

Berglund, B., Berglund, B., Engen, T. and Ekmna, G. (1973). Multidimensional analysis of 21 odours. *Scandinavian Journal of Psychology*, 14, 131–137.

Berglund, H., Lindstrom, P., Dhejne-Helmy, C. and Savic, I. (2008). Male-to-female transsexuals show sex-atypical hypothalamus activation when smelling odorous steroids. *Cerebral Cortex*, 18, 1900–1908.

Berglund, H., Lindstrom, P. and Savic, I. (2006). Brain response to putative pheromones in lesbian women. *PNAS*, 103, 8269–8274.

Bernard, J.F., Peschanski, M. and Besson, J.M. (1989). A possible spino (trigemino)-ponto-amygdaloid pathway for pain. *Neuroscience Letters*, 100, 83–88.

Bernhardt, S.J., Naim, M., Zehavi, U. and Lindemann, B. (1996). Changes in IP3 and cytosolic Ca2+ in response to sugars and non-sugar sweeteners in transduction of sweet taste in the rat. *Journal of Physiology*, 490, 325–336.

Bertino, M., Beauchamp, G.K. and Engelman, K. (1982). Long-term reduction in dietary sodium alters the taste of salt. *American Journal of Clinical Nutrition*, 36, 1134–1144.

Bertollo, D.N., Cowen, M.A. and Levy, A.V. (1996). Hypometabolism in olfactory cortical projection areas of male patients with schizophrenia: an initial positron emission tomography study. *Psychiatry Research*, 60, 113–116.

Betchen, S.A. and Doty, R.L. (1988). Bilateral detection thresholds in dextrals and sinistrals reflect the more sensitive side of the nose, which is not lateralized. *Chemical Senses*, 23, 453–457.

Birch, L.L. and Marlin, D.W. (1982). I don't like it, never tried it: effects of exposure on two-year-old children's food preferences. *Appetite*, 3, 353–360.

Bitter, T., Bruderle, J., Gudziol, H., Burmeister, H.P., Gaser, C. and Guntinas-Lichius, O. (2010a). Grey and white matter reduction in hyposmic subjects. *Brain Research*, 1347, 42–47.

Bitter, T., Gudziol, H., Burmeister, H.P., Mentzel, H.J. and Guntinas-Lichius, O. (2010b). Anosmia leads to a loss of grey matter in cortical brain areas. *Chemical Senses*, 35, 407–415.

Bitter, T., Siegert, F., Gudziol, H., Burmeister, H.P., Mentzel, H.J., Hummel, T., Gaser, C. and Guntinas-Lichius, O. (2011). Grey matter alterations in parosmia. *Neuroscience*, 177, 177–182.

Blakeslee, A.F. (1931). Genetics of sensory thresholds: taste for phenylthiocarbamide. *PNAS*, 18, 120–130.

Blansjaar, B.A., Thomassen, R. and Van Schaick, H.W. (2000). Prevalence of dementia in centenarians. *International Journal of Geriatric Psychiatry*, 15, 219–225.

Blass, E.M. and Watt, L.B. (1999). Suckling- and sucrose-induced analgesia in human newborns. *Pain*, 83, 611–623.

Boesveldt, S., Frasnelli, J., Gordon, A.R. and Lundstrom, J.N. (2010). The fish is bad: negative food odours elicit faster and more accurate reactions than other odours. *Biological Psychology*, 84, 313–317.

Boesveldt, S., Verbaan, D., Knol, D.L., Visser, M., van Rooden, S.M., van Hilten, J.J. and Berendse, H.W. (2008). A comparative study of odour identification and odour discrimination deficits in Parkinson's Disease. *Movement Disorder*, 23, 1984–1990.

Bohnen, N.I., Muller, M.L.T.M., Kotagal, V., Koeppe, R.A., Kilbourn, M.A., Albin, R.L. and Frey, K.A. (2010). Olfactory dysfunction, central cholinergic integrity and cognitive impairment in Parkinson's Disease. *Brain*, 133, 1747–1754.

Bojsen-Møller, F. and Fahrenkrug, J. (1971). Nasal swell bodies and cyclic changes in the air passages in the rat and rabbit nose. *Journal of Anatomy*, 110, 25–27.

Bonfils, P., Avan, P., Faulcon, P. and Malinvaud, D. (2005). Distorted odorant perception: analysis of a series of 56 patients with parosmia. *Archives of Otolaryngology*, 131, 107–112.

Bonnans, S. and Noble, A.C. (1993). Effect of sweetener type and of sweetener and acid level on temporal perception of sweetness, sourness and fruitiness. *Chemical Senses*, 18, 273–283.

Boyle, J.A., Djordjevic, J., Olsson, M.J., Lundstrom, J.N. and Jones-Gotman, M. (2009). The human brain distinguishes between single odorants and binary mixtures. *Cerebral Cortex*, 19, 66–71.

Braak, H. and Braak, E. (1991). Neuropathological staging of Alzheimer-related changes. *Acta Neuropathologica*, 82, 239–259.

Braak, H. and Del Tredici, K. (2008). Nervous system pathology in sporadic Parkinson Disease. *Neurology*, 70, 1916–1925.

Braak, H., Del Tredici, K., Hamm-Clement, J., Sandmann-Keil, D. and Rub, U. (2002). Staging of the intracerebral inclusion body pathology associated with idiopathic Parkinson's Disease. *Journal of Neurology*, 249, 1–5.

Braak, H., Del Tredici, K., Rub, U., de Vos, R.A., Jansen Steur, E.N. and Braak, E. (2003). Staging of brain pathology related to sporadic Parkinson's Disease. *Neurobiology of Aging*, 24, 197–211.

Bradley, R.M. (2000). Sensory receptors of the larynx. *The American Journal of Medicine*, 108(4), 47–50.

Bradley, R.M. and Beidler, L.M. (2003). Saliva: its role in taste function. In Doty, R.L. (ed.) *Handbook of Olfaction and Gustation*, 2nd edn. New York: Marcel Dekker.

Bragulat, V., Dzemidzic, M., Bruno, C., Cox, C.A., Talavage, T., Considine, R.V. and Kareken, D.A. (2010). Food-related odour probes of brain reward circuits during hunger: a pilot fMRI study. *Obesity*, 18, 1566–1571.

Bragulat, V., Dzemidzic, M., Talavage, T., Davidson, D., O'Connor, S.J. and Kareken, D.A. (2008). Alcohol sensitizes cerebral responses to the odors of alcoholic drinks: an fMRI study. *Alcohol: Clinical and Experimental Research*, 32, 1124–1134.

Brand, G. (2006). Olfactory/trigeminal interactions in nasal chemoreception. *Neuroscience and Biobehavioural Reviews*, 30, 908–917.

Brand, G. and Jacquot, L. (2002). Sensitization and desensitization to allyl isothiocyanate (mustard oil) in the nasal cavity. *Chemical Senses*, 27, 593–598.

Brand, G., Millot, J.-L. and Henquell, D. (1999). Olfaction and hemispheric asymmetry: unilateral stimulation and bilateral electrodermal recordings. *Neuropsychobiology*, 39, 160–164.

Bremner, E., Mainland, J. and Khan, R. (2003). The prevalence of androstenone anosmia. *Chemical Senses*, 28, 423–432.

Brewer, W.J., Wood, S.J., McGorry, P.D. (2003). Impairment of olfactory identification ability in individuals at ultra-high risk for psychosis who later develop schizophrenia. *American Journal of Psychiatry*, 160, 1790–1794.

Bromley, S.M. and Doty, R.L. (1995). Odour recognition memory is better under bilateral than unilateral test conditions. *Cortex*, 31, 25–40.

Bromley, S.M. and Doty, R.L. (2003). Clinical disorders affecting taste: evaluation and management. In Doty, R.L. (ed.) *Handbook of Olfaction and Gustation*, 2nd edn. New York: Marcel Dekker.

Bronshtein, A.A. and Minor, A.V. (1977). Regeneration of olfactory flagella and restoration of the electroolfactogram following application of Triton X-100 to the olfactory mucosa of frogs. *Tsitologiia*, 19, 33–39.

Brunjes, P.C., Illig, K.R. and Meyer, E.A. (2005). A field guide to the anterior olfactory nucleus (cortex). *Brain Research Reviews*, 50, 305–335.

Buchsbaum, M.S., Kesslak, J.P., Lynch, G., Chui, H., Wu, J.C., Sicotte, N., Hazlet, E., Teng, E. and Cotman, C.W. (1991). Temporal and hippocampal metabolic rate during an olfactory memory task assessed by PET in patients with dementia of the Alzheimer type and controls. *Archives of General Psychiatry*, 48, 840–847.

Buck, L. and Axel, R. (1991). A novel multigene family may encode odorant receptors: a molecular basis for odour recognition. *Cell*, 65, 175–187.

Bujas, Z., Szabo, S., Ajdukovic, D. and Mayer, D. (1991). Time course of recovery from gustatory adaptation to NaCl. *Perception and Psychophysics*, 49, 517–521.

Bulsing, P.J., Smeets, M.A.M., Gemenhardt, C., Laverman, M., Schuster, B., van den Hout, M.A. and Hummel, T. (2010). Irritancy expectancy alters odour perception: evidence from olfactory event-related potential research. *Journal of Physiology*, 104, 2749–2756.

Bult, J.H.F., de Wijk, R.A. and Hummel, T. (2007). Investigations on multimodal sensory integration: texture, taste and ortho-nasal and retronasal olfactory stimuli in concert. *Neuroscience Letters*, 411, 6–10.

Burdach, K.J. and Doty, R.L. (1987). The effects of mouth movements, swallowing and spitting on retronasal odor perception. *Physiology & Behavior*, 41(4), 353–356.

Bushteva, K.A., Polezhaev, E.F. and Semenenko, A.D. (1960). The influence of subliminal olfactory stimulations on reflex activity. *Fiziol. zh. SSSR*, 46(4), 452–457.

Caicedo, A., Kim, K.N. and Roper, S.D. (2002). Individual mouse taste cells respond to multiple chemical stimuli. *Journal of Physiology*, 544, 501–509.

Caicedo, A. and Roper, S.D. (2001). Taste receptor cells that discriminate between bitter stimuli. *Science*, 291, 1557–1560.

Cain, W.S. (1977). Differential sensitivity for smell: 'noise' at the nose. *Science*, 195, 796–798.

Cain, W.S. (1979). To know with the nose: keys to odour identification. *Science*, 203, 467–470.

Cain, W.S. (1982). Odour identification by males and females: predictions vs performance. *Chemical Senses*, 7, 129–142.

Cain, W.S. and Gent, J.F. (1991). Olfactory sensitivity: realiability, generality and association with ageing. *Journal of Experimental Psychology: Human Perception and Performance*, 17, 382–391.

Cain, W.S., Schmidt, R. and Wolkoff, P. (2007). Olfactory detection of ozone and D-limolene: reactants in indoor spaces. *Indoor Air*, 17, 337–347.

Cain, W.S., Stevens, J.C., Nickou, C., Giles, A., Johnston, I. and Garcia-Medina, M.R. (1995). Life-span development of odour identification, learning and olfactory sensitivity. *Perception*, 24, 1457–1472.

Calder, A.J., Beaver, J.D., Davis, M.H., van Ditzhuijzen, J., Keane, J. and Lawrence, A.D. (2007). Disgust sensitivity predicts the insula and pallidal response to pictures of disgusting foods. *European Journal of Neuroscience*, 25, 3422–3428.

Calder, A.J., Keane, J., Manes, F., Antoun, N. and Young, A.W. (2000). Impaired recognition and experience of disgust following brain injury. *Nature Neuroscience*, 3, 1077–1078.

Callahan, C.D. and Hinkebein, J. (1999). Neuropsychologial significance of anosmia following traumatic brain injury. *Journal of Head Trauma Rehabilitation*, 14, 581–587.

Callahan, C.D. and Hinkebein, J. (2002). Assessment of anosmia after traumatic brain injury: performance characteristics of the UPSIT. *Journal of Head Trauma Rehabilitation*, 17, 251–256.

Campbell, I.M. and Gregson, R.A.M. (1972). Olfactory short term memory in normal, schizophrenic and brain-damaged cases. *Australian Journal of Psychology*, 24, 179–185.

Carmichael, S.T., Clugnet, M.C. and Price, J.L. (1994). Central olfactory connections in the macaque monkey. *Journal of Comparative Neurology*, 346, 413–434.

Carmichael, S.T. and Price, J.L. (1994). Architectonic subdivision of the oribital and medial prefrontal cortex in the Macaque monkey. *Journal of Comparative Neurology*, 346, 366–402.

Carroll, B., Richardson, J.T.E. and Thompson, P. (1993). Olfactory information processing and temporal lobe epilepsy. *Brain and Cognition*, 22, 230–243.

Castillo, M. and Mukherjii, S.K. (1996). Magnetic resonance imaging of the olfactory apparatus. *Topics in Magnetic Resonance Imaging*, 8, 80–86.

Castriota-Scanderbeg, A., Hagberg, G.E., Cerasa, A., Committeri, G., Galati, G., Patria, F., Pitzali, S., Calatagirone, C. and Frackowiak, R. (2005). The appreciation of wine by sommeliers: a functional magnetic resonance study of sensory integration. *Neuroimage*, 25, 570–578.

Cereda, C., Ghika, J., Maeder, P. and Bogousslavsky, J. (2002). Strokes restricted to the insular cortex. *Neurology*, 59, 1950–1955.

Cerf, B., LeBihan, D., van de Moortelez, P., MacLeod, P. and Faurion, A. (1998). Functional lateralization of human gustatory cortex related to handedness disclosed by fMRI. *Annals of the New York Academy of Sciences*, 855, 575–578.

Cerf-Ducastel, B. and Murphy, C. (2003). FMRI brain activation in response to odours is reduced in primary olfactory areas of elderly subjects. *Brain Research*, 986, 39–53.

Cerf-Ducastel, B. and Murphy, C. (2006). Neural substrates of cross-modal olfactory recognition memory: an fMRI study. *Neuroimage*, 31, 386–396.

Cerf-Ducastel, B. and Murphy, C. (2009). Age-related differences in the neural substrates of cross-modal olfactory recognition memory: an fMRI investigation. *Brain Research*, 1285, 88–98.

Cerf-Ducastel, B., van der Moortele, P.F., MacLeod, P., LeBihan, D. and Faurion, A. (2001). Interaction of gustatory and lingual somatosensory perceptions at the cortical level in the human. *Chemical Senses*, 26, 371–383.

Cernoch, J.M. and Porter, R.H. (1985). Recognition of maternal auxillary odours by infants. *Child Development*, 56, 1593–1598.

Chalke, H.D. and Dewhurst, J.R. (1957). Accidental coal-gas poisoning. *BMJ*, 19, 915–917.

Chandrashekar, J., Mueller, K.L., Hoon, M.A., Adler, E., Feng, L., Guo, W., Zuker, C.S. and Ryba, N.J. (2000). T2Rs function as bitter taste receptors. *Cell*, 100, 703–711.

Chaudhari, N., Landin, A.M. and Roper, S.D. (2000). A metabotropic glutamate receptor variant functions as a taste receptor. *Nature Neuroscience*, 3, 113–119.

Chauhan, J. (1989). Pleasantness perception of salt in young vs elderly adults. *Journal of the American Diatetic Association*, 89, 834–836.

Chazal, G., Durbec, P., Jankovski, A., Rougon, G. and Cremer, H. (2000). Consequences of neural cell adhesion molecule deficiency on cell migration in the rostral migratory stream of the mouse. *Journal of Neuroscience*, 20, 1446–1457.

Chen, C., Shih, Y.H. and Yen, D.J. (2003). Olfactory auras in patients with temporal lobe epilepsy. *Epilepsia*, 44, 257–260.

Chen, D. and Dalton, P. (2005). The effect of emotion and personality on olfactory perception. *Chemical Senses*, 30, 345–351.

Cherinskii, A.A., Piskorskaya, N.G., Zima, I.G., Krizhanovskii, S.A. and Makarchouk, N. (2010). Individual/typological peculiarities of EEG rearrangements in humans related to the analysis of olfactory information: role of extroversion/introversion. *Neurophysiology*, 42, 70–77.

Chertkow, H., Whatmough, C., Saumier, D. and Duong, A. (2008). Cognitive neuroscience studies of semantic memory in Alzheimer's Disease. *Progress in Brain Research*, 169, 393–407.

Chiavaras, M.M. and Petrides, M. (2000). Orbitofrontal sulci of the human and macaque monkey brain. *Journal of Comparative Neurology*, 422, 35–54.

Chopra, A., Baur, A. and Humme, T. (2008). Thresholds and chemosensory event-related potentials to malodors before, during and after puberty: differences related to sex and age. *Neuroimage*, 40, 1257–1263.

Chou, K.L. and Bohnen, N.I. (2009). Performance on an Alzheimer-selective odour identification test in patients with Parkinson's Disease and its relationship with cerebral dopamine transporter activity. *Parkinsonism Related Disorders*, 15, 640–643.

Choudhury, E., Moberg, P. and Doty, R. (2003). Influences of age and sex on a microencapsulated odour memory test. *Chemical Senses*, 28, 799–805.

Chrea, C., Valentin, D., Sulmont-Rosse, C., Ly Mai, H., Hoang Nguyen, D. and Abdi, H. (2004). Culture and odour categorization: agreement between cultures depends upon the odours. *Food Quality and Preference*, 15, 669–679.

Christensen, C.M. (1977). Texture-taste interactions. *Cereal Food World*, 22, 243–256.

Christensen, C.M. (1980). Effect of taste quality and intensity on oral perception of viscosity. *Perception and Psychophysics*, 28, 315–320.

Chu, C.C., Tranel, D., Damasio, A.R. and van Hoesen, G.W. (1997). The autonomic-related cortex: pathology in Alzheimer's Disease. *Cerebral Cortex*, 7, 86–95.

Ciumas, C., Lindstrom, P., Aoun, B. and Savic, I. (2008). Imaging of odour perception delineates functional disintegration of the limbic circuits in mesial temporal lobe epilepsy. *Neuroimage*, 39, 578–592.

Clark, C., Kopala, L. and Hurwitz, T. (1991). Regional metabolism in microsmic patients with schizophrenia. *Canadian Journal of Psychiatry*, 36, 645–650.

Clark, C.C. and Lawless, H.T. (1994). Limiting response alternatives in time-intensity scaling. *Chemical Senses*, 19, 583–594.

Clark, E.C. and Dodge, H.W. (1955). Effect of anosmia on the appreciation of flavour. *Neurology*, 5, 671–674.

Cleland, T.A. and Linster, C. (2003). Central olfactory structures. In Doty, R.L. (ed.) *Handbook of Olfaction and Gustation*, 2nd edn. New York: Marcel Dekker.

Clepce, M., Gossler, A., Reich, K., Kornuber, J. and Thuerauf, N. (2010). The relation between depression, anhedonia and olfactory hedonic estimates: a pilot study in major depression. *Neuroscience Letters*, 471, 139–143.

Colebach, J.G., Adams, L., Murphy, K., Martin, A.J., Lammertsma, A.A. and Tochon-Danguy, H.J. (1991). Regional cerebral blood flow during volitional breathing in man. *Journal of Physiology*, 443, 91–103.

Collet, S., Grulois, V., Bertrand, B. and Rombaux, P. (2009). Post-traumatic olfactory dysfunction: a cohort study and update. *B-ENT*, 13, 97–107.

Connelly, T., Farmer, J.M., Lynch, D.R. and Doty, R.L. (2003). Olfactory dysfunction in degenerative ataxias. *Journal of Neurology, Neurosurgery and Psychiatry*, 74, 1435–1437.

Conover, K.L. and Shizgal, P. (1994). Competition and summation between rewarding effects of sucrose and lateral hypothalamic stimulation in the rat. *Behavioural Neuroscience*, 108, 537–548.

Constanzo, R.M. and Zasler, N.D. (1992). Epidemiology and pathophysiology of olfactory and gustatory dysfunction in head trauma. *Journal of Head Trauma Rehabilitation*, 7, 15–24.

Corcoran, C., Whitaker, A. and Coleman, E. (2005). Olfactory deficits, cognition and negative symptoms in early onset psychosis. *Schizophrenia Research*, 80, 283–293.

Corlis, R., Splaver, G., Wisecup, P. and Fischer, R. (1967). Myers-Briggs type personality scales and their relation to taste acuity. *Nature*, 216, 91–92.

Corso, J.F. (1971). Sensory processes and age effects in normal adults. *Journal of Gerontology*, 26, 90–115.

Corwin, J., Loury, M. and Gilbert, A.N. (1995). Workplace, age and sex as mediators of olfactory function: data from the National Geographic Smell Survey. *Journal of Gerontology B*, 50, 179–186.

Cowart, B.J. (1989). Relationships between taste and smell across the adult life span. *Annals of the New York Academy of Sciences*, 561, 39–55.

Cowart, B.J., Young, I.M., Feldman, R.S. and Lowry, L.D. (1997). Clinical disorders of smell and taste. *Occupational Medicine*, 12, 465–483.

Craig, A.D. (2009). How do you feel – now? The anterior insula and human awareness. *Nature Reviews*, 10, 59–70.

Crino, P.B., Martin, J.A., Hill, W.D., Greenberg, B., Lee, V.M. and Trojanowski, J.Q. (1995). Beta-amyloid peptide and amyloid precursor proteins in olfactory mucosa of patients with Alzheimer's Disease, Parkinson's Disease and Down's Syndrome. *Annals of Otolaryngology, Rhinology and Laryngology*, 104, 655–661.

Critchley, H.D. and Rolls, E.T. (1996). Hunger and satiety modify the responses of olfactory and visual neurons in the primate orbitofrontal cortex. *Journal of Neurophysiology*, 75, 1659–1672.

Croy, I., Springborn, M., Lotsch, J., Johnston, A.N.B. and Hummel, T. (2011). Agreeable smellers and sensitive neurotics: correlations among personality traits and sensory thresholds. *PLoS ONE*, 6(4), e18701, 1–9.

Cruz, A. and Green, B.G. (2000). Thermal stimulation of taste. *Nature*, 403, 889–892.

Crysdale, W.S., Cole, P. and Emery, P. (1985). Cephalometric radiographs, nasal airway resistance and the effect of adenoidectomy. *Journal of Otolaryngology*, 14, 92–94.

Crystal, S.R. and Bernstein, I.L. (1998). Infant salt preference and mother's morning sickness. *Appetite*, 30, 297–307.

Cui, L. and Evans, W.J. (1997). Olfactory event-related potentials to amyl acetate in congenital anosmia. *Electroencephalography and Clinical Neurophysiology*, 102, 303–306.

Cupchik, G.C. and Phillips, K. (2005). The scent of literature. *Cognition and Emotion*, 19, 101–119.

Dade, L.A., Zatorre, R.J., Evans, A.C. and Jones-Gotman, M. (2001). Working memory in another dimension: functional imaging of human olfactory working memory. *Neuroimage*, 14, 650–660.

Dade, L.A., Zatorre, R.J. and Jones-Gotman, M. (2002). Olfactory learning: convergent findings from lesion and brain imaging studies in humans. *Brain*, 125, 96–101.

Dalton, P., Doolittle, N. and Breslin, P.A.S. (2002). Gender-specific induction of enhanced sensitivity to odours. *Nature Neuroscience*, 5, 199–200.

Dalton, P., Doolittle, N. and Nagata, H. (2000). The merging of the senses: integration of subthreshold taste and smell. *Nature Neuroscience*, 3, 431–432.

Dalton, P., Mennella, J.A., Cowart, B.J., Maute, C., Pribitkin, E.A. and Reilly, J.S. (2009). Evaluating the prevalence of olfactory dysfunction in a pediatric population. *Annals of the New York Academy of Sciences*, 1170, 537–542.

Dalton, P. and Wysocki, C.J. (1996). The nature and duration of adaptation following long-term odour exposure. *Perception and Psychophysics*, 58, 781–792.

Daly, D. (1958). Uncinate fits. *Neurology*, 8, 250–260.

Damak, S., Rong, M., Yasumatsu, K., Kokrashvili, Z., Varadarajan, V., Zou, S., Jiang, P., Ninomiya, Y. and Margolskee, R.F. (2003). Detection of sweet and umami taste in the absence of taste receptor T1r3. *Science*, 301, 850–853.

Damasio, A., Grabowski, T.J. and Bechara, A. (2000). Subcortical and cortical brain activity during the feeling of self-generated emotions. *Nature Neuroscience*, 3, 1049–1056.

Daniel, S.E. and Hawkes, C.H. (1992). Prelimary diagnosis of Parkinson's Disease using olfactory bulb pathology. *Lancet*, 340, 186.

Daniels, S.K. and Foundas, A.L. (1997). The role of the insular cortex in dysphagia. *Dysphagia*, 12, 146–156.

Davidson, R.J., Shackman, A.J. and Maxwell, J.S. (2004). Asymmetries in face and brain related to emotion. *Trends in Cognitive Science*, 8, 389–391.

Davidson, T.M. and Murphy, C. (1997). Rapid clinical evaluation of anosmia: the alcohol sniff test. *Archives of Otolaryngology*, 123, 591–594.

Davies, D.C., Brooks, J.W. and Lewis, D.A. (1993). Axonal loss from olfactory tracts in Alzheimer's Disease. *Neurobiology of Ageing*, 14, 353–357.

Davies, J.T. (1953). L'odeur et la morphologie des molecules. *Ind. Parfum*, 8, 74.

De, B.G. (1990). Auriculotemporal syndrome presenting as tic douloureux: report of two cases. *Journal of Neurosurgery*, 72, 955–958.

de Araujo, I.E.T., Kringelbach, M.L., Rolls, E.T. and Hobden, P. (2003a). Represention of umami taste in the human brain. *Journal of Neurophysiology*, 90, 313–319.

de Araujo, I.E.T., Kringelbach, M.L., Rolls, E.T. and McGlone, F. (2003b). Human cortical responses to water in the mouth and the effects of thirst. *Journal of Neurophysiology*, 90, 1865–1876.

de Araujo, I.E.T., Rolls, E.T., Kringelbach, F., McGlone, F. and Phillips, N. (2003c). Taste-olfactory convergence and the representation of the pleasantness of flavour in the human brain. *European Journal of Neuroscience*, 18, 2059–2068.

de Araujo, I.E.T., Rolls, E.T., Velazco, M.I., Margot, C. and Cayeux, I. (2005). Cognitive modulation of olfactory processing. *Neuron*, 46, 671–679.

Deeb, J., Shah, M., Muhammed, N., Gunasekera, R., Gannon, K., Findley, L.J. and Hawkes, C.H. (2010). A basic smell test is as sensitive as a dopamine transporter scan: comparison of olfaction, taste and DaTSCAN in the diagnosis of Parkinson's Disease. *Quarterly Journal of Medicine*, 103, 941–952.

Deems, D.A., Doty, R.L., Settle, R.G., Moore-Gillon, V., Shaman, P., Mester, A.F., Kimmelman, C.P., Brightman, V.J. and Snow, J.B. (1991). Smell and taste disorders:

a study of 750 patients from the University of Pennsylvania Smell and Taste Center. *Archives of Otolaryngology, Head Neck Surgery*, 117, 508–528.

DeGraaf, C., Van Staveren, W. and Burema, J. (1996). Psychophysical and psychohedonic functions of four common food flavours in elderly subjects. *Chemical Senses*, 21, 293–302.

De Jong, N., De Graaf, C. and Van Steveren, W.A. (1996). Effect of sucrose in breakfast items on pleasantness and food intake in the elderly. *Physiology & Behavior*, 60, 1453–1462.

Del Tredici, K., Rub, U., de Vos, R.A., Bohl, J.R.E. and Braak, H. (2002). Where does Parkinson's Disease pathology begin in the brain? *Journal Neuropathology and Experimental Neurology*, 61, 413–426.

Delwiche, J. (2004). The impact of perceptual intercations on perceived flavour. *Food Quality and Preference*, 15, 137–146.

Delwiche, J.F., Lera, M.F. and Breslin, P.A. (2000). Selective removal of a target stimulus localized by taste in humans. *Chemical Senses*, 25, 181–187.

Demarquay, G., Royet, J.P., Mick, G. and Ryvlin, P. (2008). Olfactory hypersensitivity in migraineurs: a H2 15 PET study. *Cephalalgia*, 28, 1069–1080.

Desgranges, B., Ramirez-Amaya, V., Ricano-Cornejo, I., Levy, F. and Ferreira, G. (2010). Flavour preference learning increases olfactory and gustatory convergence onto single neurons in the basolateral amygdala but not in the insular cortex in rats. *PLoS ONE*, 5, e10097.

Desor, J.A. and Beauchamp, G.K. (1987). Longitudinal changes in sweet preferences in humans. *Physiology & Behavior*, 39, 639–641.

Desor, J.A. and Finn, J. (1989). Effects of amiloride on salt taste in humans. *Chemical Senses*, 14, 793–803.

Desor, J.A., Greene, L.S. and Maller, O. (1975). Preferences for sweet and salty in nine to 15-year-old and adult humans. *Science*, 190, 686–687.

Devanand, D.P., Michaels-Marston, K.S., Liu, X.H., Pelton, G.H., Padilla, M. and Marder, K. (2000). Olfactory deficits in patients with mild cognitive impairment predict Alzheimer's Disease at follow-up. *American Journal of Psychiatry*, 157, 1399–1405.

de Wijk, R.A. and Cain, W.S. (1994). Odour identification by name and by edibility: lifespan development and safety. *Human Factors*, 36, 182–187.

de Wijk, R.A., Polet, I.A. and Engelen, L. (2004). Amount of ingested custard dessert as affected by its color, odour and texture. *Physiology & Behavior*, 82, 397–403.

Ding, E.L., Huftless, S.M., Ding, X. and Girotra, S. (2006). Chocolate and prevention of cardiovascular disease: a systematic review. *Nutrition & Metabolism*, 3, 1–12.

Dittrich, W.H., Johansen, T., Landro, N.I., Fineberg, N.A and Goder, Y.M. (2010). Cognitive performance and specific deficits in OCD symptom dimensions: I. Olfactory perception and impaired recognition of disgust. *German Journal of Psychiatry*, 13, 127–139.

Djordjevic, J. and Jones-Gotman, M. (2006). Olfaction and the temporal lobes. In Brewer W., Castle D. and Pantelis C. (eds) *Olfaction and the Brain*. Cambridge: CUP, pp. 28–49.

Djordjevic, J., Jones-Gotman, M., De Sousa, K. and Chertkow, H. (2008). Olfaction in patients with mild cognitive impairment and Alzheimer's Disease. *Neurobiology of Aging*, 29, 693–706.

Djordjevic, J., Zatorre, R.J. and Jones-Gotman, M. (2004a). Effects of perceived and imagined odours on taste detection. *Chemical Senses*, 29, 199–208.

Djordjevic, J., Zatorre, R.J., Petrides, M. and Jones-Gotman, M. (2004b). The mind's nose: effects of odour and visual imagery on odour detection. *Psychological Science*, 15, 143–148.

Doetsch, F., Garcia-Verdugo, J.M. and Alvarez-Buylla, A. (1997). Cellular composition and three-dimensional organization of the subventricular germinal zone in the adult mammalian brain. *Journal of Neuroscience*, 17, 5046–5061.

Dominguez, P.R. (2011). The study of postnatal and later development of the taste and olfactory systems using the human brain mapping approach. *Brain Research Bulletin*, 84, 118–124.

Doty, R.L. (1995). *The Smell Identification Test Administration Manual*, 3rd edn. Haddon Heights: Sensonics Ltd.

Doty, R.L. (2003). Odour perception in neurodegenerative diseases. In Doty, R.L. (ed.) *Handbook of Olfaction and Gustation*, 2nd edn. New York: Marcel Dekker.

Doty, R.L., Agrawal, U. and Frye, R. (1989). Evaluation of the internal consistency reliability of the fractionated and whole UPSIT. *Perception and Psychophysics*, 45, 381–384.

Doty, R.L., Applebaum, S. and Zusho, H. (1985). Sex differences in odour identification ability: a cross-cultural analysis. *Neuropsychologia*, 23, 667–672.

Doty, R.L., Deems, D.A. and Stellar, S. (1988). Olfactory dysfunction in parkinsonism: a general deficit unrelated to neurologic signs, disease stage or disease duration. *Neurology*, 38, 1237–1244.

Doty, R.L. and Ferguson-Segall, M. (1989). Influence of adult castration on the olfactory sensitivity of the male rat: a signal detection analysis. *Behavioural Neuroscience*, 103, 691–694.

Doty, R.L., Fernandez, A.D., Levine, M.A., Moses, A. and McKeown, D.A. (1997). Olfactory dysfunction in type I pseudohyporathyroidism: dissociation from Gs alpha protein deficiency. *Journal of Clinical Endocrinology and Metabolism*, 82, 247–250.

Doty, R.L. and Laing, D.G. (2003). Psychophysical measurement of human olfactory function, including odorant mixture assessment. In Doty, R.L. (ed.) *Handbook of Olfaction and Gustation*, 2nd edn. New York: Marcel Dekker.

Doty, R.L., Marcus, A. and Lee, W.W. (1996). Development of the 12-item cross-cultural smell identification test. *Laryngoscope*, 106, 353–356.

Doty, R.L. and Mishra, A. (2001). Influences of nasal obstruction, rhinitis and rhinosinusitis on the ability to smell. *Laryngoscope*, 111, 409–423.

Doty, R.L., Reyes, P.F. and Gregor, T. (1987). Presence of both odour identification and detection deficits in Alzheimer's Disease. *Brain Research Bulletin*, 18, 597–600.

Doty, R.L., Shaman, P., Applebaum, S., Giberson, R., Sikorski, L. and Rosenberg, L. (1984a). Smell identification ability: changes with age. *Science*, 226, 1441–1443.

Doty, R.L., Shaman, P. and Dann, M. (1984b). Development of the University of Pennsylvania Smell Test: standardized microencapsulated test for olfactory function. *Physiology & Behavior*, 32, 489–502.

Doty, R.L., Smith, R., McKeown, D.A. and Raj, J. (1994). Tests of human olfactory function: principal components analysis suggests that most measure a common source of variance. *Perception & Psychophysics*, 56, 701–707.

Doty, R.L., Snyder, P., Huggins, G, and Lowry, L.D. (1981). Endocrine, cardiovascular and psychological correlates of olfactory sensitivity changes during the menstrual cycle. *Journal of Comparative Physiology and Psychology*, 96, 46–80.

Doty, R.L., Stern, M.B., Pfeiffer, C., Gollomp, S.M. and Hurtig, H.I. (1992). Bilateral olfactory dysfunction in early stage treated and untreated idiopathic Parkinson's Disease. *Journal of Neurology, Neurosurgery and Psychiatry*, 55, 138–142.

Double, K.L., Rowe, D.B., Hayes, M., Chan, D.K., Blackie, J., Corbett, A., Joffe, R., Fung, V.S., Morris, J. and Halliday, G. (2003). Identifying the pattern of olfactory deficits in Parkinson's Disease using the brief identification test. *Archives of Neurology*, 60, 545–549.

Douek, E., Bannister, L.H. and Dodson, H.C. (1975). Recent advances in the pathology of olfaction. *Proceedings of the Royal Society of Medicine*, 68, 467–470.

Douglass, B. (1901). Nasal condition observed in the aged. *New York Medical Journal*, 73, 885–890.

Drewnowski, A. (2003). Genetics of human taste perception. In Doty, R.L. (ed) *Handbook of Olfaction and Gustation*, 2nd edn. New York: Marcel Dekker.

Drewnowski, A., Ahlstrom Henderson, S., Driscoll, A. and Rolls, B.J. (1986). Salt taste perceptions and preferences are unrelated to sodium consumption in healthy older adults. *Journal of the American Diatetic Association*, 96, 471–474.

Drewnowski, A. and Gomez-Carenos, C. (2000). Bitter taste, phytonutrients and the consumer: a review. *American Journal of Clinical Nutrition*, 72, 1424–1435.

duBose, C.N., Cardello, A.V. and Maller, O. (1980). Effects of colorants and flavorants on identification, perceived flavour intensity and hedonic quality of fruit-flavored beverages and cake. *Journal of Food Science*, 45, 1393–1415.

Duda, J.E., Shah, U., Arnold, S.E., Lee, V.M. and Trojanowski, J.Q. (1999). The expression of alpha, beta and gamma synucleins in olfactory mucosa from patients with and without neurodegenerative disease. *Experimental Neurology*, 160, 515–522.

Eichenbaum, H., Morton, T.H., Potter, H. and Corkin, S. (1983). Selective olfactory deficits in case HM. *Brain*, 106, 459–472.

Eichenbaum, H., Shedlack, K.J. and Eckmann, K.W. (1980). Thalamocortical mechanisms in odour-guided behaviour: I. Effects of lesions to the mediodorsal thalamic nucleus and frontal cortex on olfactory discrimination in the rat. *Brain Behaviour and Evolution*, 17, 255–275.

Eldeghaidy, L., Marciani, F., McGlone, T., Hollowood, J., Hort, K., Head, J., Taylor, R.C., Spiller, P.A. and Francis, S.T. (2011). The cortical response to oral perception of fat emulsions and the effect of taster status. *Journal of Physiology*, 105, 2572–2581.

Elfhag, K. and Erlanson-Albertsson, C. (2006). Sweet and fat taste preference in obesity have different associations with personality and eating behaviour. *Physiology & Behavior*, 88, 61–66.

Elsberg, C.A. and Spotnitz, H. (1942). Value of quantitative olfactory tests for lovalization of supratentorial disease: analysis of one thousand cases. *Archives of Neurology and Psychiatry*, 48, 1–12.

Elsberg, C.A. and Stewart, J. (1938). Quantitative olfactory tests: value in localization and diagnosis of tumors of the brain, with analysis of results in three hundred patients. *Archives of Neurology and Psychiatry*, 40, 471–481.

Engelen, L., de Wijk, R.A. and Prinz, J.F. (2002). The effect of oral temperature on the temperature perception of liquids and semisolids in the mouth. *European Journal of Oral Science*, 110, 412–416.

Engen, T. (1982). *The Perception of Odors*. New York: Academic Press.

Engen, T. and Bosack, T.N. (1969). Facilitation in olfactory detection. *Journal of Comparative Physiology & Psychology*, 68, 320–326.

Enns, M.P., Van Itallie, T.B. and Grinker, J.A. (1979). Contributions of age, sex and degree of fatness on preferences and magnitude estimations for sucrose in humans. *Physiology & Behavior*, 22, 999–1003.

Epstein, J.B. and Barasch, A. (2010). Taste disorders in cancer patients. *Oral Oncology*, 46, 77–81.

Escada, P.A., Lima, C. and da Silva, J.M. (2009). The human olfactory mucosa. *European Archives of Otorhinolaryngology*, 266, 1675–1680.

Eskenazi, B., Cain, W.B., Novelly, R.A. and Friend, K.B. (1983). Olfactory functioning in temporal lobectomy patients. *Neuropsychologia*, 21, 365–374.

Eskenazi, B., Cain, W.B., Novelly, R.A. and Mattson, R. (1986). Odour perception in temporal lobe epilepsy patients with and without temporal lobectomy. *Neuropsychologia*, 24, 553–562.

Eskine, K.J., Kacinik, N.A. and Prinz, J.J. (2011). A bad taste in the mouth: gustatory disgust influences moral judgment. *Psychological Science*, 22, 295–299.

Evans, D.A., Funkenstein, H.H., Albert, M.S., Scherr, P.A., Cook, N.R., Chown, M.J., Herbert, L.E., Hennekens, C.H. and Taylor, J.O. (1989). Prevalence of Alzheimer's Disease in a community population of older persons: higher than previously reported. *Journal of the American Medical Association*, 262, 2551–2556.

Evans, W., Cui, L. and Starr, A. (1995). Olfactory event-related potentials in normal human subjects: effect of age and gender. *Electroencephalography and Clinical Neurophysiology*, 95, 293–301.

Evans, W.J., Kobal, G., Lorig, T.S. and Prah, J.D. (1993). Suggestions for collection and reporting of chemosensory (olfactory) event-related potentials. *Chemical Senses*, 18, 751–756.

Eyman, R.K., Kim, P.J. and Call, T. (1975). Judgment error in category vs magnitude scales. *Perceptual and Motor Skills*, 40, 415–423.

Farrer, L.A., Cuppies, A., Haines, J.L., Hyman, B., Kukull, W.A., Mayeux, R., Myers R.H., Pericak-Vance, M.A., Risch, N. and van Duijn, C.M. (1997). Effects of age, sex and ethnicity on the association between apolipoprotein E genotype and Alzheimer Disease. *Journal of the American Medical Association*, 278, 1349–1356.

Faurion, A., Cerf, B., LeBihan, D. and Pillasa, A.-M. (1998). fMRI taste study of taste cortical areas in humans. *Annals of the New York Academy of Sciences*, 855, 535–545.

Faurion, A., Cerf, B., van de Moortele, P.-F., Lobe, E., MacLeod, P. and LeBihan, D. (1999). Human taste cortical areas studied with fMRI: evidence of functional lateralization related to handedness. *Neuroscience Letters*, 277, 189–192.

Fearnley, J.M. and Lees, A. (1991). Ageing and Parkinson's Disease: substantia nigra regional selectivity. *Brain*, 114, 2283–2301.

Ferreira, G., Gutierrez, R., De La Cruz, V. and Bermudez-Rattoni, F. (2002). Differential involvement of cortical muscarinic and NMDA receptors in short- and long-term taste aversion memory. *European Journal of Neuroscience*, 16, 1139–1145.

Fenger, M.M., Gade, A., Adams, K.H., Hansen, E.S., Bolwig, T.G. and Knudsen, G.M. (2005). Cognitive deficits in obsessive-compulsive disorder on tests of frontal lobe function. *Nordic Journal of Psychiatry*, 59, 39–44.

Fenu, S., Bassareo, V. and Di Chara, G. (2001). A role for dopamine D1 receptors of the nucleus accumbens shell in conditioned taste aversion learning. *Journal of Neuroscience*, 21, 6897–6904.

Ferdenzi, C., Coureaud, G., Camos, V. and Schaal, B. (2008). Human awareness and uses of odour cues in everyday life: results from a questionnaire study in children. *International Journal of Behavioural Development*, 32, 422–431.

Ferdon, S. and Murphy, C. (2003). The cerebellum and olfaction in the ageing brain: an fMRI study. *Neuroimage*, 20, 12–21.

Ferri, C.P., Prince, M., Brayne, C., Brodaty, H., Fratiglioni, L., Ganguli, M., Hall, K., Hasegawa, K., Hendrie, H., Huang, Y., Jorm, A., Mathers, C., Menezes, P.R., Rimmer, E., Scazufca, M. and Alzheimer's Disease International (2005). Global prevalence of dementia: a Delphi consensus study. *The Lancet*, 366, 2112–2117.

Ferrier, D. (1876). The hemispheres considered physiologically. *The Functions of the Brain.* New York: Putnam.

Filsinger, E.E., Fabes, R.A. and Hughston, G. (1987). Introversion–extraversion and dimensions of olfactory perception. *Perceptual and Motor Skills*, 64, 696–699.

Finkenzeller, P. (1966). Gemittelte EEG-Potentiale bei olfaktorischer Reizung. *Pfugers Arch. Ges. Physiol.*, 292, 76–80.

Fiore, A.M. (1992). Effect of composition of olfactory cues on impressions of personality. *Social Behavior and Personality*, 20(3), 149–162.

Forestell, C.A. and Mennella, J.A. (2007). Early determinants of fruit and vegetable acceptance. *Pediatrics*, 120, 1247–1254.

Forster, S., Vaitl, A., Teipel, S.J., Yakushev, I., Mustafa, M., la Fougere, C., Rominger, A., Cumming, P., Bartenstein, P., Hampel, H., Hummel, T., Buerger, K., Hundt, W. and Steinbach, S. (2010). Functional representation of olfactory impairment in early Alzheimer's Disease. *Journal of Alzheimer's Disease*, 22, 581–591.

Fox, A.F. (1932). The relationship between chemical constitution and taste. *PNAS*, 18, 115–120.

Frank, G.K., Kaye, W.H., Carter, C.S., Brooks, S., May, C., Fissell, K. and Stenger, V.A. (2003). The evaluation of brain activity in response to taste stimuli: a pilot study and method for central taste activation as assessed by event-related fMRI. *Journal of Neuroscience Methods*, 131, 99–105.

Frank, G.K., Oberndorfer, T.A., Simmons, A.N., Paulus, M.P., Fudge, J.L., Yang, T.T. and Kaye, W.H. (2008). Sucrose activates human taste pathways differently from artificial sweetner. *Neuroimage*, 39, 1559–1569.

Frank, R.A. (2002). Response context affects judgements of flavour components in foods and beverages. *Food Quality and Preferences*, 14, 139–145.

Frank, R.A. and Byram, J. (1988). Taste–smell interactions are tastant and odorant dependent. *Chemical Senses*, 13, 445–455.

Frank, R.A., Dulay, M.F., Niergarth, K.A. and Gesteland, R.C. (2004). A comparison of the sniff magnitude test and the UPSIT in children and normative English speakers. *Physiology & Behavior*, 81, 475–480.

Frank, R.A., van der Klaauw, N.J. and Schifferstein, H.N. (1993). Both perceptual and conceptual factors influence taste–color and taste–taste interactions. *Perception and Psychophysics*, 54, 343–354.

Frankmann, S.P. and Green, B.G. (1987). Differential effects of cooling on the intensity of taste. *Annals of the New York Academy of Sciences*, 510, 300–303.

Frasnelli, J., Lundstrom, J.N., Boyle, J.A., Djordjevic, J., Zatorre, R.J. and Jones-Gotman, M. (2010). Neuroanatomical correlates of olfactory performance. *Experimental Brain Research*, 201, 1–11.

Frey, S. and Petrides, M. (1999). Re-examination of the human taste region: a PET study. *European Journal of Neuroscience*, 11, 2985–2988.

Fried, I., Spencer, D.D. and Spencer, S.S. (1995). The anatomy of epileptic auras: focal pathology and surgical outcome. *Journal of Neurosurgery*, 83, 60–66.

Frijters, J.E.R., Kooistra, A. and Vereijken, P.F.G. (1980). Tables of d' for the triangular method and the 3-AFC signal detection performance. *Perception and Psychophysics*, 27, 176–178.

Fulbright, R.K., Skudlarski, P. and Lacadie, C.M. (1998). Functional MR imaging of regional brain responses to pleasant and unpleasant odours. *American Journal of Neuroradiology*, 19, 1721–1726.

Fuller, G.N. and Guiloff, R.J. (1987). Migrainous olfactory hallucinations. *Journal of Neurosurgery and Psychiatry*, 50, 1688–1690.

Galindo-Cuspinera, V. and Breslin, P. (2006). The liaison of sweet and savoury. *Chemical Senses*, 31, 221–225.

Galindo-Cuspinera, V., Winnig, M., Bufe, B., Meyerhof, W. and Breslin, P.A.S. (2006). A TAS1R receptor-based explanation of sweet 'water-taste'. *Nature*, 441, 354–357.

Gamble, E.M. (1898). The applicability of Weber's law to smell. *American Journal of Psychology*, 10, 82–142.

Gangestad, S.W. and Thornhill, R. (1998). Menstrual cycle variation in women's preferences for the scent of symmetrical men. *Proceedings of the Royal Society B*, 265, 927–933.

Garcia-Falgueras, A., Junque, C., Gimenez, M., Caldu, X., Segovia, S. and Guilamon, A. (2006). Sex differences in the human olfactory system. *Brain Research*, 1116, 103–111.

Garcia-González, D.L., Vivancos, J. and Aparicio, R. (2011). Mapping brain activity induced by olfaction of virgin olive oil aroma. *Journal of Agricultural and Food Chemistry*, 201(59), 102–110.

Getchell, T.V., Krishna, N.S., Dhooper, N., Sparks, D.L. and Getchell, M.L. (1995). Human olfactory receptor neurons express heat shock protein 70: age-related trends. *Annals of Otolaryngology, Rhinology and Laryngology*, 104, 47–56.

Getchell, T.V., Margolis, F.L. and Getchell, M.L. (1984). Perireceptor and receptor events in vertebrate olfaction. *Progress in Neurobiology*, 23, 317–345.

Ghadami, M., Majidzadeh-A, K., Morovvati, S., Damavandi, E., Nishimura, G., Komatsu, K., Kinoshita, A., Najafi, M.-T., Niikawa, N. and Yoshiura, K. (2004). Isolated congenital anosmia with morphologically normal olfactory bulb in two Iranian families. *American Journal of Medical Genetics A*, 127, 307–309.

Giduck, A., Threatte, R.M. and Kare, M.R. (1987). Cephalic reflexes: their role in digestion and possible roles in absorption and metabolism. *Journal of Nutrition*, 117, 1191–1196.

Gilbert, A.N., Greenberg, M.S. and Beauchamp, G.K. (1999). Sex, handedness and side of nose modulate human odour perception. *Neuropsychologia*, 27, 505–511.

Gillan, D.J. (1983). Taste–taste, odour–odour and taste–odour mixtures: greater suppression within than between modalities. *Perception & Psychophysics*, 33, 183–185.

Gilmore, M.M. and Green, B.G. (1993). Sensory irritation and taste produced by NaCl and citric acid: effects of capsaicin desensitization. *Chemical Senses*, 18, 257–272.

Gilmore, M.M. and Murphy, C. (1989). Ageing is associated with increased Weber ratios for caffeine, but not for sucrose. *Perception & Psychophysics*, 46, 555–559.

Glaser, O. (1918). Hereditary deficiencies in the sense of smell. *Science*, 48, 647–648.

Glusman, G., Yanai, I., Rubin, I. and Lancet, D. (2001). The complete human olfactory subgenome. *Genome Research*, 11, 685–702.

Goldman-Rakic, P. (1996). Multiple working memory domains and the concept of the central executive. In A.C. Roberts, T.W. Robbins and L. Weiskrantz (eds) *The Prefrontal Cortex: Executive and Cognitive Functions*. London: Royal Society.

Gonzalez, J., Barros-Loscertales, A., Pulvermuller, F., Meseguer, V., Sanjuan, A., Belloch, V. and Avila, C. (2006). Reading cinnamon activates olfactory brain regions. *Neuroimage*, 32, 906–912.

Goodale, M.A. and Milner, A.D. (1992). Separate visual pathways for perception and action. *Trends in Neuroscience*, 15, 20–25.

Gordon, H.W. (1974). Olfaction and cerebral separation. In Kinsbourne, K. and Smith, W.L. (eds) *Hemispheric Disconnection and Cerebral Function*. Springfield: Charles Thomas, pp. 137–154.

Gordon, H.W. and Sperry, R.W. (1969). Lateralization of olfactory perception in the surgically separated hemispheres of man. *Neuropsychologia*, 7, 111–120.

Gorno-Tempini, M.L., Rankin, K.P., Woolley, J.D., Rosen, H.J., Phengrasamy, L. and Miller, B.L. (2004). Cognitive and behavioural profile in a case of right anterior temporal lobe neurodegeneration. *Cortex*, 40, 631–644.

Gottfried, J.A., Deichmann, R., Winston, J.S. and Dolan, R.J. (2002). Functional heterogeneity in human olfactory cortex: an event-related functional magnetic resonance imaging study. *Journal of Neuroscience*, 22, 10819–10828.

Gottfried, J.A. and Dolan, R.J. (2003). The nose smells what the eye sees: cross-modal facilitation of human olfactory perception. *Neuron*, 39, 375–386.

Gottfried, J.A. and Zald, D.H. (2005). On the scent of human olfactory orbitofrontal cortex: meta-analysis and comparison to non-human primates. *Brain Research Reviews*, 50, 287–304.

Goubet, N., Rattaz, C., Pierrat, V., Bullinger, A. and Lequien, P. (2003). Olfactory experience mediates response to pain in preterm newborns. *Developmental Psychobiology*, 42, 171–180.

Gould, A. and Martin, G.N. (2001). 'A good odour to breathe?' The effect of pleasant ambient odour on human visual vigilance. *Applied Cognitive Psychology*, 15, 225–232.

Gozal, D., Hathout, G.M., Kirlew, K.A., Tang, H., Woo, M.S. and Zhang, J. (1994). Localization of putative neural respiratory regions in the human by fMRI. *Journal of Applied Physiology*, 76, 2076–2083.

Gozal, D., Omidvar, O., Kirlew, K.A., Hathout, G.M., Hamilton, R. and Lufkin, R.B. (1995). Identification of human brain regions underlying responses to resistive inspiratory loading with fMRI. *PNAS*, 92, 6607–6611.

Grabenhorst, F. and Rolls, E.T. (2009). Different representations of relative and absolute subjective value in the human brain. *Neuroimage*, 48, 258–268.

Grabenhorst, F., Rolls, E.T., Margot, C., da Silva, M.A.A. and Velazco, M.I. (2007). How pleasant and unpleasant stimuli combine in different brain regions: odour mixtures. *The Journal of Neuroscience*, 27, 13532–13540.

Graff-Radford, N.R., Lin, S.C., Brazis, P.W., Bolling, J.P., Liesegang, T.J. and Lucas, J.A. (1997). Tropicamide eyedrops cannot be used for reliable diagnosis of Alzheimer's Disease. *Mayo Clinic Proceedings*, 72, 495–504.

Graves, A.B., Bowen, J.D. and Rajaram, L. (1999). Impaired olfaction as a marker for cognitive decline: interaction with apolipoprotein E episilon4 status. *Neurology*, 53, 1480–1487.

Graziadei, P.P.C. and Monti Graziadei, G.A. (1986). Principles of organisation of the vertebrate olfactory glomerulus: an hypothesis. *Neuroscience*, 19, 1025–1035.

Green, B.G. and Schullery, M.T. (2004). Stimulation of bitterness by capsaicin and menthol. *Chemical Senses*, 28, 45–55.

Greenfield, H., Smith, A.M. and Wills, R.B.H. (1984). Influence of multi-holed shakers on salting food. *Human Nutrition*, 38a, 199–201.

Griffiths, N.M. and Patterson, R.L.S. (1970). Human olfactory responses to 5-alpha-androst-516-en-3-one principal component of boar taint. *Journal of Science, Food and Agriculture*, 21, 4–6.

Gross-Isseroff, R., Luca-Haimovici, K., Sasson, Y., Kindler, S., Kotler, M. and Zohar, J. (1994). Olfactory sensitivity in major depressive disorder and obsessive compulsive disorder. *Biological Psychiatry*, 35, 798–802.

Gruzelier, J., Seymour, K., Wilson, L., Idley, A. and Hirsch, S. (1988). Impairments on neuropsychological tests of temporohippocampal and frontohippocampal function and word fluency in remitting schizophrenia and affective disorders. *Archives of General Psychiatry*, 45, 623–629.

Grzegorczyk, P.B., Jones, S.W. and Mistretta, C.M. (1979). Age-related differences in salt taste acuity. *Journal of Gerontology*, 34, 834–840.

Gudziol, V., Paech, I. and Hummel, T. (2010). Unilateral reduced sense of smell is an early indicator for global olfactory loss. *Journal of Neurology*, 257, 959–963.

Guittérez, R., Rodriguez-Ortiz, C.J., De la Cruz, V., Nunez-Jaramillo, L. and Bermudez-Rattoni, F. (2003). Cholinergic dependence of taste memory formation: evidence of two distinct processes. *Neurobiology of Learning and Memory*, 80, 323–331.

Gulyas, A.I., Toth, K., McBain, C.J. and Freund, T.F. (1998). Stratum radiatum giant cells: a type of principal cell in the rat hippocampus. *European Journal of Neuroscience*, 10, 3813–3822.

Guo, S.W. and Reed, D.R. (2001). The genetics of phenylthiocarbamide perception. *Annals of Human Biology*, 28, 111–142.

Gupta, A.K., Jeavons, P.M. and Hughes, R.C. (1983). Aura in temporal lobe epilepsy: clinical and electroencephalographic correlation. *Journal of Neurology, Neurosurgery and Psychiatry*, 46, 1079–1083.

Gur, R.E., Turetsky, B.I. and Cowell, P.E. (2000). Temporolimbic volume reductions in schizophrenia. *Archives of General Psychiatry*, 57, 769–775.

Haase, L., Cerf-Ducastel, B., Buracas, G. and Murphy, C. (2007). On-line psychophysical data acquisition and event-related fMRI protocol optimized for the investigation of brain activation in response to gustatory stimuli. *Journal of Neuroscience Methods*, 159, 98–107.

Haase, L., Cerf-Ducastel, B. and Murphy, C. (2009). Cortical activation in response to pure taste stimuli during the physiological states of hunger and satiety. *Neuroimage*, 44, 1008–1021.

Haase, L., Green, E. and Murphy, C. (2011). Males and females show differential brain activation to taste when hungry and sated in gustatory and reward areas. *Appetite*, 57, 421–434.

Habel, U., Loch, K., Pauly, K., Kellermann, T., Reske, M., Backes, V., Seiferth, N.Y., Stocker, T., Kircher, T., Amunts, K., Shah, N.J. and Schneider, F. (2007). The influence of olfactory-induced negative emotion on verbal working memory: individual differences in neurobehavioural findings. *Brain Research*, 1152, 158–170.

Haberly, L.B. (2001). Parallel-distributed processing in olfactory cortex: new insights from morphological and physiological analysis of neuronal circuitry. *Chemical Senses*, 26, 551–576.

Haddad, R., Weiss, T., Khan, R., Nadler, B., Mandairon, N., Bensafi, M., Schneidman, E. and Sobel, N. (2010). Global features of neural activity in the olfactory system form a parallel code that predicts olfactory behavior and perception. *The Journal of Neuroscience*, 30, 9017–9026.

Haehner, A., Hummel, T., Hummel, C., Sommer, U., Jughanns, S. and Reichmann, H. (2007). Olfactory loss may be a first sign of idiopathic Parkinson's Disease. *Movement Disorders*, 22, 839–842.

Haehner, A., Mayer, A.-M., Landis, B.N., Pournaras, I., Lill, K., Gudziol, V. and Hummel, T. (2009). High test–retest reliability of the extended version of the 'Sniffin' Sticks' test. *Chemical Senses*, 34, 705–711.

Hahn, I., Scherer, P.W. and Mozell, M.M. (1993). Velocity profiles measured for airflow through a large-scale model of the human nasal cavity. *Journal of Applied Physiology*, 75, 2273–2287.

Haller, R., Rummel, C., Henneberg, S., Pollmer, U. and Koster, E.P. (1999). The influence of early experience with vanillin on food preference later in life. *Chemical Senses*, 24, 465–467.

Halpern, B.P. (1985). Time as a factor in gustation. In Pfaff, D.W. (ed.) *Taste, Olfaction and the Central Nervous System*. New York: Rockerfeller University Press, pp. 181–209.

Halpern, M., Shapiro, L.S. and Jia, C. (1998). Heterogeneity in the accessory olfactory system. *Chemical Senses*, 23, 477–481.

Hamada, T. and Yamaguch, M. (2001). Evoked and oscillatory neuromagnetic responses to sniffing odour in human subjects. *Behavioural Brain Research*, 123, 219–223.

Harada, H., Shiraishi, K., Kato, T. and Soda, T. (1996). Coherence analysis of EEG changes during odour stimulation in humans. *The Journal of Laryngology and Otology*, 110, 652–656.

Harding, A.J., Stimson, E., Henderson, J.M. and Halliday, G.M. (2002). Clinical correlates of selective pathology in the amygdala of patients with Parkinson's Disease. *Brain*, 125, 2431–2445.

Hardy, S.L., Brennand, C.P. and Wyse, B.W. (1981). Taste thresholds of individuals with diabetes and control subjects. *Journal of American Dietetic Association*, 79, 286–289.

Harper, H.W., Jay, J.R. and Erickson, R.P. (1966). Chemically evoked sensations from single human taste papillae. *Physiology & Behavior*, 1, 319–325.

Harris, H. and Kalmus, H. (1949). The measurement of taste sensitivity to phenylthiourea. *Annals of Eugenics*, 15, 24–31.

Hasan, K.S., Reddy, S.S. and Barsony, N. (2007). Taste perception in Kallmann Syndrome: a model of congenital anosmia. *Endocrine Practice*, 13, 716–720.

Hasegawa, M. and Kern, E. (1977). The human nasal cycle. *Mayo Clinic Proceedings*, 52, 28–34.

Hausser-Hauw, C.J.B. and Bancaud, J. (1987). Gustatory hallucinations in epileptic seizures. *Brain*, 110, 339–359.

Hawkes, C.H. (2006). Olfaction in parkinsonian syndromes. In Brewer, W., Castle, D. and Pantelis, C. (eds) *Olfaction and the Brain*. Cambridge: CUP.

Hawkes, C.H., Del Tredici, K. and Braak, H. (2009). Parkinson's Disease: the dual hit theory revisited. *Annals of the New York Academy of Sciences*, 1170, 615–622.

Hawkes, C.H. and Shephard, B.C. (1998). Olfactory evoked responses and identification tests in neurological disease. *Annals of the New York Academy of Science*, 855, 608–615.

Hawkes, C.H., Shephard, B.C. and Daniel, S.E. (1997). Olfactory dysfunction in Parkinson's Disease. *Journal of Neurology, Neurosurgery and Psychiatry*, 62, 436–446.

Haxel, B.R., Grant, L. and Mackay-Sim, A. (2008). Olfactory dysfunction after head injury. *Journal of Head Trauma Rehabilitation*, 23, 407–413.

Hayes, J.E. and Duffy, V.B. (2008). Oral sensory phenotype identifies level of sugar and fat required for maximal liking. *Physiology & Behavior*, 95, 77–87.

Heck, G.L., Mierson, S. and DeSimone, J.A. (1984). Salt taste transduction occurs through an amiloride-sensitive sodium transport pathway. *Science*, 223, 43–45.

Hedner, M., Larsson, M., Arnold, N., Zucco, G.M. and Hummel, T. (2010). Cognitive factors in odour detection, odour discrimination and odour identification tasks. *Journal of Clinical and Experimental Neuropsychology*, 32, 1062–1067.

Heining, M., Young, A.W., Ioannou, G., Andrew, C.M., Brammer, M.J., Gray, J.A. and Phillips, M.L. (2003). Disgusting smells activate human anterior insula and ventral striatum. *Annals of the New York Academy of Sciences*, 1000, 380–384.

Hemdal, P., Corwin, J. and Oster, H. (1993). Olfactory identification deficits in Down's Syndrome and idiopathic mental retardation. *Neuropsychologia*, 31, 977–984.

Henkin, R.I. Comiter, H., Fedio, P. and O'Doherty, D. (1977). Defects in taste and smell recognition following temporal lobectomy. *Transactions of the American Neurological Association*, 102, 146–150.

Henkin, R.I., Gill, J.R. and Bartter, F.C. (1963). Studies on taste thresholds in normal man and in patients with adrenal cortical insufficiency: the role of adrenal cortical steroids and of serum sodium concentration. *Journal of Clinical Investment*, 42, 727–735.

Henkin, R.I. and Levy, L.M. (2001). Lateralization of brain activation to imagination and smell of odours using fMRI: left hemispheric localization of pleasant and right hemispheric localization of unpleasant odours. *Journal of Computer Assisted Tomography*, 25, 493–514.

Henning, H. (1916). *Der Geruch*. Leipzig: Barth.

Henning, H. (1924). *Zweite Auflage*. Leipzig: Barth.

Herbener, E.S., Kagan, J. and Cohen, M. (1989). Shyness and olfactory threshold. *Personality and Individual Differences*, 10(11), 1159–1163.

Herlitz, A., Airaksinen, E. and Nordstrom, E. (1999). Sex differences in episodic memory: the impact of verbal and visuospatial ability. *Neuropsychology*, 13, 1–8.

Herrick, C.J. (1933). The functions of the olfactory parts of the cerebral cortex. *Proc Natl Acad Sci USA*., 19, 7–14.

Herz, R.S. (1997. The effects of cue-distinctiveness on odour-based context dependent memory. *Memory and Cognition*, 25, 375–380.

Herz, R.S. and Cupchik, G. (1993). The effect of hedonic context on evaluations and experience of paintings. *Empirical Studies of the Arts*, 11, 147–166.

Herz, R.S., Eliassen, J., Beland, S. and Souza, T. (2004). Neuroimaging evidence for the emotional potency of odour-evoked memory. *Neuropsychologia*, 42, 371–378.

Herz, R.S., McCall, C. and Cahill, L. (1999). Hemispheric lateralization in the processing of odour pleasantness versus odour names. *Chemical Senses*, 24, 691–695.

Hock, C., Golombowski, S., Muller-Spahn, F., Peschel, O., Riederer, A., Probst, A., Mandelknow, E. and Unger, J. (1998). Histological markers in nasal mucosa of patients with Alzheimer's Disease. *Experimental Neurology*, 40, 31–36.

Hodges, J.R. and Patterson, K. (2007). Semantic dementia: a unique clinicopathological syndrome. *Lancet Neurology*, 6, 1004–1014.

Hoffman, H.J., Cruickshanks, K.J. and Davis, B. (2009). Perspectives on population-based epidemiological studies of olfactory and taste impairment. *Annals of the New York Academy of Sciences*, 1170, 514–530.

Hoffman, H.J., Ishii, E.K. and MacTurk, R.H. (1998). Age-related changes in the prevelance of smell–taste problems in the US adult population. *Annals of the New York Academy of Sciences*, 855, 716–722.

Holbrook, E.H. and Leopold, D.A. (2006). An updated review of clinical olfaction. *Current Opinion in Otolaryngology*, 14, 23–28.

Hollingworth, H.L. and Poffenberger, A.T. (1917). *The Sense of Taste*. New York: Moffat.

Homewood, J. and Stevenson, R.J. (2001). Differences in naming accuracy of odours presented to the left and right nostrils. *Biological Psychology*, 58, 65–73.

Hoon, M.A., Adler, E., Lindemeier, J., Battey, J.F., Ryba, N.J. and Zuker, C.S. (1999). Putative mammalian taste receptors. *Cell*, 96, 541–551.

Hornung, D. and Leopold, D. (1999). Relationship between uninasal anatomy and uninasal olfactory ability. *Archives of Otolaryngology, Head and Neck Surgery*, 125, 53–58.

Howard, J.D., Plailly, J., Grueschow, M., Haynes, J.-D. and Gottfried, J.A. (2009). Odour quality coding and categorization in human posterior piriform cortex. *Nature Neuroscience*, 12, 932–938.

Hübener, F., Laska, M., Kobayakawa, T. and Saito, S. (in press) Cross-cultural comparison of olfactory perception in Japanese and Germans. *Food Quality and Preference*.

Hudson, R. (1999). From molecule to mind: the role of experience in shaping olfactory function. *Journal of Comparative Physiology A*, 185, 297–304.

Hughlings-Jackson, J.H. (1899). On asphyxia in slight epileptic paroxysms: on the symptomatology of slight epileptic fits supposed to depend on discharge-lesions in the uncinate gyrus. *Lancet*, 1, 79.

Hulshoff Pol, H., Hijman, R. and Baare, W. (2000). Odour discrimination and task duration in young and older adults. *Chemical Senses*, 25, 461–464.

Hummel, T., Barz, S., Lotsch, J., Roscher, S., Kettenmann, B. and Kobal, G. (1996). Loss of olfactory function leads to a decrease of trigeminal sensitivity. *Chemical Senses*, 21, 75–79.

Hummel, T., Doty, R.L. and Yousem, D.M. (2005a). Functional MRI of intranasal chemosensory trigeminal activation. *Chemical Senses*, 30, 205–206.

Hummel, T., Fliessbach, K., Abele, M., Okulla, T., Reden, J., Reichmann, H., Wullner, U. and Haehner, A. (2010a). Olfactory fMRI in patients with Parkinson's Disease. *Frontiers in Integrative Neuroscience*, 4, 1–7.

Hummel, T., Frasnelli, J., Gerber, J. and Hummel, T. (2007). Cerebral processing of gustatory stimuli in patients with taste loss. *Behavioural Brain Research*, 185, 59–64.

Hummel, T., Futschik, T., Frasnelli, J. and Huttenbrink, K.B. (2003). Effects of olfactory function, age and gender on trigeminally mediated sensations: a study based on the lateralization of chemosensory stimuli. *Toxicology Letter*, 140, 273–280.

Hummel, T., Genow, A. and Landis, B.N. (2010b). Clinical assessment of human gustatory function using ERPs. *Journal of Neurology, Neurosurgery and Psychiatry*, 81, 459–464.

Hummel, T., Gollisch, R. and Wildt, G. (1991). Changes in olfactory perception during the menstrual cycle. *Experientia*, 47, 712–715.

Hummel, T., Guel, H. and Delank, W. (2004). Olfactory sensitivity of subjects working in odorous environments. *Chemical Senses*, 29, 533–536.

Hummel, T. and Kobal, G. (1992). Differences in human evoked potentials related to olfactory or trigeminal chemosensory stimuli. *Electroencephalography and Clinical Neurophysiology*, 84, 84–89.

Hummel, T., Krone, F., Lundstrom, J.N. and Bartsch, O. (2005b). Androstadienone odour threshold in adolescents. *Hormones and Behavior*, 47, 306–310.

Hummel, T., Mohammadian, P. and Kobal, G. (1998). Handedness is a determining factor in lateralized olfactory discrimination. *Chemical Senses*, 23, 541–544.

Hummel, T., Nesztler, C., Kallert, S., Kobal, G., Bende, M. and Nordin, S. (2001). Gustatory sensitivity in patients with anosmia. *Chemical Senses*, 26, 118.

Hummel, T., Sekinger, B., Wolf, S., Pauli, E. and Kobal, G. (1997). 'Sniffin' Sticks': olfactory performance assessed by the combined testing of odour identification, odour discrimination and olfactory threshold. *Chemical Senses*, 22, 39–52.

Huttenen, J., Kobal, G., Kaukoranta, E. and Hari, R. (1986). Cortical responses to painful CO_2 stimulation of nasal mucosa: an MEG study in man. *Electroencephalography and Clinical Neurophysiology*, 64, 347–349.

Hvastja, L. and Zanuttini, L. (1991). Recognition of nonexplictly presented odours. *Perceptual and Motor Skills*, 72, 883–892.

Hyman, B.T., Phelps, C.H., Beach, T.G., Bigio, E., Cairns, N.J., Carrillo, M.C., Dickson, D.W., Duyckaerts, C., Frosch, M.P., Masliah, E., Mirra, S.S., Nelson, P.T., Schnieder, J.A., Thal, D.R., Thies, B., Trojanowski, J.Q., Vinters, H.V. and Montine, T.J. (2012). National Institue on Aging: Alzheier's Association guidelines for the neuropathologic assessment of Alzheimer's Disease. *Alzheimer's and Dementia*, 8, 1–13.

Ilmberger, J., Heuberger, E., Mahrhofer, C., Dessovic, H., Kowarik, D. and Buchbauer, G. (2001). The influence of essential oils on human attention. 1: alertness. *Chemical Senses*, 26, 239–245.

Insausti, R., Tunon, T. and Sobreviela, T. (1995). The human entorhinal cortex. *Journal of Comparative Neurology*, 355, 171–198.

Issel-Tarver, L. and Rine, J. (1997). The evolution of mammalian olfactory receptor genes. *Genetics*, 145, 185–195.

Jabbi, M., Swart, M. and Keysers, C. (2007). Empathy for positive and negative emotions in the gustatory cortex. *Neuroimage*, 34, 1744–1753.

Jackson, J.C. and Benjamin, R.M. (1974). Unit discharges in the mediodorsal nucleus of the rabbit evoked by electrical stimulation of the olfactory bulb. *Brain Research*, 26, 193–201.

Jacob, S., Kinnunen, L.H., Metz, J., Cooper, M. and McClintock, M.K. (2001). Sustained human chemosignal unconsciously alters brain function. *Neuroreport*, 12, 2391–2394.

Jacobson, A., Green, E. and Murphy, C. (2010). Age-related functional changes in gustatory and reward processing regions. *Neuroimage*, 53, 602–610.

Jacquot, L., Monnin, J. and Brand, G. (2004). Influence of nasal trigeminal stimuli on olfactory sensitivity. *Comptes Rendus Biologies*, 327, 305–311.

Jacquot, L., Monnin, J., Lucarz, A. and Brand, G. (2005). Trigeminal sensitization and desentitization in the nasal cavity: the study of cross interactions. *Rhinology*, 43, 93–98.

Jafek, B.W., Moran, D.T., Eller, P.M., Rowley, J.C. and Jafek, T.B. (1987). Steroid-dependent anosmia. *Archives of Otolaryngology*, 113, 547–549.

James, C.E., Laing, D.G. and Oram, N. (1997). A comparison of the ability of eight to nine-year-old children and adults to detect taste stimuli. *Physiology & Behavior*, 62, 193–197.

Jeannerod, M. (1997). *The Cognitive Neuroscience of Action.* Oxford: Blackwell.

Jehl, C., Royet, J.-P. and Holley, A. (1995). Odour discrimination and recognition memory as a function of familiarisation. *Perception and Psychophysics*, 57, 1002–1011.

Jella, S.A. and Shannahoff-Khalsa, D.S. (1993). The effects of unilateral forced nostril breathing on cognitive performance. *International Journal of Neuroscience*, 73, 61–68.

Johnson, B.A., Farahbod, H., Saber, S. and Leon, M. (2005). Effects of functional group position on spatial representations of aliphatic odorants in the rat olfactory bulb. *Journal of Comparative Neurology*, 483, 192–204.

Johnson, B.A., Farahbod, H., Xu, Z., Saber, S. and Leon, M. (2004). Local and global chemotopic organization: general features of the glomerular representations of aliphatic odorants differing in carbon number. *Journal of Comparative Neurology*, 480, 234–249.

Johnson, J.L., Dzendolet, E. and Clydesdale, F.M. (1983). Psychophysical relationship between sweetness and redness in strawberry-flavoured drinks. *Journal of Food Prot.*, 46, 21–28.

Johnson, J.L., Dzendolet, E. and Damon, E. (1982). Psychophysical relationship between perceived sweetness and colour in cherry-flavoured beverages. *Journal of Food Prot.*, 45, 601–606.

Johnson, T.M. (1996). Ataxic hemiparesis and hemiageusia from an isolated post-traumatic midbrain lesion. *Neurology*, 47, 1348–1349.

Jones, C.L., Ward, J. and Critchley, H.D. (2010). The neuropsychological impact of insular cortex lesions. *Journal of Neurology, Neurosurgery and Psychiatry*, 81, 611–618.

Jones-Gotman, M. and Zatorre, R.J. (1988a). Olfactory identification deficits in patients with focal cerebral excision. *Neuropsychologia*, 26, 387–400.

Jones-Gotman, M. and Zatorre, R.J. (1988b). Contribution of the right temporal lobe to odour memory. *Epilepsia*, 29, 661.

Jones-Gotman, M. and Zatorre, R.J. (1993). Odour recognition memory in humans: role of right temporal and orbitofrontal regions. *Brain and Cognition*, 22, 182–198.

Jones-Gotman, M., Zatorre, R.J., Cendes, F., Oliver, A., Andermann, F., McMackin, D., Staunton, H., Siegel, A.M. and Wieser, H.G. (1997). Contribution of medial versus lateral temporal lobe structures to human odour identification. *Brain*, 120, 1845–1856.

Joyce, E.M., Collinson, S.L. and Crichton, P. (1996). Verbal fluency in schizophrenia: relationship with executive function, semantic memory and clinical alogia. *Psychological Medicine*, 26, 39–49.

Kadohisa, M., Rolls, V. and Verhagen, J.V. (2005). The primate amygdala: neuronal representation of the viscosity, fat texture, grittiness and taste of foods. *Neuroscience*, 132, 33–48.

Kaijura, H., Cowart, B.J. and Beauchamp, G.K. (1992). Early developmental change in bitter taste responses in human infants. *Developmental Psychobiology*, 25, 375–386.

Kami, Y.N., Goto, T.K., Tokumori, K., Yoshiura, T., Kobayashi, K., Nakamura, Y., Honda, H., Ninomiya, Y. and Yoshiura, K. (2008). The development of a novel automated taste stimulus delivery system for fMRI studies on the human cortical segregation of taste. *Journal of Neuroscience Methods*, 172, 48–53.

Kampov-Polevoy, A.B., Garbutt, J.C. and Janowsky, D.S. (1999). Association between preference for sweets and excessive alcohol intake: a review of animal and human studies. *Alcohol*, 34, 386–395.

Kareken, D.A., Mosnik, D., Doty, R.L., Dzemidzic, M. and Hutchins, G.D. (2003). Functional anatomy of human odour sensation, discrimination and identification in health and ageing. *Neuropsychology*, 17, 482–495.

Kareken, D.A., Mosnik, D., Doty, R.L., Moberg, P.J., Chen, S.H., Farlow, M.R. and Hutchins, G.D. (2001). Olfactory-evoked regional cerebral blood flow in Alzheimer's Disease. *Neuropsychology*, 15, 18–29.

Kareken, D.A., Sabri, M., Radnovich, A.J., Claus, E., Foresman, B., Hector, D. and Hutchins, G.D. (2004). Olfactory system activation from sniffing: effects in piriform and orbitofrontal cortex. *Neuroimage*, 22, 456–465.

Karlson, P. and Luscher, M. (1959). 'Pheromones': a new term for a class of biologically active substances. *Nature*, 183, 55–56.

Karp, J.R., Johnston, J.D., Tecklenburg, S., Mickleborough, T.D., Fly, A.D. and Stager, J.M. (2006). Chocolate milk as a sport-exercise recovery aid. *International Journal of Sport Nutrition and Exercise Metabolism*, 16, 78–91.

Kassab, A., Schaub, F., Vent, J., Huttenbrink, K.-B. and Damm, M. (2009). Effects of short inter-stimulus intervals on olfactory and trigeminal event-related potentials. *Acta Oto-laryngologica*, 129, 1250–1256.

Katata, K., Sakai, N., Doi, K., Kawamitsu, H., Fuiji, M., Sugimura, K. and Nibu, K.-I. (2009). Functional MRI of regional brain responses to 'pleasant' and 'unpleasant' odours. *Acta Oto-laryngologica*, 129, 85–90.

Katz, D.B., Nicoletis, M.A.L. and Simon, S.A. (2002). Gustatory processing is dynamic and distributed. *Current Opinion in Neurobiology*, 12, 448–454.

Kawagoe, T., Tamura, R., Uwano, T., Asahi, T., Nishijo, H., Eifuku, S. and Ono, T. (2007). Neural correlates of stimulus-reward association in the rat mediodorsal thalamus. *Neuroreport*, 18, 683–688.

Kay, L.M. and Sherman, S.M. (2006). An argument for an olfactory thalamus. *Trends in Neurosciences*, 30, 47–53.

Keller, A., Zhuang, H., Chi, Q., Vosshall, L.B. and Matsunami, H. (2007). Genetic variation in a human odorant receptor alters odour perception. *Nature*, 449, 468–472.

Kenneth, J.H. (1927). An experimental study of affects and associations due to certain odours. *Psychological Monographs*, 37, 1–64.

Keshavan, M.S., Tandon, R., Boutros, N.N. and Nasrallah, H.A. (2008). Schizophrenia, 'just the facts': what we know in 2008. Part 3: Neurobiology. *Schizophrenia Research*, 100, 89–107.

Kesslak, J.P., Nalcioglu, O. and Cotman, C.W. (1991). Quantification of MRI scans for hippocampal atrophy in Alzheimer's Disease. *Neurology*, 41, 51–54.

Kettenmann, B., Hummel, C., Stefan, H. and Kobal, G. (1997). Multiple olfactory activity in the human neocortex identified by magnetic source imaging. *Chemical Senses*, 22, 493–502.

Kettenmann, B., Jousmaki, V., Portin, K., Salmelin, R., Kobal, G. and Hari, R. (1996). Odorants activate the human superior temporal sulcus. *Neuroscience Letters*, 203, 143–145.

Keuning, J. (1968). On the nasal cycle. *International Rhinology*, 6, 99–136.

Keverne, E.B. (1999). The vomeronasal organ. *Science*, 286, 716–720.

Khan, N.L., Katzenschlager, R. and Watt, H. (2004). Olfaction differentiates parkin disease from early-onset parkinsonism and Parkinson's Disease. *Neurology*, 62, 1224–1226.

Khan, R., Luk, C., Flinker, A., Aggarwal, A., Lapid, H., Haddad, R. and Sobel, N. (2007). Predicting odour pleasantness from odorant structure: pleasantness as a reflection of the physical world. *Journal of Neuroscience*, 27, 10015–10023.

Kida, A. (1974). Relations between the taste evoked potential and the background brain wave. *Journal of Nihon University Medical Association*, 34, 43–52.

Kikuchi, S., Kubota, F., Nisijima, K., Washiya, S. and Kato, S. (2005). Cerebral activation focusing on strong tasting food: an fMRI study. *Neuroreport*, 16, 281–283.

Kikuchi, Y. and Watanabe, S. (2000). Personality and dietary habits. *Journal of Epidemiology*, 10, 191–198.

Killgore, W.S., Young, A.D., Femia, L.A., Bogorodzki, P., Rogowska, J. and Yurgelun-Todd, D.A. (2003). Cortical and limbic activation during viewing of high versus low-calorie foods. *Neuroimage*, 19, 1381–1394.

Killgore, W.S. and Yurgelun-Todd, D.A. (2005). Developmental changes in the functional brain responses of adolescents to images of high and low-calorie foods. *Developmental Psychobiology*, 47, 377–397.

Killgore, W.S. and Yurgelun-Todd, D.A. (2010). Sex differences in cerebral responses to images of high and low-calorie food. *Neuroreport*, 31, 354–358.

Kinomura, S., Kawashima, R., Yamada, K., Ona, S., Itoh, S., Yoshioka, S., Yamaguchi, T., Matsui, H., Miyazawa, H., Itoh, H., Goto, R., Fujiwara, T., Satoh, K. and Fukuda, H. (1994). Functional anatomy of taste perception in the human brain studied with PET. *Brain Research*, 659, 263–266.

Kipps, C.M., Duggins, A.J., McCusker, E.A. and Calder, A.J. (2007). Disgust and happiness recognition correlate with anteroventral insula and amygdala volume respectively in preclinical Huntington's Disease. *Journal of Cognitive Neuroscience*, 19, 1206–1217.

Kishikawa, M., Iseki, M., Nishimura, M., Sekine, I. and Fujii, H. (1990). A histopathological study on senile changes in the human olfactory bulb. *Acta Pathologica Japan*, 40, 255–260.

Klein, R., Pilon, D., Prosser, S. and Shannahoff-Khalsa, D. (1986). Nasal airflow asymmetries and human performance. *Biological Psychology*, 23, 127–137.

Kleitman, N. (1963). *Sleep and Wakefulness*. Chicago: University of Chicago Press.

Klemm, W.R., Lutes, S.D., Hendrix, D.V. and Warrenberg, S. (1992). Topographical EEG maps of human responses to odours. *Chemical Senses*, 17, 347–361.

Kline, N.A. and Rausch, J.L. (1985). Olfactory precipitants of flashbacks in PTSD: case reports. *Journal of Clinical Psychiatry*, 46, 383–384.

Kline, J.P., Blackhart, G.C., Woodward, K.M., Williams, S.R. and Schwartz, G.E.R. (2000a). Anterior electroencephalographic asymmetry changes in elderly women in response to a pleasant and an unpleasant odour. *Biological Psychology*, 52, 241–250.

Kline, J.P., Schwartz, G.E., Dikman, Z.V. and Bell, I.R. (2000b). Electroencephalographic registration of low concentrations of isoamyl acetate. *Consciousness and Cognition*, 9, 50–65.

Kloek, J. (1961). The smell of some steroid sex-hormones and their metabolites: reflections and experiments concerning the significance of smell for the mutual relation of the sexes. *Psychiatry, Neurology and Neurochemistry*, 64, 309–344.

Knapp, M., Comas-Herrera, A., Somani, A. and Banerjee, S. (2007). *Dementia: International Comparisons*. 2418. London: Personal Social Services Research Unit.

Knecht, M. and Hummel, T. (2004). Recording of the human electro-olfactogram. *Physiology & Behavior*, 83, 13–19.

Knecht, M., Witt, M., Huttenbrink, K.-B., Heilman, S. and Hummel, T. (2003). Assessment of olfactory function and androstenone odour thresholds in humans with or without functional occlusion of the vomeronasal duct. *Behavioural Neuroscience*, 117, 1135–1141.

Kobal, G. (1981). *Electrophysiologische Untersuchungen des menschlichen Geruchssinns*. Stuttgart: Verlag.

Kobal, G. (1982). A new method for the determination of the olfactory and the trigeminal nerve's dysfunction. In Rothenberger, A. (ed.) *Event-related Potentials in Children*. Amsterdam: Elsevier.

Kobal, G. and Hummel, C. (1988). Cerebral chemosensory evoked potentials elicited by chemical stimulation of the human olfactory and respiratory mucosa. *Electroencephalography and Clinical Neurophysiology*, 71, 241–250.

Kobal, G. and Hummel, C. (1991). Human electro-olfactograms and brain responses to olfactory stimulation. In Laing, D.G., Doty, R.L. and Breipohl, W. (eds) *The Human Sense of Smell*. New York: Springer Verlag.

Kobal, G. and Hummel, C. (1998). Olfactory and intranasal trigeminal event-relaed potentials in anosmic patients. *Laryngoscope*, 108, 1033–1035.

Kobal, G., Hummel, T. and Van Toller, C. (1992). Differences in human chemosensory evoked potentials to olfactory and somatosensory chemical stimuli presented to left and right nostrils. *Chemical Senses*, 17, 233–244.

Kobayakawa, T., Endo, H., Ayabe-Kanamura, S., Kumagi, T., Yamaguchi, Y., Kikuchi, Y., Takeda, T., Saito, S. and Ogawa, H. (1996b). The primary gustatory area in human cerebral cortex studied by MEG. *Neuroscience Letters*, 212, 155–158.

Kobayakawa, T., Endo, H., Saito, S., Ayabe-Kanamura, S., Kikuchi, Y., Yamaguchi, Y., Ogawa, H. and Takeda, T. (1996a). Trial measurements of gustatory-evoked magnetic fields. *Elecreoencephalographyand Clinical Neurophysiology*, 47, 133–141.

Kobayakawa, T., Ogawa, H., Kaneda, H., Ayabe-Kanamura, S., Endo, H. and Saito, S. (1999). Spatiotemporal analysis in cortical activity evoked by gustatory stimulation in humans. *Chemical Senses*, 24, 201–209.

Kobayashi, M., Sasabe, T., Shigihara, Y., Tanaka, M. and Watanabe, Y. (2011). Gustatory imagery reveals functional connectivity from the prefrontal to insula cortices traced with MEG. *PLoS ONE*, 6, e21736.

Kobayashi, M., Takeda, M., Hattori, N., Fukunga, M., Sasabe, T. and Inoue, N. (2004). Functional imaging of gustatory perception and imagery: 'top down' processing of gustatory signals. *Neuroimage*, 23, 1271–1282.

Koelega, H.S. (1970). Extraversion, sex arousal and olfactory sensitivity. *Acta Psychologica*, 34, 51–66.

Koelega, H.S. (1979). Olfaction and sensory asymmetry. *Chemical Senses and Flavour*, 4, 89–95.

Koelega, H.S. (1994). Sex differences in olfactory sensitivity and the problem of generality of smell acuity. *Perceptual and Motor Skills*, 78, 203–213.

Koelega, H.S. and Koster, E.P. (1974). Some experiments on sex differences in odour perception. *Annals of the New York Academy of Sciences*, 237, 234–246.

Kohler, C.G., Moberg, P.J. and Gur, R.E. (2001). Olfactory dysfunction in schizophrenia and temporal lobe epilepsy. *Neuropsychiatry, Neuropsychology and Behavioural Neurology*, 14, 83–88.

Kojima, Y. and Hirano, T. (1999). A case of gustatory disturbance caused by ipsilateral pontine haemorrhage. *Rinsho Shinkeigaku*, 39, 979–981.

Konstantinidis, I., Hummel, T. and Larsson, M. (2006). Identification of unpleasant odours is independent of age. *Archives of Clinical Neuropsychology*, 21, 615–621.

Kopala, L., Clark, C. and Hurwitz, T.A, (1989). Sex differences in olfactory function in schizophrenia. *American Journal of Psychiatry*, 146, 1320–1322.

Koritnik, B., Azam, S., Andrew, C.M., Leigh, P.N. and Williams, S.C.R. (2009). Imaging the brain during sniffing: a pilot fMRI study. *Pulmonary Pharmacology & Therapeutics*, 22, 97–101.

Koss, E., Weiffenbach, J.M., Haxby, J.V. and Friedland, R.P. (1988). Olfactory detection and identification performance are dissociated in early Alzheimer's Disease. *Neurology*, 38, 1228–1232.

Koster, E.P. (1965). Olfactory sensitivity and the menstrual cycle. *International Rhinology*, 3, 57–64.

Koster, E.P. (1968). Olfactory sensitivity and ovulatory cycle duration. *Olfactogia*, 1, 43–51.

Kovacs, T., Cairns, N.J. and Lantos, P.I. (2001). Olfactory centres in Alzheimer's Disease: olfactory bulb is involved in early Braak's stages. *Neuroreport*, 12, 285–288.

Krimer, L.S., Herman, M.M., Saunders, R.C., Boyd, J.C., Hyde, T.M. and Carter, J.M. (1997). A qualitative and quantitative analysis of the entorhinal cortex and schizophrenia. *Cerebral Cortex*, 7, 732–739.

Kringelbach, M.L., de Araujo, I.E.T. and Rolls, E.T. (2004). Taste-related activity in the human dorsolateral prefrontal cortex. *Neuroimage*, 21, 781–788.

Kringelbach, M.L., O'Doherty, J., Rolls, E.T. and Andrews, C. (2003). Activation of the human orbitofrontal cortex to a liquid food stimulus is correlated with its subjective pleasantness. *Cerebral Cortex*, 13, 1064–1071.

Krolak-Salmon, P., Henaff, M.A. and Insard, J. (2003). An attention modulated response to disgust in human ventral anterior insula. *Annals of Neurology*, 53, 446–453.

Kromer Vogt, L.J., Hyman, B.T., van Hoesen, G.W. and Damasio, A.R. (1990). Pathological alterations in the amygdala in Alzheimer's Disease. *Neuroscience*, 37, 377–385.

Kronenburger, M., Sobel, S., Ilgner, J., Finkelmeyer, A., Reinacher, P., Coenen, V.A., Wilms, H., Kloss, M., Kiening, K., Daniels, C., Falk, D., Schultz, J.B., Deuschl, G. and Hummel, T. (2010). Effects of deep brain stimulation of the cerebellothalamic pathways on the sense of smell. *Experimental Neurology*, 222, 144–152.

Kruger, R., Kuhn, W., Muller, T., Woitalla, D., Graeber, M., Kosel, S., Przuntek, H., Epplen, J.T., Schols, L. and Riess, O. (1998). Ala30Pro mutation in the gene encoding alpha-synuclein in Alzheimer's Disease. *Nature Genetics*, 18, 106–108.

Kuehn, M., Welsch, H., Zahnert, T. and Hummel, T. (2008). Changes of pressure and humidity affect olfactory function. *European Archives of Otorhinolaryngology*, 265, 299–302.

Kulkarni-Narla, A., Getchell, T.V., Schmitt, F.A. and Getchell, M.L. (1996). Manganese and copper-zinc superoxide dismutases in the human olfactory mucosa: increased immunoreactivity in Alzheimer's Disease. *Experimental Neurology*, 140, 115–125.

Kurtovic, A., Widmer, A. and Dickson, B.J. (2007). A single class of olfactory neurons mediates behavioural responses to a Drosophilia sex pheromone. *Nature*, 446, 542–546.

Kveton, J.F. and Bartoshuk, L.M. (1994). The effect of unilateral chorda tympani damage on taste. *Laryngoscope*, 104, 25–29.

Laing, D.G. (1983). Natural sniffing gives optimum odour perception for humans. *Perception*, 12, 99–117.

Laing, D.G., Eddy, A., Francis, G.W. and Stephens, L. (1994). Evidence for temporal processing of odour mixtures in humans. *Brain Research*, 651, 317–328.

Laing, D.G. and Francis, G.W. (1989). The capacity of humans to identify odours in mixtures. *Physiology & Behavior*, 46, 809–814.

Laing, D.G. and Glemarec, A. (1992). Selective attention and the perceptual analysis of odour mixtures. *Physiology & Behavior*, 52, 1047–1053.

Lara, A.H., Kennerley, S.W. and Wallis, J.D. (2009). Encoding of gustatory working memory by orbitofrontal neurons. *The Journal of Neuroscience*, 29, 765–774.

Larsson, M. (1997). Semantic factors in episodic recognition of common odours in early and late adulthood: a review. *Chemical Senses*, 22, 623–633.

Larsson, M., Finkel, D. and Pedersen, N.L. (2000). Odour identification: influences of age, gender, cognition and personality. *Journal of Gerontology*, 55B, 304–310.

Larsson, M., Louden, M. and Nilsson, L.-G. (2003). Sex differences in recollective experience for olfactory and verbal information. *Acta Psychologica*, 112, 89–103.

Larsson, M., Oberg, C. and Backman, L. (2006). Recollective experience in odour recognition: influences of adult age and familiarity. *Psychological Research*, 70, 68–75.

Lascano, A.M., Hummel, T., Lacroix, J.-S., Landis, B.N. and Michel, C.M. (2010). Spatio-temporal dynamics of olfactory processing in the human brain: an event-related source imaging study. *Neuroscience*, 167, 700–708.

Lasiter, P.S., Deems, D.A. and Garcia, J. (1985). Involvement of the anterior insular gustatory neocortex in taste potentiated odour aversion learning. *Physiology & Behavior*, 34, 71–77.

Laudien, J.H., Kuster, D., Sojka, B., Fertl, R. and Pause, B.M. (2006). Central odour processing in subjects experiencing helplessness. *Brain Research*, 1120, 141–150.

Laudien, J.H., Wencker, S., Fertl, R. and Pause, B.M. (2008). Context effects on odour processing: an event-related potential study. *Neuroimage*, 41, 1426–1436.

Lauer, R.M., Filer, L.J., Reiter, M.A. and Clarke, W.R. (1976). Blood pressure, salt preference, salt threshold and relative weight. *American Journal of Disability in Childhood*, 130, 493–497.

Lawless, H. (1978). Recognition of common, odours, pictures and simple shapes. *Perception & Psychophysics*, 24, 493–495.

Lawless, H. (1991). The sense of smell in food quality and sensory evaluation. *Journal of Food Quality*, 14, 33–60.

Lawless, H. and Clark, C.C. (1992). Psychological biases in time-intensity scaling. *Food Technology*, 46, 81–90.

Lawless, H. and Stevens, D.A. (1984). Effects of oral chemical irritation on taste. *Physiology & Behavior*, 32, 995–998.

Lawrence, G., Salles, C., Palicki, O., Septier, C., Busch, J. and Thomas-Sanguin, T. (2011). Using cross-modal interactions to counterbalance salt reduction in sold foods. *International Dairy Journal*, 21, 103–110.

Lawrence, G., Salles, C., Septier, C., Busch, J. and Thomas-Sanguin, T. (2009). Odour-taste interactions: a way to enhance saltiness in low-salt content solutions. *Food Quality and Preference*, 20, 241–248.

Lee, G.P. (2004). Differential diagnosis in epilepsy. In Ricker, J.H. (ed.) *Differential Diagnosis in Adult Neuropsychological Assessment.* New York: Springer.

Lee, J.H., Goedert, M., Hill, W.D., Lee, V.M. and Trojanowski, J.Q. (1993). Tau proteins are abnormally expressed in olfactory epithelium of Alzheimer patients and developmentally regulated in human fetal spinal cord. *Experimental Neurology*, 121, 93–105.

Lee, R.S., Koob, G.F. and Heinriksen, S.J. (1998). Electrophysiological responses of nucleus accumbens neurons to novelty stimuli and exploratory behaviour in the awake, unrestrained rat. *Brain Research*, 799, 317–322.

Lee, S.H., Lim, H.H., Lee, H.M., Park, H.J. and Choi, J.O. (2000). Olfactory mucosal findings in patients with persistent anosmia after endoscopic sinus surgery. *Annals of Otolaryngology, Rhinology and Laryngology*, 109, 720–725.

Lehmann, D. and Knauss, T.A. (1976). Respiratory cycle and EEG in man and cat. *Electroencephalography and Clinical Neurophysiology*, 40, 187.

Lehrner, J.P. (1993). Gender differences in long-term odour recognition memory: verbal versus sensory influences and the consistency of label use. *Chemical Senses*, 18, 17–26.

Lehrner, J., Eckersberger, C., Walla, P., Potsch, G. and Deecke, L. (2000). Ambient odour of orange in a dental office reduces anxiety and improves mood in female patients. *Physiology & Behavior*, 71, 83–86.

Lehrner, J.P., Walla, P., Laska, M. and Deecke, L. (1999). Different forms of human odour memory: a developmental study. *Neuroscience Letters*, 272, 17–20.

LeMagnen, J. (1945/1946). Etude des facteurs dynamiques de l'excitation olfactive. *L'Annee Psychologique*, 77–89.

LeMagnen, J. (1952). Les phenomenes olfact-sexueles chex l'homme. *Arch. Sci Physiol.*, 6, 295–332.

Lemon, C.H., Brasser, S.M. and Smith, D.V. (2004). Alcohol activates a sucrose-dependent gustatory neural pathway. *Journal of Neurophysiology*, 92, 536–544.

Leon, M. and Johnson, B.A. (2003). Olfactory coding in the mammalian olfactory bulb. *Brain Research Review*, 42, 23–32.

Leshem, M., Maroun, M. and Weintraub, Z. (1998). Neonatal diuretic therapy may not alter children's preference for salt taste. *Appetite*, 30, 53–64.

Lesne, S., Koh, M.T., Kotilinek, L., Kayed, R., Glabe, C.G., Yang, A., Gallagher, M. and Ashe, K.H. (2006). A specific amyloid-B protein assembly in the brain impairs memory. *Nature*, 440, 352–357.

Levy, L.M., Henkin, R.I., Hutter, A., Lin, C.S. and Schellinger, D. (1999a). Mapping brain activation to odorants in patients with smell loss by functional MRI. *Journal of Computer Assisted Tomography*, 22, 96–103.

Levy, L.M., Henkin, R.I., Lin, C.S., Finley, A. and Schellinger, D. (1999b). Taste memory induces brain activation as revealed by fMRI. *Journal of Computer Assisted Tomography*, 23, 499–505.

Levy, L.M., Henkin, R.I., Lin, C.S., Hutter, A. and Schellinger, D. (1999c). Odour memory induces brain activation as measured by functional MRI. *Journal of Computer Assisted Tomography*, 23, 487–498.

Li, W., Moallem, I., Paller, K.A. and Gottfried, J.A. (2007). Subliminal smells can guide social preferences. *Psychological Science*, 18, 1044–1049.

Li, X., Staszewski, L., Xu, H., Durick, K., Zoller, M. and Adler, E. (2002). Human receptors for sweet and umami taste. *PNAS*, 99, 4692–4696.

Liem, D.G. and de Graaf, C. (2004). Sweet and sour preferences in young children and adults: role of repeated exposure. *Physiology & Behavior*, 83, 421–429.

Liljenquist, K., Zhong, C.-B. and Galinsky, A.D. (2010). The smell of virtue: clean scents promote reciprocity and charity. *Psychological Science*, 21, 381–383.

Livermore, A., Hummel, T. and Kobal, G. (1992). Chemosensory evoked potentials in the investigation of interactions between the olfactory and the somatosensory (trigeminal) systems. *Electroencephalography and Clinical Neurophysiology*, 83, 201–210.

Livermore, A. and Laing, D.G. (1996). Influence of training and experience on the perception of multicomponent odour mixtures. *Journal of Experimental Psychology: Human Perception and Performance*, 22, 267–277.

Locatelli, M., Belloudi, L., Grassi, B. and Scarone, S. (1996). EEG power modifications in obsessive-compulsive disorder during olfactory stimulation. *Biological Psychiatry*, 39, 326–331.

Loo, A.T., Youngentob, S.L., Kent, P.F. and Schwob, J.E. (1996). The ageing olfactory epithelium: neurogenesis, response to damage and odorant induced activity. *International Journal of Developmental Neuroscience*, 14, 881–900.

Lord, T. and Kasprzak, M. (1989). Identification of self through olfaction. *Perceptual and Motor Skills*, 69, 219–224.

Lorig, T.S., Herman, K.B., Schwartz, G.E. and Cain, W.S. (1990). EEG activity during administration of low-concentration odours. *Bulletin of the Psychonomic Society*, 28, 405–408.

Lorig, T.S., Huffman, E., DeMartino, A. and DeMarco, J. (1991). The effects of low concentration odours on EEG activity and behaviour. *Journal of Psychophysiology*, 5: 69–77.

Lorig, T.S., Matia, D.C., Peszka, J.J. and Bryant, D.N. (1996). The effects of active and passive stimulation on chemosensory event-related potentials. *International Journal of Psychophysiology*, 23, 199–205.

Lorig, T.S., Sapp, A.C., Campbell, J. and Cain, W. (1993). Event-related potentials to odour stimuli. *Bulletin of the Psychonomic Society*, 31, 131–134.

Lorig, T.S. and Schwartz, G.E. (1988). Brain and odour: I. Alteration of human EEG by odour administration. *Psychobiology*, 16(3), 281–284.

Lorig, T.S., Schwartz, G.E., Herman, K.B. and Lane, R.D. (1988). Brain and odour: II. EEG activity during nose and mouth breathing. *Psychobiology*, 16, 285–287.

Lotsch, J., Reichmann, H. and Hummel, T. (2008). Different odour tests contribute differently to the evaluation of olfactory loss. *Chemical Senses*, 33, 17–21.

Louilot, A. and Besson, C. (2000). Specificity of amygdalostriatal interactionsin the involvement of mesencephalic dopaminergic neurons in affective perception. *Neuroscience*, 96, 73–82.

Luders, E., Narr, K.L., Thompson, P.M., Woods, R.P., Rex, D.E., Jancke, L., Steinmetz, H. and Toga, A.W. (2005). Mapping cortical grey matter in the young adult brain: effects of gender. *Neuroimage*, 26, 493–501.

Lundstrom, J.N., Boyle, J.A., Zatorre, R.J. and Jones-Gotman, M. (2008). Functional neuronal processing of body odours from that of similar common odours. *Cerebral Cortex*, 18, 1466–1474.

Lundstrom, J.N. and Hummel, T. (2006). Sex-specific hemispheric differences in cortical activation to a bimodal odour. *Behavioural Brain Research*, 166, 197–203.

Lundstrom, J.N., Olsson, M.J., Schaal, B. and Hummel, T. (2006a). A putative social chemosignal elicits faster cortical responses than perceptually similar odorants. *Neuroimage*, 30, 1340–1346.

Lundstrom, J.N., Seven, S., Olsson, M.J., Schaal, B. and Hummel, T. (2006b). Olfactory event-related potentials reflect individual differences in odour valence perception. *Chemical Senses*, 31, 705–711.

Luzzi, S., Snowden, J.S., Neary, D., Coccia, M., Provinciali, L. and Lambon Ralph, M.A. (2007). Distinct patterns of olfactory impairment in Alzheimer's Disease: semantic dementia, frontotemporal dementia and corticobasal degeneration. *Neuropsychologia*, 45, 1823–1831.

Lygonis, C.S. (1969). Familial absence of olfaction. *Hereditas*, 61, 413–416.

Lyman, B.J. and McDaniel, M.A. (1986). Effects of encoding strategy on long-term memory for odours. *Quarterly Journal of Experimental Psychology*, 38A, 753–765.

Lyman, B.J. and McDaniel, M.A. (1990). Memory for odours and odour names: modalities of elaboration and imagery. *Journal of Experimental Psychology: Learning, Memory and Cognition*, 16, 656–664.

Macht, M. and Dettmer, D. (2006). Everyday mood and emotions after eating a chocolate bar or an apple. *Appetite*, 46, 332–336.

Mackay-Sim, A. and Royet, J.-P. (2006). Structure and function of the olfactory system. In Brewer, W., Castle, D. and Pantelis C. (eds) *Olfaction and the Brain*. Cambridge: CUP, pp. 3–27.

Mackenzie, R.A., Burke, D., Skuse, N.F. and Lethlean, A.K. (1975). Fibre function and perception during cutaneous nerve block. *Journal of Neurology, Neurosurgery and Psychiatry*, 38, 865–873.

Mackey, A. and Valassi, K. (1956). The discernment of primary tastes in the presence of different food textures. *Food Technology*, 10, 238–240.

Macrides, F., Eichenbaum, H.B. and Forbes, W.B. (1982). Temporal relationship between sniffing and the limbic theta rhythm during odour discrimination reversal learning. *Journal of Neuroscience*, 2, 1705–1717.

Mainen, Z.F. (2006). Behavioural analysis of olfactory coding and computation in rodents. *Current Opinion in Neurobiology*, 16, 429–434.

Mainland, J.D., Bremner, E.A., Young, N., Johnson, B.N. and Khan, R.M. (2002). Olfactory pasticity: one nostril knows what the other learns. *Nature*, 419, 802.

Makin, J. and Porter, R. (1989). Attractiveness of lactating females' breast odours to neonates. *Child Development*, 60, 803–810.

Mallet, P. and Schaal, B. (1998). Rating and recognition of peers' personal odours by nine-year-old children. *Journal of General Psychology*, 125, 47–64.

Marella, S., Fischler, W., Kong, P., Asgarian, S., Rueckert, E. and Scott, K. (2006). Imaging taste responses in the fly: brain reveals a functional map of taste category and behavior. *Neuron*, 49, 285–295.

Maresh, A., Rodriguez Gil, D., Whitman, M.C. and Greer, C.A. (2008). Principles of glomerular organization in the human olfactory bulb. *PLoS ONE*, 3, e2640.

Mark, G.P., Blander, D.S. and Hoebel, B.G. (1991). A conditioned stimulus decreases extracellular dopamine in the nucleus accumbens after the development of a learned taste aversion. *Brain Research*, 551, 308–310.

Markovic, K., Reulbach, U., Vassiliadu, A., Lunkenheimer, J., Lunkenheimer, B., Spannenberger, R. and Thueraf, N. (2007). Good news for elderly persons: olfactory pleasure increases at later stages of the life span. *Journal of Gerontology: Series A*, 62, 1287–1293.

Markley, E.J., Mattes-Kulig, D.A. and Henkin, R.I. (1985). A classification of dysgeusia. *Journal of the American Dietetic Association*, 83, 578–580.

Markopoulou, K., Larsen, K.W. and Wszolek, E.K. (1997). Olfactory dysfunction in familial parkinsonism. *Neurology*, 49, 1262–1267.

Marlier, L. and Schaal, B. (2005). Human newborns prefer human milk. *Child Development*, 76, 155–168.

Marlier, L., Schaal, B. and Soussignan, R. (1997). Orientation responses to biological odours in the human newborn. *Comptes Rendus de l'Academie des Sciences*, 320, 999–1005.

Marlier, L., Schaal, B. and Soussignan, R. (1998). Neonatal responsiveness to the odour of amniotic and lacteal fluids: a test of perinatal chemosensory continuity. *Child Development*, 69, 611–623.

Martin, A., Haxby, J.V., Lalonde, F.M., Wiggs, C.L. and Ungeleider, L.G. (1995). Discrete cortical regions associated with knowledge of colour and knowledge of action. *Science*, 270, 102–105.

Martin, C., Beshel, J. and Kay, L.M. (2007). An olfacto-hippocampal network is dynamically involved in odour-discrimination learning. *Journal of Neurophysiology*, 98, 2196–2205.

Martin, C., Gervais, R., Hugues, E., Messaoudi, B. and Ravel, N. (2004). Learning modulation of odour-induced oscillatory responses in the rat olfactory bulb: a correlate of odour recognition? *Journal of Neuroscience*, 24, 389–397.

Martin, G.N. (1996). Olfactory remediation: current evidence and possible applications. *Social Science and Medicine*, 43(1), 63–70.

Martin, G.N. (1998). Human electroencephalographic (EEG) response to olfactory stimulation: two experiments using the aroma of food. *International Journal of Psychophysiology*, 30, 287–302.

Martin, G.N. (2006). The effect of exposure to odour on the perception of pain. *Psychosomatic Medicine*, 68, 613–616.

Martin, G.N. (2013). *Beyond Smell and Taste: Psychology, flavour and our response to the multisensory aspects of food*. Institute of Cultural Research Monograph Series.

Martin, G.N., Apena, F., Chaudry, Z., Mulligan, Z. and Nixon, C. (2001a). The development of an attitude towards the sense of smell questionnaire (SoSQ) and a comparison of different professions' responses. *North American Journal of Psychology*, 3(3), 491–502.

Martin, R.E., Goodyear, B.G., Gati, J.S. and Menon, R.S. (2001b). Cerebral cortical representation of automatic and volitional swallowing in humans. *Journal of Physiology*, 85, 938–950.

Martin, R.E., Kemppainen, P., Masuda, Y., Yao, D., Murray, G.M. and Sessle, B.J. (1999). Cortically evoked swallowing in the awake primate. *Journal of Neurophysiology*, 82, 1529–1541.

Martinez-Marcos, A. (2009). On the organization of olfactory and vomeronasal cortices. *Progress in Neurobiology*, 87, 21–30.

Martinez-Marcos, A. and Halpern, M. (1999). Differential projections from the anterior and posterior divisions of the accessory olfactory bulb to the medial amygdala in the opossum. *The European Journal of Neuroscience*, 11, 3789–3799.

Martins, Y. and Pliner, P. (2006). 'Ugh! That's disgusting!' Identification of the characteristics of foods underlying rejections based on disgust. *Appetite*, 46, 75–85.

Martins, Y., Preti, G., Crabtree, C.R., Runyan, T., Vainius, A.A. and Wysocki, C.J. (2005). Preference for human body odours is influenced by gender and sexual orientation. *Psychological Science*, 16, 694–701.

Masago, R., Shimomura, Y., Iwanaga, K. and Katsuura, T. (2001). The effects of hedonic properties of odours and attentional modulation to the olfactory event-related potentials. *Journal of Physiological Anthropology*, 20, 7–13.

Mast, T.G. and Samuelson, C.L. (2009). Human pheromone detection by the vomeronasal organ: unnecessary for mate selection? *Chemical Senses*, 34, 529–531.

Matsuda, T. and Doty, R.L. (1995). Regional taste sensitivity to NaCl: relationship to subject age, tongue locus and area of stimulation. *Chemical Senses*, 20, 283–290.

Matsunami, H., Montmayeur, J.P. and Buck, L.B. (2000). A family of candidate taste receptors in human and mouse. *Nature*, 404, 601–604.

Mattes, R.D. (1990). Discretionary salt and compliance with reduced sodium diet. *Nutrition Research*, 10, 1337–1344.

Mattes, R.D. (1997). The taste for salt in humans. *American Journal of Clinical Nutrition*, 65, 692–697.

Mattes, R.D., Cowart, B.J., Schiavo, M.A., Arnold, C., Garrison, B., Kare, M.R. and Lowry, L.D. (1990). Dietary evaluation of patients with smell and/or taste disorders. *American Journal of Clinical Nutrition*, 51, 233–240.

Mattes, R.D. and Donnelly, D. (1991). Relative contributions to dietary sodium sources. *Journal of American College of Nursing*, 10, 383–393.

Mattes-Kulig, D.A. and Henkin, R.I. (1985). Energy and nutrient consumption of patients with dysgeusia. *Journal of the American Dietetic Association*, 85, 822–826.

Mattila, P.M., Rinne, J.O., Helenius, H., Dickson, D.W. and Roytta, M. (2000). Alpha-synuclein-immunoreactive cortical Lewy bodies are associated with cognitive impairment in Parkinson's Disease. *Acta Neuropathologica*, 100, 285–290.

Mattila, P.M., Rinne, J.O., Helenius, H. and Roytta, M. (1999). Neuritic degeneration in the hippocampus and amygdala in Parkinson's Disease in relation to Alzheimer pathology. *Acta Neuropathologica*, 98, 157–164.

Maurage, P., Callot, C., Phillpott, P., Rombaux, P. and de Timary, P. (2011). Chemosensory event-related potentials in alcoholism: a specific impairment for olfactory function. *Biological Psychology*, 88, 28–36.

McBurney, D.H. (1986). Taste, smell and flavour terminology: taking the confusion out of fusion. In Meiselman, H.L. and. Rivkin, R.S. (eds) *Clinical Measurement of Taste and Smell*. New York: Macmillan.

McCabe, C. and Rolls, E.T. (2007). Umami: a delicious flavour formed by convergence of taste and olfactory pathways in the human brain. *European Journal of Neuroscience*, 25, 1855–1864.

McCaughey, S.A. (2007). Taste-evoked response to sweeteners in the nucleus of the solitary tract differ between C57BL/6Bnj and 129P3/J mice. *Journal of Neuroscience*, 27, 35–45.

McCaughey, S.A. (2008). The taste of sugars. *Neuroscience and Biobehavioral Reviews*, 32, 1024–1043.

McCaughey, S.A., Giza, B.K., Nolan, L.J. and Scott, T.R. (1997). Extinction of a conditioned taste aversion in rates II: neural effects in the nucleus of the solitary tract. *Physiology & Behavior*, 61, 373–379.

McClintock, M.K. (1971). Menstrual synchrony and suppression. *Nature*, 229, 244–245.

McKeown, D.A., Doty, R.L., Perl, D.P., Frye, R.E., Simms, I. and Mester, A.F. (1996). Olfactory function in young adolescents with Down's Syndrome. *Journal of Neurology, Neurosurgery and Psychiatry*, 61, 412–414.

McLaughlin, N.C.R. and Westervelt, H.J. (2008). Odour identification defictis in frontotemporal dementia: a preliminary study. *Archives of Clinical Neuropsychology*, 23, 119–123.

McShane, R.H., Nagy, Z., Esiri, M.M., King, E., Joachim, C., Sullivan, N. and Smith, A.D. (2001). Anosmia in dementia is associated with Lewy bodies rather than Alzheimer's pathology. *Journal of Neurology, Neurosurgery and Psychiatry*, 70, 739–743.

Melrose, D.R., Reed, H.C.B. and Patterson, R.L.S. (1971). Androgen steroids associated with boar odour as an aid to the detection of oestrus in pig artificial insemination. *British Vetinary Journal*, 127, 497–502.

Mennella, J.A. and Beauchamp, G.K. (1991a). Maternal diet alters the sensory qualities of human milk and the nursling's behavior. *Pediatrics*, 88, 737–744.

Mennella, J.A. and Beauchamp, G.K. (1991b). The transfer of alcohol to human milk: effects on flavour and the infant's behavior. *New England Journal of Medicine*, 325, 981–985.

Mennella, J.A. and Beauchamp, G.K. (1993). The effects of repeated exposure to garlic-flavoured milk on the nursling's behavior. *Pediatric Research*, 34, 805–808.

Mennella, J.A. and Beauchamp, G.K. (1996). The human infants' response to vanilla flavours in mother's milk and formula. *Infant Behavior and Development*, 19, 13–19.

Mennella, J.A. and Beauchamp, G.K. (1998). Smoking and the flavour of breast-milk. *New England Journal of Medicine*, 339, 1559–1560.

Mennella, J.A. and Beauchamp, G.K. (1999). Experience with a flavour in mother's milk modifies the infant's acceptance of flavoured cereal. *Developmental Psychobiology*, 35, 197–203.

Mennella, J.A., Pepino, M.Y. and Reed, D.R. (2005). Genetic and environmental determinants of bitter perception and sweet preferences. *Pediatrics*, 115, 216–222.

Mesholam, R.I., Moberg, P.J., Mahr, R.N. and Doty, R.L. (1998). Olfaction in neurogenerative disease. *Archives of Neurology*, 55, 84–90.

Mesulam, M.M. and Mufson, E.J. (1982a). Insula of the old world monkey I. *Journal of Comparative Neurology*, 212, 1–22.

Mesulam, M.M. and Mufson, E.J. (1982b). Insula of the old world monkey III. *Journal of Comparative Neurology*, 212, 38–52.

Miller, A., Barr, R.G. and Young, S.N. (1994). The cod pressor test in children: methodological aspects of the analgesic effect of intraoral sucrose. *Pain*, 56, 175–183.

Millot, J.-L. and Brand, G. (2000). Behavioral lateralization during spontaneous smelling tasks. *Perceptual and Motor Skills*, 90, 444–450.

Min, B.-C., Jin, S.-H., Kang, I.-H., Lee, D.H., Kang, J.K., Lee, S.T. and Sakamoto, K. (2003). Analysis of mutual information content for EEG responses to odour stimulation for subjects classified by occupation. *Chemical Senses*, 28, 741–749.

Mingo, S.A. and Stevenson, R.J. (2007). Phenomenological differences between familiar and unfamiliar odours. *Perception*, 36, 931–947.

Minor, K.L., Wright, B.D. and Park, S. (2004). The Smell Identification Test as a measure of olfactory identification ability in schizophrenia and healthy populations. *Journal of Abnormal Psychology*, 113, 207–216.

Mirza, N., Kroger, H. and Doty, R.L. (1997). Influence of age on the nasal cycle. *Laryngoscope*, 107, 62–66.

Miyanari, A., Kaneoke, Y., Noguchi, Y., Honda, M., Sadato, N., Sagara, Y. and Kakigi, R. (2007). Human brain activation in response to olfactory stimulation by intravenous administration of odorants. *Neuroscience Letters*, 423, 6–11.

Moberg, P. and Turetsky, B. (2006). Olfaction in psychosis. In Brewer, W., Castle, D. and Pantelis, C. (eds) *Olfaction and the Brain*. Cambridge: CUP, pp. 296–321.

Moberg, P.J., Agrin, R., Gur, R.E., Gur, R.C., Turetsky, B.I. and Doty, R.L. (1999). Olfactory dysfunction in schizophrenia: a qualitative and quantitative review. *Neuropsychopharmacology*, 21, 325–340.

Moberg, P.J., Arnold, S.E. and Doty, R.L. (2003). Impairment of odour hedonics in men with schizophrenia. *American Journal of Psychiatry*, 160, 1784–1789.

Moberg, P.J., Doty, R.L., Mahr, R.N., Mesholam, R.I., Arnold, S.E., Turestky, B.I. and Gur, R.E. (1997a). Olfactory identification in elderly schizophrenia and Alzheimer's Disease. *Neurobiology of Ageing*, 18, 163–167.

Moberg, P.J., Doty, R.L., Turetsky, B.I., Arnold, S.E., Mahr, R.N. and Gur, R.C. (1997b). Olfactory identification deficits in schizophrenia: correlation with duration of illness. *American Journal of Psychiatry*, 154, 1016–1018.

Moberg, P., Roalf, D.R., Gur, R.E. and Turetsky, B.I. (2004). Smaller nasal volumes as stigmata of aberrant neurodevelopment in schizophrenia. *American Journal of Psychiatry*, 161, 2314–2316.

Mohr, C., Rohrenbach, C.M., Laska, M. and Brugger, P. (2001). Unilateral olfactory perception and magical ideation. *Schizophrenia Research*, 47, 255–264.

Mojet, J., Christ-Hazelhof, E. and Heidema, J. (2001). Taste perception with age: generic or specific losses in threshold sensitivity to the five basic tastes? *Chemical Senses*, 26, 845–860.

Mojet, J., Christ-Hazelhof, E. and Heidema, J. (2003). Taste perception with age: generic or specific losses in suprathreshold sensitivity to the five taste qualities? *Chemical Senses*, 28, 397–413.

Mojet, J., Heidema, J. and Christ-Hazelhof, E. (2004). Effect of concentration on taste–taste interactions in foods for the elderly and young subjects. *Chemical Senses*, 29, 671–681.

Mombaerts, P., Wang, F. and Dulac, C. (1996). Visualising an olfactory sensory map. *Cell*, 87, 675–686.

Moncrieff, R.W. (1962). Effect of odours on EEG records. *Perfumery and Essential Oil Records*, 53, 757–760.

Moncrieff, R.W. (1966). *Odor Preferences*. New York: Wiley.

Moncrieff, R.W. (1967). *The Chemical Senses*. Cleveland: CRC Press.

Monnery-Patris, S., Rouby, C., Nicklaus, S. and Issanchou, S. (2009). Development of olfactory ability in children: sensitivity and identification. *Developmental Psychobiology*, 51, 268–276.

Montgomery, E.B., Baker, K.B. and Lyons, K. (1999). Abnormal performance on the PD test battery as asymptomatic first-degree relatives. *Neurology*, 52, 757–762.

Montmayeur, J.P., Liberles, S.D., Matsunami, H. and Buck, L.B. (2001). A candidate taste receptor gene near a sweet taste locus. *Nature Neuroscience*, 4, 492–498.

Moran, D.T., Rowley, J.C., Jafek, B.W. and Lovell, M.A. (1982). The fine structure of the olfactory mucosa in man. *Journal of Neurocytology*, 11, 721–746.

Morgan, C.D. and Murphy, C. (2010). Differential effects of active attention and age on event-related potentials to visual and olfactory stimuli. *International Journal of Psychophysiology*, 78, 190–199.

Morgan, C.D., Nordin, S. and Murphy, C. (1995). Odour identification as an early marker for Alzheimer's Disease. *Journal of Clinical and Experimental Neuropsychology*, 17, 793–803.

Mori, K. and Yoshihara, Y. (1995). Molecular recognition and olfactory processing in the mammalian olfactory system. *Progress in Neurobiology*, 45, 585–619.

Moriceau, S. and Sullivan, R.M. (2004). Unique neural circuitry for neonatal olfactory learning. *The Journal of Neuroscience*, 24, 1182–1189.

Morrot, G., Brochet, F. and Dubourdieu, D. (2001). The color of odours. *Brain and Language*, 79, 309–320.

Mozell, M.M. (1970). Evidence for a chromatographic model of olfaction. *Journal of General Physiology*, 56, 46–63.

Mozell, M.M., Kent, P. and Murphy, S. (1991). The effect of flow rate upon the magnitude of the olfactory response differs for different odorants. *Chemical Senses*, 16, 631–649.

Mozell, M.M., Smith, B.P., Smith, P.E., Sullivan, R.L. and Swender, P. (1969). Nasal chemoreception in flavour identification. *Archives of Otolaryngology*, 90, 367–373.

Mueller, A., Rodewald, A., Reden, J., Gerber, J., von Kummer, R. and Hummel, T. (2005). Reduced olfactory bulb volume in post-traumatic and post-infectious olfactory dysfunction. *Neuroreport*, 16, 475–478.

Mueller, C.A. and Hummel, T. (2009). Recovery of olfactory function after nine years of post-traumatic anosmia: a case study. *Journal of Medical Case Reports*, 3, 1–3.

Murakami, M., Kashiwadani, H., Kirino, Y. and Mori, K. (2005). State-dependent sensory gating in olfactory cortex. *Neuron*, 46, 285–296.

Murphy, C. (1985). Cognitive and chemosensory influences on age-related changes in the ability to identify blended foods. *Journal of Gerontology*, 40, 47–52.

Murphy, C. (1999). Loss of olfactory function in dementing disease. *Physiology & Behavior*, 66, 177–182.

Murphy, C., Cain, W.S. and Bartoshuk, L.M. (1977). Mutual action of taste and olfaction. *Sensory Processes*, 1, 204–211.

Murphy, C., Cain, W., Gilmore, M.M. and Skinner, R.B. (1991). Sensory and semantic factors in recognition memory for odours and graphic stimuli: elderly vs young persons. *American Journal of Psychology*, 104, 161–192.

Murphy, C., Doty, R.L. and Duncan, H.J. (2003). Clinical disorders of olfaction. In Doty, R.L. (ed.) *Handbook of Olfaction and Gustation*, 2nd edn. New York: Marcel Dekker.

Murphy, C., Gilmore, M.M., Seery, C.S., Salmon, D.P. and Lasker, B.R. (1990). Olfactory thresholds are associated with degree of dementia in Alzheimer's Disease. *Neurobiology of Aging*, 11, 465–469.

Murphy, C., Nordin, S., de Wijk, R., Cain, W. and Polich, J. (1994). Olfactory-evoked potentials: assessment of young and elderly and comparison to psychophysical threshold. *Chemical Senses*, 19, 47–56.

Murphy, C., Schubert, C.R., Cruickshanks, K.J., Klein, B.E.K., Klein, R. and Nondahl, D.M. (2002). Prevalence of olfactory impairment in older adults. *JAMA*, 288, 2307–2312.

Murphy, C. and Withee, J. (1986). Age-related differences in the pleasantness of chemosensory stimuli. *Psychology and Ageing*, 1, 312–318.

Nagata, Y. and Takeuchi, N. (1990). Measurement of odour threshold by triangle odour bag method. *Bulletin of the Japanese Environmental Sanitation Centre*, 17, 77–89.

Naik, S., Shetty, N. and Maben, E.V.S. (2010). Drug-induced taste disorders. *European Journal of Internal Medicine*, 21, 240–243.

Naka, A., Riedl, M., Luger, A., Hummel, T. and Mueller, C. (2010). Clinical significance of smell and taste disorders in patients with diabetes mellitus. *European Journal of Oto-Rhino-Laryngology*, 267, 547–550.

Nakamura, T. and Gold, G. (1987). A cyclic nucleotide-gated conductance in olfactory receptor cilia. *Nature*, 325, 442–444.

Nasri, N., Beno, N., Septier, C., Salles, C. and Thomas-Danguin, T. (2011). Cross-modal interactions between taste and smell: odour-induced saltiness enhancement depends on salt level. *Food Quality and Preference*, 22, 678–682.

Nelson, G., Hoon, M.A., Chandrashekar, J., Zhang, Y., Ryba, N.J. and Zuker, C.S. (2002). An amino-acid taste receptor. *Nature*, 416, 199–202.

Nelson, P.T., Abner, E.L., Schmitt, F.A., Kryscio, R.J., Kicha, G.A., Smith, C.D., Davis, D.G., Posuska, J.W., Patel, E., Mendiondo, M.S. and Markesbery, W.R. (2010). Modelling the association between 43 different clinical and pathological variables and the severity of cognitive impairment in a large autopsy cohort of elderly persons. *Brain Pathology*, 20, 66–79.

Nicell, J.A. (2009). Assessment and regulation of odour impacts. *Atmospheric Environment*, 43, 196–206.

Nishijo, H., Uwano, T., Tamura, R. and Ono, T. (1998). Gustatory and multimodal neuronal responses in the amygdala during licking and discrimination of sensory stimuli in awake rats. *Journal of Neurophysiology*, 79, 21–36.

Nitschke, J.B., Dixon, G.E., Sarinopoulous, I., Short, S.J., Cohen, J.D., Smith, E.E., Kosslyn, S.M., Rose, R.M. and Davidson, R.J. (2006). Altering expectancy dampens neural response to aversive taste in primary taste cortex. *Nature Neuroscience*, 9, 435–442.

Nordin, S., Andersson, L., Olofsson, J.K., McCormack, M. and Polich, J. (2011). Evaluation of auditory, visual and olfactory event-related potentials for comparing interspersed and single-stimulus paradigms. *International Journal of Psychophysiology*, 81, 252–262.

Nordin, S., Bramerson, L.E. and Bende, M. (1998). The Scandinavian odour-identification test: development, reliability, validity and normative data. *Acta Otolaryngology*, 118, 226–234.

Nordin, S. and Murphy, C. (1996). Impaired sensory and cognitive olfactory function in questionable Alzheimer's Disease. *Neuropsychology*, 10, 113–119.

Nordin, S., Razani, L.J., Markinson, S. and Murphy, C. (2003). Age-associated increases in intensity discrimination for taste. *Experimental Aging Research*, 29, 371–381.

Norgren, R. (1976). Taste pathways to hypothalamus and amygdala. *Journal of Comparative Neurology*, 166, 17–30.

Norgren, R., Hajnal, A. and Nungarndee, S.S. (2006). Gustatory reward and the nucleus accumbens. *Physiology & Behavior*, 89, 531–535.

Nunez-Jaramillo, L., Ramirez-Lugo, L., Herrera-Morales, W. and Miranda, M.I. (2010). Taste memory formation: latest advances and challenges. *Behavioral and Brain Research*, 207, 232–248.

Oberg, C., Larsson, M. and Backman, L. (2002). Differential sex effects in olfactory functioning: the role of verbal processing. *International Journal of the Neuropsychological Society*, 8, 691–698.

O'Doherty, R.M., Rolls, E.T., Francis, S., Bowtell, R. and McGlone, F. (2001). Representation of pleasant and aversive taste in the human brain. *Journal of Neurophysiology*, 85, 1315–1321.

O'Doherty, R.M., Rolls, E.T., Francis, S., Bowtell, R., McGlone, F., Kobal, G., Renner, B. and Ahne, G. (2000). Sensory-specific satiety-related olfactory activation of the human orbitofrontal cortex. *Neuroreport*, 11, 399–403.

Ogawa, H., Wakita, M., Hasegawa, K., Kobayakawa, T., Sakai, N., Hirai, T., Yamashita, Y. and Saito, S. (2005). Functional MRI detection of activation in the primary gustatory cortices in humans. *Chemical Senses*, 30, 583–592.

Ohla, K., Toepel, U., le Coutre, J. and Hudry, J. (2010). Electrical neuroimaging reveals intensity-dependent activation of human cortical gustatory and somatosensory areas by electric taste. *Biological Psychology*, 85, 446–455.

Okamoto, M., Dan, H., Singh, A.K., Hayakawa, F., Jurcak, V., Suzuki, T., Kohyama, K. and Dan, I. (2006). Prefrontal activity during flavour difference test. *Appetite*, 47, 220–232.

Okubo, Y., Suhara, T., Suzuki, K., Kobayashi, K., Inoue, O., Terasaki, O., Someya, Y., Sassa, T., Sudo, Y., Matsushima, E., Iyo, M., Tateno, Y. and Toru, M. (1997). Decreased prefrontal dopamine D1 receptors in schizophrenia revealed by PET. *Nature*, 385, 634–636.

Olofsson, J.K., Broman, D.A., Gilbert, P.E., Dean, P., Nordin, S. and Murphy, C. (2006). Laterality of the olfactory event-related potential response. *Chemical Senses*, 31, 699–704.

Olofsson, J.K., Ericsson, E. and Nordin, S. (2008). Comparison of chemosensory, auditory and visual event-related potential amplitudes. *Scandinavian Journal of Psychology*, 49, 231–237.

Olofsson, J.K. and Nordin, S. (2004). Gender differences in chemosensensory perception and event-related potentials. *Chemical Senses*, 29, 629–637.

Olofsson, J.K., Ronnlund, M., Nordin, S., Nyberg, L., Nilsson, L.-G. and Larsson, M. (2009). Odour identification deficit as a predictor of five-year global cognitive change: interactive effects with age and ApoE-e4. *Behaviour Genetics*, 39, 496–503.

Olson, J.M., Boehnke, M., Neiswanger, K., Roche, A.F. and Siervogel, R.M. (1989). Alternative genetic models for the inheritance of the phenylthiocarbamide taste deficiency. *Genetics and Epidemiology*, 6, 423–434.

Olsson, M.J. and Fridén, M. (2001). Evidence of odour priming: edibility judgements are primed differently between the hemispheres. *Chemical Senses*, 26, 117–123.

Olsson, M.J. and Jonsson, F.U. (2008). Is it easier to match a name to an odour than vice versa? *Chemosensory Perception*, 1, 184–189.

O'Mahony, M. (1979). Short-cut signal detection measurements for sensory analysis. *Journal of Food Science*, 44, 302–303.

Ongur, D., Ferry, A.T. and Price, J.L. (2003). Architectonic subdivision of the human orbital and medial prefrontal cortex. *Journal of Comparative Neurology*, 460, 425–449.

Ongur, D. and Price, J.L. (2000). The organisation of networks within the orbital and medial prefrontal cortex of rats, monkeys and humans. *Cerebral Cortex*, 10, 206–219.

Ono, T. (2004). Cortical activation during tongue movement revealed by fMRI: fact and myth on chewing-side preference. *International Congress Series*, 1270, 99–104.

Onoda, K. and Ikeda, M. (1999). Gustatory disturbance due to cerebrovascular disorder. *Laryngoscope*, 109, 123–128.

Onoda, K., Kobayakawa, T., Ikeda, M., Saito, S. and Kida, A. (2005). Laterality of human primary gustatory cortex studied by MEG. *Chemical Senses*, 30, 657–666.

Onozuka, M., Fujita, M., Watanabe, K., Hirano, Y., Niwa, M., Nishiyama, K. and Saito, S. (2002). Mapping brain regional activity during chewing: a Functional Magnetic Resonance Imaging study. *Journal of Dental Research*, 81, 743–746.

Onozuka, M., Fujita, M., Watanabe, K., Hirano, Y., Niwa, M., Nishiyama, K. and Saito, S. (2003). Age-related changes in brain regional activity during chewing: a Functional Magnetic Resonance Imaging study. *Journal of Dental Research*, 82, 657–600.

Osterhammel, P., Terkildsen, K. and Zilstorff, K. (1969). Electro-olfactograms in man. *The Journal of Laryngology & Otology*, 83, 731–733.

Owen, C.M. and Patterson, J. (2002). Odour liking physiological indices: a correlation of sensory and electrophysiological responses to odour. *Food Quality and Preference*, 13, 307–316.

Padiglia, A., Zonza, A., Atzori, E., Chilloti, C., Calo, C. and Tepper, B.J. (2010). Sensitivity to 6-n-propythiouracil is associated with gustin (carbonic anhydrase VI) gene polymorphism, salivary zinc and body mass index in humans. *American Journal of Clinical Nutrition*, 92, 539–545.

Pangborn, R.M., Chrisp, R.B. and Bertolero, L. (1970). Gustatory, salivary and oral thermal responses to solutions of sodium chloride at four temperatures. *Perception and Psychophysics*, 8, 69–75.

Pangborn, R.M. and Giovanni, M.E. (1984). Dietary intake of sweet foods and of dairy fats and resultant gustatory responses to sugar in lemonade and to fat in milk. *Appetite*, 5, 317–327.

Pardini, M., Huey, E.D., Cavanagh, A.L. and Grafman, J. (2009). Olfactory function in corticobasal syndrome and frontotemporal dementia. *Archives of Neurology*, 66, 92–96.

Parker, G.H. (1912). The relation of smell, taste and the common chemical sense in vertebrates. *Journal of the Academy of Natural Science Philadelphia*, 15, 221–234.

Passamonti, L., Rowe, J.B., Scwarzbauer, C., Ewbank, M.P., von dem Hagen, E. and Calder, A.J. (2009). Personality predicts the brain's response to viewing appetizing foods: the neural basis of a risk factor for overeating. *The Journal of Neuroscience*, 29, 43–51.

Pause, B.M., Ferstl, R. and Fehm-Wolfsdorf, G. (1998). Personality and olfactory sensitivity. *Journal of Research in Personality*, 32, 510–518.

Pause, B.M. and Krauel, K. (2000). Chemosensory event-related potentials as a key to the psychology of odours. *International Journal of Psychophysiology*, 36, 105–122.

Pause, B.M., Miranda, A., Göder, R., Aldenhoff, J.B. and Ferstl, R. (2001). Reduced olfactory performance in patients with major depression. *Journal of Psychiatry Research*, 35, 271–277.

Pause, B.M., Sojka, B., Krauel, K. and Ferstl, R. (1995). The nature of the late positive complex within the olfactory event-related potential. *Psychophysiology*, 33, 376–384.

Pearce, R.K.B., Hawkes, C.H. and Daniel, S.E. (1995). The anterior olfactory nucleus in Parkinson's Disease. *Movement Disorders*, 10, 283–287.

Pelletier, C.A. and Dhanaraj, G.E. (2006). The effect of taste and palatability on lingual swallowing pressure. *Dysphagia*, 21, 121–128.

Pendse, S.G. (1987). Hemispheric asymmetry in olfaction on a category judgement task. *Perceptual and Motor Skills*, 64, 495–498.

Penfield, W. and Faulk, M.E. (1955). The insula: further observations of its function. *Brain*, 78, 445–470.

Pierce, J. and Halpern, B.P. (1996). Orthonasal and retronasal odorant identification based upon vapour phase input from common substances. *Chemical Senses*, 21, 529–543.

Piwnica-Worms, K.E., Omar, R., Hailstone, J.C. and Warren, J.D. (2010). Flavour processing in semantic dementia. *Cortex*, 46, 761–768.

Plailly, J., Bensafi, M., Pachot-Clouard, M., Delon-Martin, C., Kareken, D.A., Rouby, C., Segebarth, C. and Royet, J.-P. (2005). Involvement of right piriform cortex in olfactory familiarity judgements. *Neuroimage*, 24, 1032–1041.

Plailly, J., Delon-Martin, C. and Royet, J.-P. (2012). Experience induces functional reorganization in brain regions involved in odour imagery in perfumers. *Human Brain Mapping*, 33, 224–234.

Plailly, J., Howard, J.D., Gitelman, D.R. and Gottfried, J.A. (2008). Attention to odour modulates thalamocortical connectivity in the human brain. *The Journal of Neuroscience*, 28, 5257–5267.

Plailly, J., Radnovich, A.J., Sabri, M., Royet, J.-P. and Kareken, D.A. (2007a). Involvement of the left anterior insula and frontopolar gyrus in odour discrimination. *Human Brain Mapping*, 28, 363–372.

Plailly, J., Tillmann, B. and Royet, J.-P. (2007b). The feeling of familiarity of music and odours: the same neural signature? *Cerebral Cortex*, 17, 2650–2658.

Plattig, K.-H. and Kobal, G. (1979). Spatial and temporal distribution of olfactory evoked potentials and techniques involved in their measurement. In Lehmann, D. and Callaway, E. (eds) *Human Evoked Potentials*. New York: Plenum, pp. 285–301.

Pliner, P. (1982). The effects of mere exposure on liking for edible substances. *Appetite*, 3, 283–290.

Poellinger, A., Thomas, R., Lio, P., Lee, A., Makris, N., Rosen, B.R. and Kwong, K.W. (2001). Activation and habituation on olfaction: an fMRI study. *Neuroimage*, 13, 547–560.

Poncelet, J., Rinck, F., Bourgeat, F., Schaal, B., Rouby, C., Bensafi, M. and Hummel, T. (2010). The effect of early experience on odour perception in humans: psychological and physiological correlates. *Behavioural Brain Research*, 208, 458–465.

Ponsen, M.M., Stoffers, D., Booij, J., Van Eck-Smit, W. and Berendse, H.W. (2004). Idiopathic hyposmia as a preclinical sign of Parkinson's Disease. *Annals of Neurology*, 56, 173–181.

Porter, J., Anand, T., Johnson, B., Khan, R.M. and Sobel, N. (2005). Brain mechanisms for extracting spatial information about smell. *Neuron*, 47, 581–592.

Porter, J., Craven, B., Khan, R.M., Chang, S.-J., Kang, I., Judkewicz, B., Volpe, J., Settles, G. and Sobel, N. (2007). Mechanisms of scent-tracking in humans. *Nature Neuroscience*, 10, 27–29.

Porter, R.H., Cernoch, J.M. and McLaughlin, F.J. (1983). Maternal recognition of neonates through olfactory cues. *Physiology & Behavior*, 30, 151–154.

Porter, R.H. and Moore, J.D. (1981). Human kin recognition by olfactory cues. *Physiology & Behavior*, 27, 493–495.

Porubska, K., Veit, R., Preissl, H., Fritsche, A. and Birbaumer, N. (2006). Subjective feeling of appetite modulates brain activity: an fMRI study. *Neuroimage*, 32, 1273–1280.

Postolache, T.T., Doty, R.L., Wehr, T.A., Jimma, L.A., Han, L., Turner, E.H., Matthews, J.R., Neumeister, A., No, C., Kroger, H., Bruder, G.E. and Rosenthal, N.E. (1999). Monorhinal odour identification and depression scores in patients with seasonal affective disorder. *Journal of Affective Disorders*, 56, 27–35.

Postuma, R. and Gagnon, J.-F. (2010). Cognition and olfaction in Parkinson's Disease. *Brain*, August, e1–2.

Potter, H. and Butters, N. (1980). As assessment of olfactory deficits in patients with damage to prefrontal cortex. *Neuropsychologia*, 18, 621–628.

Potter, H. and Nauta, W.J. (1979). A note on the problem of olfactory associations of the orbitofrontal cortex in the monkey. *Neuroscience*, 4, 361–367.

Prah, J.D. and Benignus, V.A. (1992). Olfactory evoked responses to odorous stimuli of different intensities. *Chemical Senses*, 17, 417–425.

Prasad, K.M., Patel, A.R., Muddasani, S., Sweeney, J. and Keshavan, M.S. (2004). The entorhinal cortex in first-episode psychotic disorders: a structural magnetic resonance imaging study. *American Journal of Psychiatry*, 161, 1612–1619.

Prehn-Kristensen, A., Wiesner, C., Bergmann, T.O., Wolff, S., Jansen, O., Mehdorn, H., Ferstl, R. and Pause, B.M. (2009). Induction of empathy by the smell of anxiety. *PLoS ONE*, 4, e5987.

Prescott, J. (1999). Flavour as a psychological construct. *Food Quality and Preference*, 10, 349–356.

Prescott, J. and Stevenson, R.J. (1995). Effects of oral chemical irritation on tastes and flavours in frequent and infrequent users of chilli. *Physiology & Behavior*, 58, 1117–1127.

Pribram, K. and Bagshaw, W. (1954). Further analysis of the temporal lobe syndrome utilizing frontotemporal ablations. *Journal of Comparative Neurology*, 98, 347–373.

Price, J.L. (1990). Olfactory system. In Paxinos, G. (ed.) *The Human Nervous System*. San Diego: Academic.

Price, J.L., Carmichael, S.T. and Carnes, K.M. (1991a). Olfactory input to the prefrontal cortex. In Davis, J. and Eichenbaum, H. (eds) *Olfaction: A model system for computational neuroscience*. Cambridge, MA: MIT Press.

Price, J.L., Davis, P.F., Morris, J.C. and White, D.L. (1991b). The distribution of tangles, plaques and related immunohistochemical markers in healthy ageing and Alzheimer's Disease. *Neurobiology of Aging*, 12, 295–312.

Pritchard, J.A. (1965). Deglutition by normal and anacephalic fetuses. *Obstetrics and Gynaecology*, 5, 289–297.

Pritchard, T., Hamilton, R., Morse, J. and Norgren, R. (1986). Projections from thalamic gustatory and lingual areas in the monkey. *Journal of Comparative Neurology*, 244, 213–228.

Pritchard, T.C., Macaluso, D.A. and Eslinger, P.J. (1999). Taste perception in patients with insular cortex lesions. *Behavioural Neuroscience*, 113, 663–671.

Punter, P.H. (1983). Measurement of human olfactory thresholds for several groups of structurally related compounds. *Chemical Senses*, 7, 215–235.

Purdon, S.E. (1998). Olfactory identification and Stroop interference converge in schizophrenia. *Journal of Psychiatry and Neuroscience*, 23, 163–171.

Purdon, S.E. and Flor-Henry, P. (2000). Asymmetrical olfactory acuity and neuroleptic treatment in schizophrenia. *Schizophrenia Research*, 44, 221–232.

Quinn, N.P., Rossor, M.N. and Marsden, C.D. (1987). Olfactory threshold in Parkinson's Disease. *Journal of Neurology, Neurosurgery and Psychiatry*, 50, 88–89.

Qureshy, A., Kawashima, R., Imran, M.B., Sugiura, M., Goto, R., Okada, K., Inoue, K., Itoh, M., Schormann, T., Zilles, K. and Fukuda, H. (2000). Functional mapping of human brain in olfactory processing: a PET study. *Journal of Neurophysiology*, 84, 1656–1666.

Rabin, M.D. (1988). Experience facilitates olfactory quality discrimination. *Perception and Psychophysics*, 44, 532–540.

Rabin, M.D. and Cain, W.S. (1984). Odour recognition: familiarity, identifiability and encoding consistency. *Journal of Experimental Psychology: Learning, Memory and Cognition*, 10, 316–325.

Rami, L., Loy, C.T., Hailstone, J. and Warren, J.D. (2007). Odour identification in frontotemporal lobar degeneration. *Journal of Neurology*, 254, 431–435.

Ramirez-Amaya, V., Alvarez-Borda, B. and Bermudez-Rattoni, F. (1998). Differential effects of NMDA-induced lesions into the insular cortex and amygdala on the acquisition and evocation of conditioned immunosuppression. *Brain, Behaviour and Immunity*, 12, 149–160.

Ramirez-Lugo, L., Nunez-Jaramillo, L. and Bermudez-Rattoni, F. (2007). Taste memory formation: role of nucleus accumbens. *Chemical Senses*, 32, 93–97.

Ramirez-Lugo, L., Zavala-Vega, S. and Bermudez-Rattoni, F. (2006). NMDA and muscarinic receptors of the nucleus accumbens have differential effects on taste memory formation. *Learning & Memory*, 13, 45–51.

Raudenbush, B., Koon, J., Smith, J. and Zoladz, P. (2003). Effects of odorant administration on objective and subjective measures of sleep quality, post-sleep mood and alertness and cognitive performance. *North American Journal of Psychology*, 5(2), 181–192.

Rausch, R., Serafetinides, E.A. and Crandall, P.H. (1977). Olfactory memory in patients with anterior temporal lobectomy. *Cortex*, 13, 445–452.

Rawson, N.E. and Li, X. (2007). The cellular basis of flavour perception: taste and aroma. In Taylor, A.J. and Roberts, D.D. (eds) *Flavour Perception*. Chichester: Wiley.

Razani, J., Nordin, S., Chan, A. and Murphy, C. (2010). Semantic networks for odours and colours in Alzheimer's Disease. *Neuropsychology*, 24, 291–299.

Rehn, T. (1978). Perceived odour intensity as a function of airflow through the nose. *Sensory Processes*, 2, 198–205.

Reichenberg, A. and Harvey, P.D. (2007). Neuropsychological impairments in schizophrenia: integration of performance-based and brain imaging findings. *Psychological Bulletin*, 133(5), 833–858.

Reske, M., Kellermann, T., Shah, N.J., Schneider, F. and Habel, U. (2010). Impact of valence and age on olfactory induced brain activation in healthy women. *Behavioral Neuroscience*, 124, 414–422.

Richman, R.A., Post, E.M. and Sheehe, P.R. (1992). Olfactory performance during childhood I. Development of an odorant identification test for children. *Journal of Pediatrics*, 121, 908–911.

Roalf, D.R., Turetsky, B.I. and Owzar, K. (2006). Unirhinal olfactory function in schizophrenia patients and first-degree relatives. *Journal of Neuropsychiatry and Clinical Neurosciences*, 18, 389–396.

Roitman, M.F., Wheeler, R.A. and Carelli, R.M. (2005). Nucleus accumbens neurons are innately tuned for rewarding and aversive taste stimuli, encode their predictors and are linked to motor output. *Neuron*, 45, 587–597.

Rolls, B.J. (1986). Sensory-specific satiety. *Nutrition Review*, 44, 93–101.

Rolls, E.T. (2008). *Memory, Attention and Decision-making*. Oxford: OUP.

Rolls, E.T. and Baylis, L.L. (1994). Gustatory, olfactory and visual convergence within the primate orbitofrontal cortex. *Journal of Neuroscience*, 14, 5437–5452.

Rolls, E.T., Critchley, H.D., Browning, A.S., Hernadi, I. and Lenard, L. (1999). Responses to the sensory properties of fat of neurons in the primate orbitofrontal cortex. *Journal of Neuroscience*, 19, 1532–1540.

Rolls, E.T., Grabenhorst, F., Margot, C., da Silva, M.A. and Velazco, M.I. (2008). Selective attention to affective value alters how the brain processes olfactory stimuli. *Journal of Cognitive Neuroscience*, 20, 1815–1826.

Rolls, E.T., Grabenhorst, F. and Parris, B.A. (2009). Neural systems underlying decisions about affective odours. *Journal of Cognitive Neuroscience*, 22, 1069–1082.

Rolls, E.T., Kringelbach, M.L. and de Araujo, I.E.T. (2003a). Different representations of pleasant and unpleasant odours in the human brain. *European Journal of Neuroscience*, 18, 695–703.

Rolls, E.T. and Rolls, J.H. (1997). Olfactory sensory-specific satiety in humans. *Physiology & Behaviour*, 61(3), 461–473.

Rolls, E.T., Scott, T.R., Sienkiewicz, Z.J. and Yaxley, S. (1988). The responsiveness of neurons in the frontal opercular gustatory cortex of the macaque monkey is independent of hunger. *Journal of Physiology*, 397, 1–12.

Rolls, E.T., Sienkiewicz, Z.J. and Yaxley, S. (1989). Hunger modulates the responses to gustatory stimuli of single neurons in the caudolateral orbitofrontal cortex of the macaque monkey. *European Journal of Neuroscience*, 1, 53–70.

Rolls, E.T., Verhagen, J.V. and Kadohisa, M. (2003b). Representations of the texture of food in the primate orbitofrontal cortex: neurons responding to viscosity, grittiness and capsaicin. *Journal of Neurophysiology*, 90, 3711–3724.

Rolls, E.T., Yaxley, S. and Sienkiewicz, Z.J. (1990). Gustatory responses of single neurons in the orbitofrontal cortex of the macaque monkey. *Journal of Neurophysiology*, 64, 1055–1066.

Rombaux, P., Bertrand, B., Keller, T. and Mouraux, A. (2007). Clinical significance of olfactory event-related potentials related to orthonasal and retronasal olfactory testing. *The Laryngoscope*, 117, 1096–1101.

Rombaux, P., Guerit, J.-M. and Mouraux, A. (2008). Lateralisation of intranasal trigeminal chemosensory event-related potentials. *Clinical Neurophysiology*, 38, 23–30.

Rombaux, P., Weitz, H., Mouraux, A., Nicolas, G., Bertrand, B., Duprez, T. and Hummel, T. (2006). Olfactory function assessed with orthonasal and retronasal testing, olfactory bulb volume and chemosensory event-related potentials. *Archives of Otolaryngology*, 132, 1346–1351.

Rosen, H.J., Allison, S.C., Ogar, J.M., Amici, S., Rose, K. and Dronkers, N. (2006). Behavioural features in semantic dementia versus other forms of progressive aphasias. *Neurology*, 67, 1752–1756.

Ross, G.W., Petrovitch, H., Abbott, R.D., Tanner, C.M., Popper, J., Masaki, K., Launer, L. and White, L.R. (2007). Association of olfactory dysfunction with risk for future Parkinson's Disease. *Annals of Neurology*, 63, 167–173.

Roth, H.A., Radle, L.J., Gifford, S.R. and Clydesdale, F.M. (1988). Psychophysical relationships between perceived sweetness and colour in lemon- and lime-flavoured drinks. *Journal of Food Science*, 53, 1116–1119.

Rotton, J. (1983). Affective and cognitive consequences of malodorous pollution. *Basic and Applied Social Psychology*, 4, 171–191.

Rotton, J., Barry, T., Frey, J. and Soler, E. (1978). Air pollution and interpersonal attraction. *Journal of Applied Social Psychology*, 8, 57–71.

Rouseff, R.L. (1990) *Bitterness in Foods and Beverages*. New York: Elsevier Science Publishers.

Rousseaux, M., Muller, P., Gahide, I., Mottin, Y. and Romon, M. (1996). Disorders of smell, taste and food intake in a patient with a dorsomedial thalamic infarct. *Stroke*, 27, 2328–2330.

Rovee, C.K., Harris, S.L. and Yopp, R. (1973). Olfactory thresholds and level of anxiety. *Bulletin of the Psychonomic Society*, 12, 76–78.

Royet, J.-P., Hudry, J., Zald, D.H., Godinot, D., Gregoire, M.C., Lavenne, F., Costes, N. and Holley, A. (2001). Functional neuroanatomy of different olfactory judgements. *Neuroimage*, 13, 506–519.

Royet, J.-P., Koenig, O., Gregoire, M.-C., Cinotti, L., Lavenne, F., Bars, D., Vigouroux, M., Farget, V., Sicard, G., Holley, A., Mauguiere, F., Comar, D. and Froment, J.-C. (1999). Functional anatomy of perceptual and semantic processing for odours. *Journal of Cognitive Neuroscience*, 11, 94–109.

Royet, J.-P., Plailly, J., Delon-Martin, C., Kareken, D.A. and Segebarth, C. (2003). fMRI of emotional responses to odours: influence of hedonic valence and judgement, handedness and gender. *Neuroimage*, 20, 713–728.

Royet, J.-P., Zald, D., Versace, R., Costes, N., Lavenne, F., Koenig, O. and Gervais, R. (2000). Emotional responses to pleasant and unpleasant olfactory, visual and auditory stimuli: a PET study. *The Journal of Neuroscience*, 20, 7752–7759.

Rozin, P. (1982). 'Taste-smell confusions' and the duality of the olfactory sense. *Perception & Psychophysics*, 21, 397–401.

Rozin, P. and Fallon, A.E. (1987). A perspective on disgust. *Psychological Review*, 94, 23–41.

Russell, M.I., Suritz, G.M. and Thompson, K. (1980). Olfactory influences on the human menstrual cycle. *Pharmacology, Biochemenstrual Behaviour*, 13, 737–738.

Sack, A.T., Camprodon, J.A., Pascual-Leone, A. and Goebel, R. (2005). The dynamics of interhemispheric compensatory processes in mental imagery. *Science*, 308, 702–704.

Saint-Eve, A., Deleris, I., Feron, G., Ibarra, D., Guichard, E. and Souchon, I. (2010). How trigeminal, taste and aroma perceptions are affected in mint-flavoured carbonated beverages. *Food Quality and Preference*, 21, 1026–1033.

Sakai, N., Kobayakawa, T., Gotow, N., Saitp, S. and Imada, S. (2001). Enhancement of sweetness ratings of aspartame by a vanilla odour presented either by orthonasal or retronasal routes. *Perceptual and Motor Skills*, 92, 1002–1008.

Sakamoto, R., Minoura, K., Usui, A., Ishizuka, Y. and Kanba, S. (2005). Effectiveness of aroma on work efficiency: lavender aroma during recesses prevents deterioration of work performance. *Chemical Senses*, 30, 683–691.

Sanchez-Lara, K., Sosa-Sanchez, R., Green-Renner, D., Rodriguez, C., Laviano, A., Motola-Kuba, D. and Arrieta, O. (2010). Influence of taste disorders on dietary behaviours in cancer patients under chemotherapy. *Nutrition Journal*, 9, 15.

Sanders, C., Diego, M., Fernandez, M., Field, T., Hernandez-Reif, M. and Roca, A. (2002). EEG asymmetry responses to lavender and rosemary aromas in adults and infants. *International Journal of Neuroscience*, 112, 1305–1320.

Sasaguri, K., Sato, S., Hirano, Y., Aoki, S., Ishikawa, T., Fujita, M., Watanabe, K., Tomida, M., Ido, Y. and Onozuka, M. (2004). Involvement of chewing in memory processes in humans: an approach using fMRI. *Frontiers in Human Brain Topology*, 1270, 111–116.

Savic, I. and Berglund, H. (2000). Right-nostril dominance in discrimination of unfamiliar, but not familiar, odours. *Chemical Senses*, 25, 517–523.

Savic, I. and Berglund, H. (2010). Androstenol: a steroid derived odour activates the hypothalamus in women. *PLoS ONE*, 5, e8651, 1–5.

Savic, I., Berglund, H., Gulyas, B. and Roland, P. (2001). Smelling of odorous sex hormone-like compounds causes sex-differentiated hypothalamic activations in humans. *Neuron*, 31, 661–668.

Savic, I., Berglund, H. and Lindstrom, P. (2005). Brain response to putative pheromones in homosexual men. *PNAS*, 102, 7356–7361.

Savic, I., Gulyas, B. and Berglund, H. (2002). Odorant differentiated pattern of cerebral activation: comparison of acetone and vanillin. *Human Brain Mapping*, 17, 17–27.

Savic, I., Gulyas, B., Larsson, M. and Roland, P. (2000). Olfactory functions are mediated by parallel and hierarchical processing. *Neuron*, 26, 735–745.

Schaal, B. (1988). Olfaction in infants and children: development and functional perspectives. *Chemical Senses*, 13, 145–190.

Schaal, B., Marlier, L. and Soussignan, R. (1998). Olfactory function in the human fetus: evidence from selective neonatal responsiveness to the odour of amniotic fluid. *Behavioural Neuroscience*, 112, 1438–1449.

Schab, F.R. (1990). Odours and the remembrance of things past. *Journal of Experimental Psychology: Learning, Memory and Cognition*, 16, 648–655.

Schaefer, A.T. and Margrie, T.W. (2007). Spatiotemporal representations in the olfactory system. *Trends in Neurosciences*, 30, 92–100.

Schecter, P.J. and Henkin, R.I. (1974). Abnormalities of taste and smell after head trauma. *Journal of Neurology, Neurosurgery and Psychiatry*, 37, 802–810.

Schieb, J.E., Gangestad, S.W. and Thornhill, R. (1999). Facial attractiveness, symmetry and cues of good genes. *Proceedings of the Royal Society B*, 266, 1913–1917.

Schiet, F.T. and Frijters, J.E.R. (1988). An investigation of the equiratio-mixture model in olfactory psychophysics: a case study. *Perception & Psychophysics*, 44, 304–308.

Schiff, B.B. and Rump, S.A. (1995). Asymmetrical hemispheric activation and emotion: the effects of unilateral forced nostril breathing. *Brain and Cognition*, 29, 217–231.

Schifferstein, H.N.J. and Verlegh, P.W.J. (1996). The role of congruency and pleasantness and odour-induced taste-enhancement. *Acta Psychologica*, 94, 87–105.

Schiffman, S.S. (1974). Physiochemical correlates of olfactory quality. *Science*, 185, 112–117.

Schiffman, S.S. (1983). Taste and smell in disease. *New England Journal of Medicine*, 308, 1275–1280.

Schiffman, S.S. (1997). Taste and smell losses in normal ageing and disease. *JAMA*, 278, 1357–1362.

Schiffman, S.S., Graham, B.G., Sattely-Miller, E.A., Zervakis, J. and Welsh-Bohmer, K. (2002). Taste, smell and neuropsychological performance of individuals at familial risk for Alzheimer's Disease. *Neurobiology of Aging*, 23, 397–404.

Schiffman, S.S., Lockhead, E. and Maes, F.W. (1983). Amiloride reduces the taste intensity of Na+ and Li+ salts and sweeteners. *PNAS*, 80, 6136–6140.

Schiffman, S.S. and Pasternak, M. (1979). Decreased discrimination of food odours in the elderly. *Journal of Gerontology*, 34, 73–79.

Schiffman, S.S., Reynolds, M.L. and Young, F.W. (1981). *Introduction to Multidimensional Scaling*. Orlando: Academic Press.

Schiffman, S.S., Zervakis, J., Heffron, S. and Heald, A.E. (1999). Effect of protease inhibitors on sense of taste. *Nutrition*, 15, 767–772.

Schleidt, M., Neumann, P. and Morishita, H. (1988). Pleasure and disgust: memories and associations of pleasant and unpleasant odours in Germany and Japan. *Chemical Senses*, 13, 279–293.

Schmidt, H.J. and Beauchamp, G.K. (1988). Adult-like odour preferences and aversions in three-year-old children. *Child Development*, 59, 1136–1143.

Schneider, R.A. and Wolf, S. (1955). Olfactory perception thresholds for citral utilizing an new style olfactorium. *Journal of Applied Physiology*, 8, 337–342.

Schubert, C.R., Cruickshanks, K.J., Klein, B.E.K., Klein, R. and Nondahl, D.M. (2011). Olfactory impairment in older adults: five-year incidence and risk factors. *The Laryngoscope*, 121, 873–878.

Schubert, C.R., Cruickshanks, K.J., Murphy, C., Huang, G.-H., Klein, B.E.K., Klein, R., Nieto, F.J., Pankow, J.S. and Tweed, T.S. (2009). Olfactory impairment in adults: the Beaver Dam Experience. *Annals of the New York Academy of Sciences*, 1170, 531–536.

Schumm, L.P., McClintock, M., Williams, S., Leitsch, S., Lundstrom, J., Hummel, T. and Lindau, S.T. (2009). Assessment of sensory function in the national social life, health and ageing project. *Journal of Gerontology: Social Sciences*, 64B, 176–185.

Schwob, J.E., Szumowski, K.E. and Stasky, A.A. (1992). Olfactory sensory neurons are trophically dependent on the olfactory bulb for their prolonged survival. *Journal of Neuroscience*, 12, 3896–3919.

Scott, T.R., Karadi, Z., Oomura, Y., Nishino, H., Plas-Salaman, C.R. and Leonard, L. (1993). Gustatory neural coding in the amygdala of the alert macaque monkey. *Journal of Neurophysiology*, 69, 1810–1820.

Scott, T.R. and Plata-Salaman, C.R. (1999). Taste in the monkey cortex. *Physiology & Behavior*, 67, 489–511.

Scott, T.R., Plata-Salaman, C.R., Smith, V.L. and Giza, B.K. (1991). Gustatory neural coding in the monkey cortex: stimulus intensity. *Journal of Neurophysiology*, 65, 76–86.

Scott, T.R., Yan, Y. and Rolls, E.T. (1995). Brain mechanisms of satiety and taste in macaques. *Neurobiology*, 3, 281–292.

Scott, T.R., Yaxley, S., Sienkiewitz, Z.J. and Rolls, E.T. (1986). Gustatory responses in the frontal opercular cortex of the alert cynomolgus monkey. *Journal of Neuophysiology*, 56, 876–890.

Scrivani, S.J., Meghan, M., Donoff, R.B. and Kaban, L.B. (2000). Taste perception after lingual nerve repair. *Journal of Oral Maxillofacial Surgery*, 58, 3–5.

Sela, L., Sacher, Y., Srfaty, C., Yeshurun, Y., Soroker, N. and Sobel, N. (2009). Spared and impaired olfactory abilities after thalamic lesions. *The Journal of Neuroscience*, 29, 12059–12069.

Sengoku, R., Saito, Y., Ikemura, M., Hatsuta, H., Sakiyama, Y., Lanemaru, K., Arai, T., Sawabe, M., Tanaka, N., Mochizuki, H., Inoue, K. and Murayama, S.J. (2008). Incidence and extent of Lewy body-related alpha-synucleinopathy in ageing human olfactory bulb. *Neuropathology & Experimental Neurology*, 67, 1072–1083.

Serby, M., Corwin, J., Novatt, A., Conrad, P. and Rotrosen, J. (1985). Olfaction in dementia. *Journal of Neurology, Neurosurgery and Psychiatry*, 48, 848–849.

Serby, M., Larson, P. and Kalkstein, D. (1990). Olfactory sense in psychoses. *Biological Psychiatry*, 28, 829–830.

Seubert, J., Kellermann, T., Loughead, J., Boers, F., Brensinger, C., Schneider, D. and Habel, U. (2010). Processing of disgusted faces is facilitated by odour primes: a fMRI study. *Neuroimage*, 53, 746–756.

Seubert, J., Rea, A.F., Loughead, J. and Habel, U. (2009). Mood induction with olfactory stimuli reveals differential affective responses in males and females. *Chemical Senses*, 34, 77–84.

Sewards, T.V. (2004). Dual separate pathways for sensory and hedonic aspects of taste. *Brain Research Bulletin*, 62, 271–283.

Shafer, D.M., Frank, M.E., Gent, J.F. and Fischer, M.E. (1999). Gustatory function after third molar extraction. *Oral Surgery, Medicine, Pathology, Radiology and Endodontology*, 87, 419–428.

Shah, N.J., Marshall, J.C., Zafiris, O., Schwab, A., Zilles, K., Markowitsch, H.J. and Fink, G.R. (2001). The neural correlates of person familiarity: a functional magnetic resonance imaging study with clinical implications. *Brain*, 124, 804–815.

Shankar, M.U., Levitan, C.A., Prescott, J. and Spence, C. (2009). The influence of colour and label information on flavour perception. *Chemosensory Perception*, 2, 53–58.

Shepherd, G.M. (2005). Perception without a thalamus: how does olfaction do it? *Neuron*, 46, 166–168.

Sherrington, C.S. (1906). *The Integrative Action of the Nervous System*. New Haven, CT: Yale University Press.

Shimomura, J. and Motokizawa, F. (1995). Functional equivalence of the two sides of the human nose in odour detection. *Chemical Senses*, 20, 585–587.

Ship, J.A., Pearson, J.D., Cruise, L.J., Brant, L.J. and Metter, E.M. (1996). Longitudinal changes in smell identification. *Journal of Gerontology: Medical Sciences*, 51A, 86–91.

Shu, C.-H., Yuan, B.-C. and Lin, C.-C. (2007). Cross-cultural application of the Sniffin' Sticks odour identification test. *American Journal of Rhinology*, 21, 570–573.

Siep, N., Roefs, A., Roebroeck, A., Havermans, R., Bonte, M.L. and Jansen, A. (2009). Hunger is the best spice: an fMRI study of the effects of attention, hunger and calorie content on food reward processing in the amygdala and orbitofrontal cortex. *Behavioural Brain Research*, 198, 149–158.

Sigurdardottir, S., Andelic, N., Roe, C., Jerstad, T. and Schanke, A.-K. (2010). Olfactory dysfunction, gambling task performance and intracranial lesions after traumatic brain injury. *Neuropsychology*, 24, 504–513.

Silberstein, S.D., Miknam, R. and Rozen, T.D. (2000). Cluster headache with aura. *Neurology*, 54, 219–221.

Silveira-Moriyama, L., Holton, J.L., Kingsbury, A., Ayling, H., Petrie, A. and Sterlacci, W. (2009). Regional differences in the severity of Lewy body pathology across the olfactory cortex. *Neuroscience Letters*, 453, 77–80.

Sim, K., DeWitt, I. and Ditman, T. (2006). Hippocampal and parahippocampal volumes in schizophrenia: a structural MRI sudy. *Schizophrenia Bulletin*, 32, 332–340.

Simmons, W.K., Martin, A. and Barsalou, L.W. (2005). Pictures of appetizing foods activate gustatory cortices for taste and reward. *Cerebral Cortex*, 15, 1602–1608.

Simons, C.T., O'Mahoney, M. and Carstens, E. (2002). Taste suppression following lingual capsaicin pre-treatment in humans. *Chemical Senses*, 27, 353–365.

Singh, N., Grewal, M.S. and Austin, J.H. (1970). Familial anosmia. *Archives of Neurology*, 22, 40–44.

Small, D.M., Bernasconi, N., Bernasconi, A., Sziklas, V. and Jones-Gotman, M. (2005a). Gustatory agnosia. *Neurology*, 64, 311–317.

Small, D.M., Gerber, J.C., Mak, Y.E. and Hummel, T. (2005b). Differential neural responses evoked by orthonasal versus retronasal odorant perception in humans. *Neuron*, 47, 593–605.

Small, D.A., Gregory, M.D., Mak, Y.E., Gitleman, D., Mesulam, M.M. and Parrish, T. (2003). Dissociation of neural representation of intensity and affective valuation in human gustation. *Neuron*, 39, 701–711.

Small, D.M., Jones-Gotman, M., Zatorre, R.J., Petrides, M. and Evans, A.C. (1997a). Flavour processing: more than the sum of its parts. *Neuroreport*, 8, 3913–3917.

Small, D.M., Jones-Gotman, M., Zatorre, R.J., Petrides, M. and Evans, A.C. (1997b). A role for the right anterior temporal lobe in taste quality recognition. *Journal of Neuroscience*, 17, 5136–5142.

Small, D.M., Veldhuizen, M.G., Felsted, J., Mak, Y.E. and McGlone, F. (2008). Separable substrates for anticipatory and comsummatory chemosensation. *Neuron*, 13, 786–797.

Small, D.M., Voss, J., Mak, Y.E., Simmons, K.B., Parrish, T. and Gitelman, D. (2004). Experience-dependent neural integration of taste and smell in the human brain. *Journal of Neurophysiology*, 92, 1892–1903.

Small, D.M., Zald, D.H., Jones-Gotman, M., Zatorre, R.J., Pardo, J.V., Frey, S. and Petrides, M. (1999). Human cortical gustatory areas: a review of functional imaging data. *Neuroreport*, 10, 7–14.

Small, D.M., Zatorre, R.J., Dagher, A., Evans, A.C. and Jones-Gotman, M. (2001a). Changes in brain activity related to eating chocolate. *Brain*, 124, 1720–1733.

Small, D.M., Zatorre, R.J. and Jones-Gotman, M. (2001b). Increased intensity perception of aversive taste following right anteromedial temporal lobe removal in humans. *Brain*, 124, 1566–1575.

Smeets, P.A.M., de Graaf, C., Stafleu, A., van Osch, M.J.P., Nievelstein, R.A.J. and van der Grond, J. (2006). Effect of satiety on brain activation during chocolate tasting in men and women. *American Journal of Clinical Nutrition*, 83, 1297–1305.

Smith, B.A. and Blass, E.M. (1996). Taste-mediated calming in premature, preterm and full-term human infants. *Developmental Psychology*, 32, 1084–1089.

Smith, D.B., Allison, T., Goff, W.R. and Principato, J.J. (1971). Human odorant evoked responses: effects of trigeminal or olfactory deficit. *Electroencephalography and Clinical Neurophysiology*, 30, 313–317.

Smith, D.V. and Shipley, M.T. (1992). Anatomy and physiology of taste and smell. *Journal of Head Trauma Rehabilitation*, 7, 1–14.

Smith, T.D., Bhatnagar, K.P., Tuladhar, P. and Burrows, A.M. (2004). Distribution of olfactory epithelium in the primate nasal cavity: are microsmia and macrosmia valid morphological concepts? *Anat Rec A Discov Mol Cell Evolutionary Biology*, 281, 1173–1181.

Smits, M., Peeters, R.R., van Hecke, P. and Sunaert, S. (2007). A 3-T event-related fMRI study of primary and secondary gustatory cortex localization using natural tastants. *Neuroradiology*, 49, 61–71.

Smutzer, G.S., Doty, R.L., Arnold, S.E. and Trojanowski, J.Q. (2003). Olfactory system neuropathology in Alzheimer's Disease, Parkinson's Disease and schizophrenia. In Doty, R.L. (ed.) *Handbook of Olfaction and Gustation*, 2nd edn. New York: Marcel Dekker, pp. 503–524.

Smutzer, G.S., Lam S., Hastings, L., Desai, H., Abarintos, A., Sobel, M. and Sayed, N. (2008). A test for measuring gustatory function. *Laryngoscope*, 118, 1411–1416.

Snowden, J.S., Bathgate, D., Varma, A., Blackshaw, A., Gibbons, Z.C. and Neary, D. (2001). Distinct behavioural profiles in frontotemporal dementia and semantic dementia. *Journal of Neurology, Neurosurgery and Psychiatry*, 70, 323–332.

Sobel, N., Khan, R.M., Hartley, C.A., Sullivan, E.V. and Gabrieli, J.D. (2000). Sniffing longer rather than stronger to maintain olfactory detection threshold. *Chemical Senses*, 25, 1–8.

Sobel, N., Khan, R.M., Saltman, A., Sullivan, E.V. and Gabrieli, J.D.E. (1999a). The world smells different to each nostril. *Nature*, 402, 35.

Sobel, N., Prabhakaran, V., Hartley, C.A., Desmond, J.E., Glover, G.H., Sullivan, E.V. and Gabrieli, J.D.E. (1999b). Blind smell: brain activation induced by an undetected air-borne chemical. *Brain*, 122, 209–217.

Sobel, N., Prabhakaran, V., Hartley, C.A., Desmond, J.E., Zhao, Z., Glover, G.H., Gabrieli, J.D.E. and Sullivan, E.V. (1998). Odorant-induced and sniff-induced activation in the cerebellum of the human. *The Journal of Neuroscience*, 18(21), 8990–9001.

Sobel, N., Thomason, M.E., Stappen, I., Tanner, C.M., Tetrud, J.W., Bower, J.M., Sullivan, E.V. and Gabrieli, J.D.E. (2001). An impairment in sniffing contributes to the olfactory impairment in Parkinson's Disease. *PNAS*, 98, 4154–4159.

Sohrabi, H.R., Bates, K.A., Rodrigues, M., Taddei, K., Laws, S.M., Lautenschlager, N.T., Dhaliwal, S.S., Johnston, A.N.B., Mackay-Sim, A., Gandy, S., Foster, J.K. and Martins, R.N. (2009). Olfactory dysfunction is associated with subjective memory complaints in community-dwelling elderly individuals. *Journal of Alzheimer's Disease*, 17, 1387–2877.

Solomon, G.S., Petrie, W.M. and Hart, J.R. (1998). Olfactory dysfunction discriminates Alzheimer's dementia from major depression. *Journal of Neuropsychiatry and Clinical Neurosciences*, 10, 64–67.

Sosulski, D.L., Bloom, M.L., Cutforth, T., Axel, R. and Datta, S.D. (2011). Distinct representations of olfactory information in different cortical centres. *Nature*, 472, 213–216.

Soubeyrand, L. (1964). Action des medicaments vaso-moteurs sur le cycle nasal et la fonction ciliare. *Revue de Laryngologie Otorhinologie*, 85, 48–113.

Soussignan, R., Schaal, B., Marlier, L. and Jiang, T. (1997). Facial and autonomic responses to biological and artificial olfactory stimuli in human neonates: re-examining early hedonic discrimination of odours. *Physiology & Behavior*, 62, 745–758.

Soussignan, R., Schaal, B., Schmit, G. and Nadel, J. (1995). Facial responsiveness to odours in normal and pervasively developmentally disordered children. *Chemical Senses*, 20, 47–59.

Spangler, R., Wittowski, K.M., Goddard, N.L., Avena, N.M., Hoebel, B.G. and Leibowitz, S.F. (2004). Opiate-like effects of sugar on gene expression in reward areas of the rat brain. *Brain Research*, 124, 134–142.

Spencer, N.A., McClintock, M.K., Sellergren, S.A., Bullivant, S., Jacob, S. and Mennella, J.A. (2004). Social chemosignals from breastfeeding women increase sexual motivation. *Hormones and Behavior*, 46, 362–370.

Steiner, J.E. (1977). Facial expressions of the neonate infant indicating the hedonics of food-related chemical stimuli. In Weiffenbach, J.M. (ed.) *Taste and Development: The Genesis of Sweet Preference*. Bethesda: US Department of Health.

Steiner, J.E., Glaser, D., Hawilo, M.E. and Berridge, K.C. (2011). Comparative expression of hedonic impact: affective reactions to taste by human infants and other primates. *Neuroscience & Biobehavioral Reviews*, 25, 53–74.

Stern, K. and McClintock, M.K. (1998). Regulation of ovulation by human pheromones. *Nature*, 392, 177–179.

Stevens, J.C., Bartoshuk, L.M. and Cain, W.S. (1984). Chemical senses and ageing: taste versus smell. *Chemical Senses*, 9, 167.

Stevens, J.C., Cain, W.S. and Burke, R.J. (1988). Variability of olfactory thresholds. *Chemical Senses*, 13, 643–653.

Stevens, J.C., Cain, W.S. and Demarque, A. (1990). Memory and identification of simulated odours in elderly and young persons. *Bulletin of the Psychonomic Society*, 28, 293–296.

Stevens, J.C., Cain, W.S. and Weinstein, D.E. (1987). Ageing impairs the ability to detect gas odour. *Fire Technology*, 23, 198.

Stevenson, R.J. and Boakes, R.A. (2004). Sweet and sour smells: the acquisition of taste-like qualities by odours. In Calvert, G., Spence, C. and Stein, B. (eds) *Handbook of Multisensory Processing*. Boston: MIT Press.

Stevenson, R.J. and Mahmut, M.K. (2011). Experience dependent changes in odour-viscosity perception. *Acta Psychologica*, 136, 60–66.

Stevenson, R.J., Miller, L.A. and Thayer, Z.C. (2008). Impairments in the perception of odour-induced tastes and their relationship to taste perception. *Journal of Experimental Psychology*, 34, 1183–1197.

Stevenson, R.J., Prescott, J. and Boakes, R.A. (1999). Confusing tastes and smells: how odours can influence the perception of sweet and sour tastes. *Chemical Senses*, 24, 627–635.

Stevenson, R.J. and Repacholi, B. (2003). Age-related changes in children's hedonic response to make body odour. *Developmental Psychology*, 36, 670–679.

Stinton, N., Atif, M.A., Barkat, N. and Doty, R.L. (2010). Influence of smell loss on taste function. *Behavioural Neuroscience*, 124, 256–264.

Stockhorst, U. and Pietrowsky, R. (2004). Olfactory perception, communication and the nose-to-brain pathway. *Physiology & Behavior*, 83, 3–11.

Stone, L.J. and Pangborn, R.M. (1990). Preferences and intake measures of salt and sugar and their relation to personality traits. *Appetite*, 15, 63–79.

Stuck, B.A., Frey, S., Freiburg, C., Hormann, K., Zahnert, T. and Hummel, T. (2006). Chemosensory event-related potentials in relation to side of stimulation, age, sex and stimulus concentration. *Clinical Neurophysiology*, 117, 1367–1375.

Su, C.-Y., Martelli, C., Emonet, T. and Carlson, J.R. (2011). Temporal coding of odour mixtures in an olfactory receptor neuron. *PNAS*, 108, 5075–5080.

Sucker, K., Both, R. and Winneke, G. (2009). Review of adverse health effects of odours in field studies. *Water Science & Technology*, 59, 1281–1289.

Sugita, M. and Shiba, Y. (2005). Genetic tracing shows segregation of taste neuronal circuitries for bitter and sweet. *Science*, 309, 781–785.

Suh, G.S., Wong, A.M., Hergarden, A.C., Wang, J.W., Simon, A.F., Benzer, S., Axel, R. and Anderson, D.J. (2004). A single population of olfactory sensory neurons mediates an innate avoidance behaviour in Drosophila. *Nature*, 431, 854–859.

Sullivan, S. and Birch, L.L. (1990). Pass the sugar, pass the salt: experience dictates preference. *Developmental Psychology*, 26, 546–551.

Sumner, D. (1964). Post-traumatic anosmia. *Brain*, 87, 107–120.

Suzuki, Y., Critchley, H.D., Suckling, J., Fukuda, R., Williams, S.C.R., Andrew, C., Howard, R., Ouldred, E., Bryant, C., Swift, C.G. and Jackson, S.H.D. (2001). Functional Magnetic Resonance Imaging of odour identification: the effect of ageing. *Journal of Gerontology A*, 56, 756–760.

Tabaton, M., Cammarata, S., Mancardi, G.L., Cordone, G., Perry, G. and Loeb, C. (1991). Abnormal tau-reactive filaments in olfactory mucosa in biopsy specimens of patients with probable Alzheimer's Disease. *Neurology*, 41, 391–394.

Tabert, M.H., Liu, X., Doty, R.L., Serby, M., Zamora, D., Pelton, G.H., Marder, K., Albers, M.W., Stern, Y. and Devanand, D.P. (2005). A 10-item smell identification scale related to risk for Alzheimer's Disease. *Annals of Neurology*, 58, 155–160.

Takeda, A., Saito, N., Baba, T., Kikuchi, A., Sugeno, N., Kobayashi, M., Hasegawa, T. and Itoyama, Y. (2010). Functional imaging studies of hyposmia in Parkinson's Disease. *Journal of the Neurological Sciences*, 289, 36–39.

Talamo, B.R., Rudel, R., Kosik, K.S., Lee, V.M., Neff, S., Adelman, L. and Kauer, J.S. (1989). Pathological changes in olfactory neurons in patients with Alzheimer's Disease. *Nature*, 337, 736–739.

Tanabe, T., Ino, M., Ooshima, Y. and Takagi, S.F. (1974). An olfactory area in the prefrontal lobe. *Brain Research*, 80, 127–130.

Tanabe, T., Ino, M. and Takagi, S.F. (1975a). Discrimination of odours in olfactory bulb, pyriform-amygdaloid areas and orbitofrontal cortex of the monkey. *Journal of Neurophysiology*, 38, 1269–1283.

Tanabe, T., Yarita, H., Iino, M., Ooshima, Y. and Takagi, S.F. (1975b). An olfactory projection area in orbitofrontal cortex of the monkey. *Journal of Neurophysiology*, 38, 1269–1283.

Tandon, R., Keshavan, M.S. and Nasrallah, H.A. (2008). Schizophrenia, 'just the facts': what we know in 2008. 2: epidemiology and etiology. *Schizophrenia Research*, 100, 1–18.

Tataranni, P.A., Gaultier, J.F., Chen, K., Uecker, A., Bandy, D. and Salbe, A.D. (1999). Neuroanatomical correlates of hunger and satiation in humans using PET. *PNAS*, 96, 4569–4574.

Taylor, D.C. and Lochery, M. (1987). Temporal lobe epilepsy: origin and significance of simple and complex auras. *Journal of Neurology, Neurosurgery and Psychiatry*, 50, 673–681.

Teghtsoonian, R., Teghtsoonian, M., Berglund, B. and Berglund, U. (1978). Invariance of odour strength with sniff vigor. *Journal of Experimental Psychology: Human Perception and Performance*, 4, 144–152.

Tham, W.W.P., Stevenson, R.J. and Miller, L.A. (2011). The impact of mediodorsal thalamic lesions on olfactory attention and flavour perception. *Brain and Cognition*, 77, 71–79.

Thesen, T. and Murphy, C. (2002). Reliability analysis of event-related brain potentials to olfactory stimuli. *Psychophysiology*, 39, 733–738.

Thomann, P.A., Dos Santos, V., Toro, P., Schonknecht, P., Essig, M. and Schroder, J. (2009). Reduced olfactory bulb and tract volume in early Alzheimer's Disease: a MRI study. *Neurobiology of Aging*, 30, 838–841.

Titchener, E.B. (1909). *A Textbook of Psychology*. London: Macmillan.

Tomita, H., Ikeda, M. and Okuda, Y. (1986). Basis and practice of clinical taste examinations. *Auris-Nasus-Larynx*, 13, S1–15.

Tonosaki, K. and Funakoshi, M. (1988). Cylic nucleotides may mediate taste transduction. *Nature*, 331, 354–356.

Toulouse, E. and Vaschide, N. (1899). Mesure de l'odorat chez l'homme et chez la femme. *C.R. Societe Biologie*, 51, 381–383.

Toulouse, E. and Vaschide, N. (1900). L'asymmetrie sensorielle olfactive. *Revue Philosophique*, 49, 176–187.

Trojanowski, J.Q., Newman, P.D., Hill, W.D. and Lee, V.M. (1991). Human olfactory epithelium in normal ageing, Alzheimer's Disease and other neurodegenerative disorders. *Journal of Comparative Neurology*, 310, 365–376.

Trotier, D., Eloit, C., Wassef, M., Talmain, G., Bensimon, J.L., Doving, K.B. and Ferrand, J. (2000). The vomeronasal cavity in adult humans. *Chemical Senses*, 25, 369–380.

Tsuboi, Y., Wszolek, Z.K., Graff-Radford, N.R. (2003). Tau pathology in the olfactory bulb correlates with Braak stage, Lewy body pathology and apolipoprotein episilon4. *Neuropathology and Applied Neurobiology*, 29, 503–510.

Turetsky, B.I., Glass, C.A., Abbazia, J., Kohler, C.G., Gur, R.E. and Moberg, P.J. (2007). Reduced posterior nasal cavity volume: a gender-specific neurodevelopmental abnormality in schizophrenia. *Schizophrenia Research*, 93, 237–244.

Turetsky, B.I., Hahn, C.G., Arnold, S.E. and Moberg, P.J. (2009a). Olfactory receptor neuron dysfunction in schizophrenia. *Neuropsychopharmacology*, 34, 767–774.

Turetsky, B.I., Hahn, C.G., Borgmann-Winter, K. and Moberg, P.J. (2009b). Scents and nonsense: olfactory dysfunction in schizophrenia. *Schizophrenia Bulletin*, 35, 1117–1131.

Turetsky, B.I., Kohler, C.G., Gur, R.E. and Moberg, P.J. (2008). Olfactory physiological impairment in first-degree relatives of schizophrenia patients. *Schizophrenia Research*, 102, 220–229.

Turetsky, B.I. and Moberg, P.J. (2009). An odour-specific detection threshold sensitivity deficit implicates abnormal intracellular cyclic AMP signalling in schizophrenia. *American Journal of Psychiatry*, 166, 226–233.

Turetsky, B.I., Moberg, P.J., Arnold, S.A., Doty, R.L. and Gur, R.E. (2003a). Low olfactory bulb volume in first-degree relatives of patients with schizophrenia. *American Journal of Psychiatry*, 160, 703–708.

Turetsky, B.I., Moberg, P.J., Roalf, D.R., Arnold, S.E. and Gur, R.E. (2003). Decrements in volume of anterior ventromedial temporal lobe and olfactory dysfunction in schizophrenia. *Archives of General Psychiatry*, 60, 1193–1200.

Turetsky, B.I., Moberg, P.J., Roalf, D.R., Arnold, S.E. and Gur, R.E. (2003b). Decrements in volume of anterior ventromedial temporal lobe and olfactory dysfunction in schizophrenia. *Archives of General Psychiatry*, 60, 1193–1200.

Turetsky, B.I., Moberg, P.J., Yousem, D., Arnold, S.E., Doty, R.L. and Gur, R.E. (2000). Olfactory bulb volume is reduced in patients with schizophrenia. *American Journal of Psychiatry*, 157, 828–830.

Turner, B.H., Mishkin, M. and Knapp, M. (1980). Organization of the amygdalopetal projections from modality-specific cortical association areas in the monkey. *Journal of Comparative Neurology*, 191, 515–543.

Ueno, Y. (1993). Cross-cultural study of odour perception in Sherpa and Japanese people. *Chemical Senses*, 18, 352–353.

van Hoesen, G.W., Augustinack, J.C., Dierking, J., Redman, S.J. and Thangavel, R. (2000). The parahippocampal gyrus in Alzheimer's Disease: clinical and preclinical neuroanatomical correlates. *Annals of the New York Academy of Sciences*, 911, 254–274.

van der Laan, L.N., de Ridder, D.T.D., Viergever, M.A. and Smeets, P.A.M. (2011). The first taste is always with the eyes: a meta-analysis on the neural correlates of processing visual food cues. *Neuroimage*, 55, 296–303.

Vanderwolf, C.H. (1992). Hippocampal activity, olfaction and sniffing: an olfactory input to the dentate gyrus. *Brain Research*, 593, 197–208.

Van Toller, S., Behan, J., Howells, P., Kendal-Reed, M. and Richardson, A. (1993). An analysis of spontaneous human cortical EEG activity to odours. *Chemical Senses*, 18, 1–16.

Varendi, H., Porter, R.H. and Winberg, J. (1994). Does the newborn baby find the nipple by smell? *Lancet*, 334, 989–990.

Varney, N.R. (1988). Prognostic significance of anosmia in patients with closed head injury. *Journal of Clinical Experimental Neuropsychology*, 10, 250–254.

Varney, N.R. and Bushnell, D. (1998). NeuroSPECT findings in patients with post-traumatic anosmia: a quantitative analysis. *Journal of Head Trauma Rehabilitation*, 13, 63–72.

Varney, N.R., Pinkston, J.B. and Wu, J.C. (2001). Quantitative PET findings in patients with post-traumatic anosmia. *Journal of Head Trauma Rehabilitation*, 16, 253–259.

Vaschide, N. (1904). L'etat de la sensibilite olfactive dans la vieillesse. *Bulletin of Laryngology, Otolaryngology and Rhinology*, 7, 323–333.

Vasquez, M., Pearson, P.B. and Beauchamp, G.K. (1982). Flavour preferences in malnourished Mexican infants. *Physiology & Behavior*, 28, 513–519.

Velakoulis, D. (2006). Olfactory hallucinations. In Brewer, W. , Castle, D. and. Pantelis, C. (eds) *Olfaction and the Brain*. Cambridge: CUP, pp. 322–333.

Veldhuizen, M.G., Bender, G., Constable, R.T. and Small, D.M. (2007). Trying to detect taste in a tasteless solution: modulation of early gustatory cortex by attention to taste. *Chemical Senses*, 32, 569–581.

Veldhuizen, M.G., Nachtigal, D., Teulings, L., Gitelman, D.R. and Small, D.A. (2010). The insular taste cortex contributes to odour quality coding. *Frontiers in Human Neuroscience*, 4(58), 1–11.

Vennemann, M.M., Hummel, T. and Berger, K. (2008). The association between smoking and smell and taste impairment in the general population. *Journal of Neurology*, 255, 1121–1126.

Ventanas, S., Mustonen, S., Puolanne, E. and Tuorila, H. (2010). Odour and flavour perception in flavoured model systems. *Food Quality and Preference*, 21, 453–462.

Verhagen, J.V. and Engelen, L. (2006). The neurocognitive bases of human multimodal food perception: sensory integration. *Neuroscience and Biobehavioral Reviews*, 30, 613–650.

Verhagen, J.V., Kadohisa, M. and Rolls, E.T. (2004). The primate insular/opercular taste cortex: neuronal representations of the viscosity, fat texture, grittiness, temperature and taste of foods. *Journal of Neurophysiology*, 92, 1685–1699.

Verhagen, J.V., Rolls, E.T. and Kadohisa, M. (2003). Neurons in the primate orbitofrontal cortex respond to fat texture independently of viscosity. *Journal of Neurophysiology*, 90, 1514–1525.

Vermetten, E. and Bremner, J.D. (2003). Olfaction as a traumatic reminder in post-traumatic stress disorder: case reports and review. *Journal of Clinical Psychiatry*, 64, 202–207.

Verron, H. and Gaultier, G. (1976). Processus olfactifs et structures relationelles. *Physiologie Française*, 21, 205–209.

Vickers, Z.M. and Wasserman, S.S. (1980). Sensory qualities of food sounds based on individual perceptions. *Journal of Texture Studies*, 10, 319–332.

Vierling, J. and Rock, J. (1967). Variations in olfactory sensitivity to Exaltolide during the menstrual cycle. *Journal of Applied Physiology*, 22, 311–315.

Villemure, C. and Bushnell, M.C. (2007). The effects of the steroid androstadienone and pleasant odorants on the mood and the pain perception of men and women. *European Journal of Pain*, 11, 181–191.

Villemure, C., Wassimi, S., Bennett, G.J., Shir, Y. and Bushnell, M.C. (2006). Unpleasant odours increase pain processing in a patient with neuropathic pain: psychophysical and fMRI investigation. *Pain*, 120, 213–220.

Voirol, E. and Daget, N. (1986). Comparative study of nasal and retronasal olfactory perception. *Food Science and Technology*, 19, 316–319.

Vowles, R.H., Bleach, N.R. and Rowe-Jones, J.M. (1997). Congenital anosmia. *International Journal of Pediatric Otorhinolaryngology*, 41, 207–214.

Wahab, N.A., Jones, R.D. and Huckabee, M.-L. (2011). Effects of olfactory and gustatory stimuli on the biomechanics of swallowing. *Physiology & Behavior*, 102, 485–490.

Walla, P. and Deecke, L. (2010). Odours influence visually induced emotion: behavior and neuroimaging. *Sensors*, 10, 8185–8197.

Walla, P., Hufnagl, B., Lehrner, J., Mayer, D., Lindinger, G., Imhof, H., Deecke, L. and Lang, W. (2002). Evidence of conscious and subconscious olfactory information processing during word encoding: an MEG study. *Cognitive Brain Research*, 14, 309–316.

Walla, P., Hufnagl, B., Lehrner, J., Mayer, D., Lindinger, G., Imhof, H., Deecke, L. and Lang, W. (2003a). Olfaction and depth of word processing: a magnetoencephalographic study. *Neuroimage*, 18, 104–116.

Walla, P., Hufnagl, B., Lehrner, J., Mayer, D., Lindinger, G., Imhof, H., Deecke, L. and Lang, W. (2003b). Olfaction and face encoding in humans: a magnetoencephalographic study. *Cognitive Brain Research*, 15, 105–115.

Wang, J., Eslinger, P.J., Smith, M.B. and Yang, Q.X. (2005). Functional magnetic resonance imaging study of human olfaction and normal ageing. *Journal of Gerontology A*, 60, 510–514.

Wang, L., Walker, V.E., Sardi, H., Fraser, C. and Jacob, T.J.C. (2002). The correlation between physiological and psychological responses to odour stimulation in human subjects. *Clinical Neurophysiology*, 113, 542–551.

Wang, L., Zhou, M., Brand, J. and Huang, L. (2009). Inflammation and taste disorders. *Annals of the New York Academy of Sciences*, 1170, 596–603.

Ward, C.D., Hess, W.A. and Calne, D.B. (1983). Olfactory impairment in Parkinson's Disease. *Neurology*, 33, 943–946.

Wardle, J., Cooke, L., Gibson, E.L., Sapochnik, M., Sheilham, A. and Lawson, M. (2003a). Increasing children's acceptance of vegetables: a randomized trial of guidance to parents. *Appetite*, 40, 155–162.

Wardle, J., Herrera, M.-L., Cook, L. and Gibson, E.L. (2003b). Modifying children's food preferences: the effects of exposure and reward on acceptance of an unfamiliar vegetable. *European Journal of Clinical Nutrition*, 57, 341–348.

Warner, M.D., Peabody, C.A. and Berger, P.A. (1988). Olfactory deficits in Down's Syndrome. *Biological Psychiatry*, 23, 836–839.

Wattendorf, E., Welge-Lussen, A., Fiedler, K., Bilecen, D., Wolfensberger, M., Fuhr, P., Hummel, T. and Westermann, B. (2009). Olfactory impairment predicts brain atrophy in Parkinson's Disease. *The Journal of Neuroscience*, 29, 15410–15413.

Wehling, E., Nordin, S., Espeseth, T., Reinvang, I. and Lundervold, A.J. (2011). Unawareness of olfactory dysfunction and its association with cognitive functioning in middle aged and old adults. *Archives of Clinical Neuropsychology*, 26, 260–269.

Weiffenbach, J.M., Baum, B.J. and Burghauser, R. (1982). Taste thresholds: quality specific variation with human ageing. *Journal of Gerontology*, 37, 372–377.

Weiffenbach, J.M., Wolf, R.O., Benheim, A.E. and Folio, C.J. (1983). Taste threshold assessment: a note on quality specific differences between methods. *Chemical Senses*, 8, 151–159.

Weiner, M.W., Veitch, D.P., Aisen, P.S., Beckett, L.A., Cairns, N.J., Green, R.C., Harvey, D., Jack, C.R., Jagust, W., Liu, E., Morris, J.C., Petersen, R.C., Saykin, A.J., Schmidt, M.E., Shaw, L., Siuciak, J.A., Soares, H., Toga, A.W. and Trojanowski, J.Q. (2012). The Alzheimer's Disease Neuroimaging Initiative: a review of papers. *Alzheimer's & Dementia*, 8, S1–68.

Welge-Lussen, A., Dorig, P., Wolfensberger, M., Krone, F. and Hummel, T. (2010). A study about the frequency of taste disorders. *Journal of Neurology*, 258, 386–392.

Welge-Lussen, A., Husner, A., Wolfensberger, M. and Hummel, T. (2009). Influence of simultaneous gustatory stimuli on orthonasal and retronasal olfaction. *Neuroscience Letters*, 454, 124–128.

Wells, D.L. and Hepper, P.G. (2000). The discrimination of dog odours by humans. *Perception*, 29, 111–115.

Werntz, D.A., Bickford, R.G., Bloom, F.E. and Shannahoff-Khalsa, D.S. (1983). Alternating cerebral hemispheric activity and the lateralization of autonomic nervous function. *Human Neurobiology*, 2, 39–43.

Werntz, D.A., Bickford, R.G. and Shannahoff-Khalsa, D.S. (1987). Selective heimispheric stimulation by unilateral forced nostril breathing. *Human Neurobiology*, 6, 165–171.

Wesson, D.W., Levy, E., Nixon, R.A. and Wilson, D.A. (2010). Olfactory dysfunction correlates with Β-amyloid plaque burden in an Alzheimer's Disease mouse model. *Journal of Neuroscience*, 30, 505–514.

Whisman, M., Goetzinger, J., Cotton, F. and Brinkman, D. (1978). Odorant evaluation: a study of ethanethiol and tetrahydrothiophene as warning agents in propane. *Environmental Science and Technology*, 12, 1285–1288.

Whissell-Buechy, D. and Amoore, J.E. (1973). Odour-blindness to musk: simple recessive inheritance. *Nature*, 242, 271.

Wicker, B., Keysers, C., Plailly, J., Royet, J.-P., Gallese, V. and Rizzolatti, G. (2003). Both of us disgusted in my insula: the common neural basis of seeing and feeling disgust. *Neuron*, 40, 655–664.

Williams, S.S., Williams, J., Combrinck, M., Christie, S., Smith, A.D. and McShane, R. (2009). Olfactory impairment is more marked in patients with mild dementia with Lewy bodies and those with mild Alzheimer's Disease. *Journal of Neurology, Neurosurgery and Psychiatry*, 80, 667–670.

Wilson, R.I. and Mainen, Z.F. (2006). Early events in olfactory processing. In Brewer, W., Castle, D. and Pantelis, C. (eds) *Olfaction and the Brain*. Cambridge: CUP, pp. 163–201.

Wilson, R.S., Arnold, S.E., Schneider, J.A., Boyle, P.A., Buchman, A.S. and Bennett, D.A. (2009). Olfactory impairment in presymptomatic Alzheimer's Disease. *Annals of the New York Academy of Sciences*, 1170, 730–735.

Wilson, R.S., Arnold, S.E., Schneider, J.A., Tang, Y. and Bennett, D.A. (2007). The relationship between cerebral Alzheimer's Disease pathology and odour identification in old age. *Journal of Neurology, Neurosurgery and Psychiatry*, 78, 30–35.

Winston, J.S., Gottfried, J.A., Kilner, J.M. and Dolan, R.J. (2004). Integrated neural representations of odour intensity and affective in human amygdala. *The Journal of Neuroscience*, 25, 8903–8907.

Wrobel, B.B. and Leopold, D.A. (2004). Clinical assessment of patients with smell and taste disorders. *Otolaryngol Clin North America*, 37, 1127–1142.

Wu, J., Buchsbaum, M.S., Moy, K., Denlea, N., Kesslak, P. and Tseng, H. (1993). Olfactory memory in unmedicated schizophrenics. *Schizophrenia Research*, 9, 41–47.

Wysocki, C. and Gilbert, A. (1989). National Geographic smell survey. *Annals of the New York Academy of Sciences*, 561, 12–28.

Wysocki, C., Dorries, K.M. and Beauchamp, G.K. (1989). Ability to perceive androstenone can be acquired by ostensibly anosmic people. *PNAS*, 4, 7976–7978.

Wysocki, C. and Preti, G. (2004). Facts, fallacies, fears and frustrations with human pheromones. *Anat Rec A Discov Mol Cell Evol Biol*, 281A, 1201–1211.

Yamada, H. and Imoto, T. (1987). Inhibitory effect of the extract from Zizyphus jujuba leaves on sweet taste responses of the chorda tympani in the rat and hamster. *Comparative Biochemistry and Physiology A*, 88, 355–360.

Yamagishi, M., Getchell, M.L., Takami, S. and Getchell, T.V. (1998). Increased density of olfactory receptor neurons immunoreactive for apolipoprotein E in patients with Alzheimer's Disease. *Annals of Otolaryngology, Rhinology and Laryngology*, 107, 421–426.

Yamagishi, M., Ishizuka, Y. and Seki, K. (1994). Pathology of olfactory mucosa in patients with Alzheimer's Disease. *Annals of Otolaryngology, Rhinology and Laryngology*, 103, 421–427.

Yamamoto, C., Nagai, H., Takahashi, K., Nakagawa, S., Yamaguchi, M., Tonoike, M. and Yamamoto, T. (2006). Cortical representation of taste-modifying action of miracle fruit in humans. *Neuroimage*, 33, 1145–1151.

Yamamoto, C., Takehara, S., Morikawa, K., Nakagawa, S., Yamaguchi, M., Iwaki, S., Tonoike, M. and Yamamoto, T. (2003). MEG study of cortical activity evoked by electrogustatory stimuli. *Chemical Senses*, 28, 245–251.

Yamamoto, T. (2006). Neural substrates for the processing of cognitive and affective aspects of taste in the brain. *Archives of Histology and Cytology*, 69, 243–255.

Yamamoto, T., Sako, N. and Maeda, S. (2000). Effects of taste discrimination on beta-endorphin levels in rat cerebrospinal fluid and plasma. *Physiology & Behavior*, 69, 345–350.

Yan, J. and Scott, T.R. (1996). The effect of satiety on responses of gustatory neurons in the amygdala of alert cynomolgus macaques. *Brain Research*, 740, 193–200.

Yanagisawa, K., Bartoshuk, L.M., Catalanotto, F.A., Karrer, T.A. and Kveton, J.F. (1998). Anesthesia of the chorda tympani nerve and taste phantoms. *Physiology & Behavior*, 63, 329–335.

Yarita, H., Iino, M., Tanabe, T., Kogure, S. and Takagi, S.F. (1980). A transthalamic olfactory pathway to orbitofrontal cortex in the monkey. *Journal of Neurophysiology*, 43, 69–85.

Yasoshima, Y., Sako, N., Senba, E. and Yamamoto, T. (2006). Acute suppression, but not chronic genetic deficiency, of c-fos gene expression impairs long-term memory in aversive taste learning. *PNAS*, 103, 7106–7111.

Yaxley, S., Rolls, E.T. and Sienkiewicz, Z.J. (1990). Gustatory responses of single neurons in the insula of the macaque monkey. *Journal of Neurophysiology*, 63, 689–700.

Yeshurun, Y. and Sobel, N. (2010). An odour is not worth a thousand words: from multidimensional odours to unidimensional odour objects. *Annual Review of Psychology*, 61, 219–241.

Youngentob, S.L., Kurtz, D.B., Leopold, D.A., Mozell, M.M. and Hornung, D.E. (1982). Olfactory sensitivity: is there laterality? *Chemical Senses*, 7, 11–21.

Yousem, D.M., Geckle, R.J., Bilker, W.B., McKeown, D.A. and Doty, R.L. (1996). Post-traumatic olfactory dysfunction: MR and clinical evaluation. *AJNR*, 17, 1171–1179.

Yousem, D.M., Maldjian, J., Hummel, T., Alsop, D.C., Geckle, R.J., Kraut, M.A. and Doty, R.L. (1999a). The effect of age on odour-stimulated functional MR imaging. *AJNR*, 20, 600–608.

Yousem, D.M., Maldjian, J. and Siddiqi, F. (1999b). Gender effects on odour-stimulated functional magnetic resonance imaging. *Brain Research*, 818, 480–487.

Yousem, D.M., Turner, W.J.D., Li, C., Snyder, P.J. and Doty, R.L. (1993). Kallmann Syndrome: MR evaluation of olfactory system. *AJNR*, 14, 839–843.

Yousem, D.M., Williams, S.C.R., Howard, R.O., Andrew, C., Simmons, A., Allin, M., Geckle, R.J., Suskind, D., Bullmore, E.T., Brammer, M.J. and Doty, R.L. (1997). Functional MR imaging during odour stimulation: preliminary data. *Radiology*, 204, 833–838.

Zald, D.H., Hagen, M.C. and Pardo, J.V. (2002). Neural correlates of tasting concentrated quinine and sugar solutions. *Journal of Neurophysiology*, 87, 1068–1075.

Zald, D.H., Lee, J.T., Fluegel, K.W. and Pardo, J.V. (1998). Aversive gustatory stimulation activates limbic circuits in humans. *Brain*, 121, 1143–1154.

Zald, D.H. and Pardo, J.V. (1997). Emotion, olfaction and the human amygdala: amygdala activation during aversive olfactory stimulation. *PNAS*, 94, 4119–4124.

Zald, D.H. and Pardo, J.V. (2000). Functional neuroimaging of the olfactory system in humans. *International Journal of Psychophysiology*, 36, 165–181.

Zatorre, R.J. and Jones-Gotman, M. (1990). Right-nostril advantage for discrimination of odour. *Perception & Psychophysics*, 47, 526–531.

Zatorre, R.J. and Jones-Gotman, M. (1991). Human olfactory discrimination after unilateral frontal or temporal lobectomy. *Brain*, 114, 71–84.

Zatorre, R.J., Jones-Gotman, M., Evans, A.C. and Meyer, E. (1992). Functional localization and lateralization of human olfactory cortex. *Nature*, 360, 339–340.

Zellner, D.A. and Durlach, P. (2003). Effect of colour on expected and experienced refreshment, intensity and liking of beverages. *American Journal of Psychology*, 116, 633–647.

Zellner, D.A., McGarry, A., Mattern-McClory, R. and Abreu, D. (2008). Masculinity/femininity of fine fragrances affects colour-odour correspondences: a case for cognitions influencing cross-modal correspondences. *Chemical Senses*, 33, 211–222.

Zernecke, R., Haegler, K., Leemann, A.M., Albrecht, J., Frank, T., Linn, J., Bruckmann, H. and Wiesmann, M. (2011). Effects of male anxiety chemosignals on the evaluation of happy facial expressions. *Journal of Psychophysiology*, 25, 116–123.

Zernecke, R., Kleemann, A.M., Haegler, K., Albrecht, J., Vollmer, B., Linn, J., Bruckmann, H. and Wiesmann, M. (2010). Chemosensory properties of human sweat. *Chemical Senses*, 35, 101–108.

Zhang, J. and Webb, D.M. (2003). Evolutionary deterioration of the vomeronasal pheromone transduction pathway in catarrhine primates. *Proceedings of the National Academy of Science*, 100, 8337–8341.

Zhang, Y., Hoon, M.A., Chandrashekar, J., Mueller, K.L., Cook, B., Wu, D., Zuker, C.S. and Ryba, N.J. (2003). Coding of sweet, bitter and umami tastes: different receptor cells sharing similar signalling pathways. *Cell*, 112, 293–301.

Zhou, W. and Chen, D. (2009). Sociochemosensory and emotional functions: behavioural evidence for shared mechanisms. *Psychological Science*, 20, 1118–1124.

Zou, Z., Li, F. and Buck, L.B. (2005). Odour maps in the olfactory cortex. *PNAS*, 102, 7724–7729.

Zucco, G. and Tressoldi, P.E. (1988). Hemispheric differences in odour recognition. *Cortex*, 25, 607–615.

Zucco, G., Zeni, M.T., Perrone, A. and Piccolo, I. (2001). Olfactory sensitivity in early stage Parkinson patients affected by more marked unilateral disorder. *Perceptual and Motor Skills*, 92, 894–898.

Zverev, Y. and Mipando, M. (2008). Extraversion and taste sensitivity. *Coll Antropol*, 32, 21–25.

INDEX